VLSI-SOC: FROM SYSTEMS TO CHIPS

IFIP – The International Federation for Information Processing

IFIP was founded in 1960 under the auspices of UNESCO, following the First World Computer Congress held in Paris the previous year. An umbrella organization for societies working in information processing, IFIP's aim is two-fold: to support information processing within its member countries and to encourage technology transfer to developing nations. As its mission statement clearly states,

> *IFIP's mission is to be the leading, truly international, apolitical organization which encourages and assists in the development, exploitation and application of information technology for the benefit of all people.*

IFIP is a non-profitmaking organization, run almost solely by 2500 volunteers. It operates through a number of technical committees, which organize events and publications. IFIP's events range from an international congress to local seminars, but the most important are:

- The IFIP World Computer Congress, held every second year;
- Open conferences;
- Working conferences.

The flagship event is the IFIP World Computer Congress, at which both invited and contributed papers are presented. Contributed papers are rigorously refereed and the rejection rate is high.

As with the Congress, participation in the open conferences is open to all and papers may be invited or submitted. Again, submitted papers are stringently refereed.

The working conferences are structured differently. They are usually run by a working group and attendance is small and by invitation only. Their purpose is to create an atmosphere conducive to innovation and development. Refereeing is less rigorous and papers are subjected to extensive group discussion.

Publications arising from IFIP events vary. The papers presented at the IFIP World Computer Congress and at open conferences are published as conference proceedings, while the results of the working conferences are often published as collections of selected and edited papers.

Any national society whose primary activity is in information may apply to become a full member of IFIP, although full membership is restricted to one society per country. Full members are entitled to vote at the annual General Assembly, National societies preferring a less committed involvement may apply for associate or corresponding membership. Associate members enjoy the same benefits as full members, but without voting rights. Corresponding members are not represented in IFIP bodies. Affiliated membership is open to non-national societies, and individual and honorary membership schemes are also offered.

VLSI-SOC: FROM SYSTEMS TO CHIPS

IFIP TC 10/ WG 10.5 Twelfth International Conference on Very Large Scale Integration of System on Chip (VLSI-SoC 2003), December 1-3, 2003, Darmstadt, Germany

Edited by

Manfred Glesner
Technische Universität, Darmstadt, DE

Ricardo Reis
Universidade Federal do Rio Grande do Sul, BR

Leandro Indrusiak
Technische Universität, Darmstadt, DE

Vincent Mooney
Georgia Tech, USA

Hans Eveking
Technische Universität, Darmstadt, DE

VLSI-SOC: From Systems to Chips
Edited by M. Glesner, R. Reis, L. Indrusiak, V. Mooney, and H. Eveking

p. cm. (IFIP International Federation for Information Processing, a Springer Series in Computer Science)

ISSN: 1571-5736 / 1861-2288 (Internet)

eISBN: 10: 0-387-33403-3

ISBN 978-1-4419-4126-8 e-ISBN 978-0-387-33403-5

Printed on acid-free paper

9 8 7 6 5 4 3 2 1
springeronline.com

CONTENTS

PREFACE

This book contains extended and revised versions of the best papers that have been presented during the twelfth edition of the IFIP TC10/WG10.5 International Conference on Very Large Scale Integration, a Global System-on-a-Chip Design & CAD Conference. The 12th edition was held at the Lufthansa Training Center in Seeheim-Jugenheim, south of Darmstadt, Germany (December 1-3, 2003). Previous conferences have taken place in Edinburgh (81), Trondheim (83), Tokyo (85), Vancouver (87), Munich (89), Edinburgh (91), Grenoble (93), Tokyo (95), Gramado (97), Lisbon (99) and Montpellier (01).

The purpose of this conference, sponsored by IFIP TC 10 Working Group 10.5, is to provide a forum to exchange ideas and show research results in the field of microelectronics design. The current trend toward increasing chip integration brings about exhilarating new challenges both at the physical and system-design levels: this conference aims to address these exciting new issues.

The 2003 edition of VLSI-SoC conserved the traditional structure, which has been successful in previous editions. The quality of submissions (142 papers) made the selection process difficult, but finally 57 papers and 14 posters were accepted for presentation in VLSI-SoC 2003. Submissions came from Austria, Bulgaria, Brazil, Canada, Egypt, England, Estonia, Finland, France, Germany, Greece, Hungary, India, Iran, Israel, Italy, Japan, Korea, Malaysia, Mexico, Netherlands, Poland, Portugal, Romania, Spain, Sweden, Taiwan and the United States of America. From 57 papers presented at the conference, 18 were selected to have an extended and revised version included in this book.

This book also includes the paper "Effect of Power Optimizations on Soft Error Rate" which one is related to a tutorial presented by *Narayanan Vijaykrishnan, Penn State University.*

VLSI-SoC 2003 was the work of many dedicated volunteers: paper authors, reviewers, session chairs, invited speakers and chairs of various aspects, especially the local arrangements by the various members of the Institut für Mikroelektronische Systeme of Technische Universität Darmstadt and the financial management by the VDE Information Technology Society. The edition of this book was done with the significant help of Ana Cristina Pinto and Jane Follmann from UFRGS at Porto Alegre. We thank them all for their contribution.

The editors

EFFECT OF POWER OPTIMIZATIONS ON SOFT ERROR RATE

Vijay Degalahal, R. Ramanarayanan, Narayanan Vijaykrishnan, Y. Xie, M. J. Irwin

Embedded and Mobile Computing Design Center
Pennsylvania State University

Abstract Due to technology scaling, devices are getting smaller, faster and operating at lower voltages. The reduced nodal capacitances and supply voltages coupled with more dense and larger chips are increasing soft errors and making them an important design constraint. As designers aggressively address the excessive power consumption problem that is considered as a major design limiter they need to be aware of the impact of the power optimizations on the soft error rates(SER). In this chapter, we analyze the effect of increasing threshold voltage and reducing the operating voltages, widely used for reducing power consumption, on the soft error rate. While reducing the operating voltage increases the susceptibility to soft errors, increasing the threshold voltages offers mixed results. We find that increasing threshold voltage (V_t) improves SER of transmission gate based flip-flops, but can adversely affect the robustness of combinational logic due to the effect of higher threshold voltages on the attenuation of transient pulses. We also show that, in certain circuits, clever use of high V_t can improve the robustness to soft errors.

Keywords: Low Power VLSI, Power Optimizations, Soft Errors, Single Event Upset, Reliability, Noise Immunity,

1. Introduction

As silicon industry enters the nanometer regime, it is facing new challenges on several fronts. In the past, aggressive technology scaling has improved performance, reduced power consumption, and helped the industry obey Moore's law. In the sub-130nm regime, supply voltage is also scaled down to reduce the power consumption. To compensate for the lower supply voltage, the threshold voltage of the device is also reduced. This increases the subthreshold leakage [8]. In addition, the ultra thin gate oxides increase the tunneling probability of the electrons, thus increasing the gate leakage. Furthermore, the dense integration of the transistors along with increased leakage currents makes power density an important concern in newer technologies. Hence power, by many, is

---

Please use the following format when citing this chapter:
Degalahal, Vijay, Ramanarayanan, R., Vijaykrishnan, Narayanan, Xie, Y., Irwin, .J, M., 2006, in IFIP International Federation for Information Processing, Volume 200, VLSI-SOC: From Systems to Chips, eds. Glesner, M., Reis, R., Indrusiak, L., Mooney, V., Eveking, H., (Boston: Springer), pp. 1-20.

regarded as the most significant road block in realizing the benefits of scaling for next generation. Consequently various optimizations for reducing power consumption have been proposed [29, 25]. This chapter will mainly examine the impact of higher threshold voltage devices and supply voltage scaling techniques, used for reducing power, on soft error rates(SER).

The direct consequence of the lower supply voltage is lower signal to noise ratio (SNR). This results in increased susceptibility of circuits to noise sources like soft errors. In contrast, the effect of the higher V_t devices is not straight forward. Soft errors are transient circuit errors that are induced by external radiations. Radiation from cosmic rays or packaging material induces a localized ionization which, can lead to spurious pulses or upset the internal data states. While these errors cause a transient pulse, the circuit itself is not damaged. In combinatorial logic, if this transient pulse is latched by a memory element like a latch or a register, the transient pulse translates to a soft error which corrupts the computation. In memory elements like SRAM, latches and registers, these errors change the stored values of the bits. In SRAM based FPGAs, the configuration is stored in the SRAM cells, which when upset causes a change in the configuration and hence leads to an error in firmware. This error, though correctable, will lead to wrong computation until the configuration is reloaded. Conventional ways of reducing the soft error rates include adding redundancy, increasing nodal capacitance and using error correcting codes. In this work, we analyze the effect of increasing the threshold voltage (V_t) of the device and supply voltage scaling on soft errors in standard memory elements like SRAMs and flip-flops and on combinational circuits like inverters, nand gates and adders, which represent the most common CMOS logic styles. We believe such an analysis is very important because it helps us in making intelligent design choices that reduce leakage power consumption and improve the reliability of the next generation circuits.

The chapter is organized as follows: Section 2 presents the background on soft errors. Section 3 presents the experimental methodology that is used to examine soft errors in circuits. Section 4 focuses on the impact of supply voltage scaling and high V_t devices on SER. It also discusses the detailed results of the experimental analysis of SER on different circuits. It also presents guidelines to implement delay balancing to reduce power and improve robustness to soft errors at the same time. Section 5 presents the conclusions.

2. Soft Errors

When energy particles hit a silicon substrate, the kinetic energy of the particle generates electron hole pairs as they pass through p-n junctions. Some of the deposited charge will recombine to form a very short duration current pulse which causes a soft error. In memory elements, these can cause bit flips, but in combinational circuits they can cause a temporary change in the output. In combinational logic such a pulse is naturally attenuated, but if a transient pulse is latched, it corrupts the logic state of the circuit [7, 10].

There are three principle sources of soft errors: alpha particles, high-energy cosmic ray induced neutrons, and neutron induced boron-10(^{10}B) fission. Alpha particles are emitted from decaying elements like Lead present in the packaging materials. Alpha particles are also created due to the interaction of cosmic ray thermal neutron with ^{10}B present in the P-type regions of the devices [6]. A single alpha particle can generate anywhere from 4 to 16fC/m over its entire range.

The galactic flux of primary cosmic rays (mainly consisting of protons) is very large, about 100,000 particles/m^2s as compared to the much lower final flux (mainly consisting of neutrons) at sea level of about 360 particles/m^2s [30]. Only few of the galactic particles have adequate energy to penetrate the earth's atmosphere. The intensity of these neutron radiations depends on altitude, geomagnetic region and solar cycles [30]. The primary reaction by which cosmic ray induced neutrons cause SER is by silicon recoil. The impinging neutrons knock off the silicon from its lattice. The displaced silicon nucleus breaks down into smaller fragments each of which generates some charge. The charge density for silicon recoils is about 25 to 150fC/m, which is more than that from alpha particle strike. So it has a higher potential to upset the circuit.

The third significant source of ionizing particles is from the neutron induced ^{10}B fission. ^{10}B, an isotope of p-type dopant (about 19.9%), is unstable and on impact from neutron it absorbs the neutrons and breaks apart with the release of an alpha particle and ^{7}Li (Lithium). Both these by-products are capable of inducing soft errors. To reduce SER due to alpha particle induced soft errors, one can use ^{10}B free materials and shield the circuit so that components with higher alpha emission rates are physically isolated from the sensitive circuits. Such solutions though effective for alpha generated soft errors, are generally ineffective against the neutrons as they are highly penetrative.

The issue of soft errors was studied in the context of scaling trends of microelectronics as early as in 1962 [28]. Interestingly, the forecast from this study that the lower limit on supply voltage reduction will be imposed by soft errors is shared by a recent work from researchers at Intel [9]. However, most works on radiation effects, since the work in 1962, focused on space applications rather than terrestrial applications. As earth's atmosphere shields most cosmic ray particles from reaching the ground and charge per circuit node used to be large, soft errors on terrestrial devices has not been important until recently. Recent works [14, 15, 26] have shown the effect of technology scaling on soft errors. In [30], a study on radiation flux noted that particles of lower energy occur far more frequently than particles of higher energy. So it can be seen that as CMOS device sizes decrease, they are more easily affected by these lower energy particles, potentially leading to a much higher rate of soft errors.

According to [15], the soft error rate of memory element is expressed as Equation 1:

$$SER \propto N_{flux} * CS * exp\frac{-Q_{critical}}{Q_s} \quad (1)$$

where,

N_{flux} is the intensity of neutron flux,
CS is the area of cross section of the node,
Q_s is the charge collection efficiency,
$Q_{critical}$ is the charge that is stored at the node and hence is equal to $VDD * C_{node}$, where VDD is the supply voltage and
C_{node} is the nodal capacitance.

Hence any reduction in supply voltage or nodal capacitance increases the soft error rate. The soft error rate is also proportional to the area of the node CS. Smaller devices have lesser area and hence are less susceptible for an upset. But lower $Q_{critical}$ coupled with higher density of the devices and larger dies ensures an increase in soft errors for each new generation [27, 26, 15].

In combinational circuits soft errors are dependent on many factors. A transient glitch due a radiation event at a node alters the execution and generates a computation error only if the resultant glitch causes a change in the output and also such a change in the output is latched by a register. These factors derate the soft errors rates in logic. Hence in logic circuits, the

$$SER \propto N_{flux} * CS * Prob_G * Prob_P * Prob_L \quad (2)$$

Where,

$Prob_G$ is the probability that an transient pulse is generated for a particle strike at that node,
$Prob_P$ is the probability that the generated transient pulse propagates through the logic network,
$Prob_L$ is the probability that the transient pulse is latched at the output.

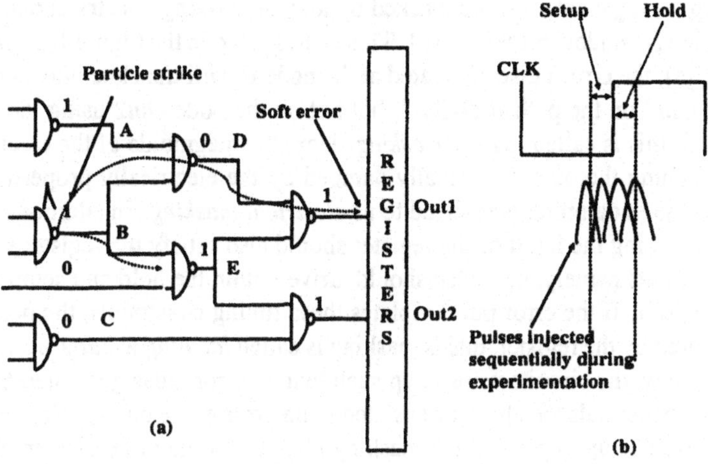

Figure 1. Circuit level evaluation of soft errors in logic circuit

For static CMOS logic, all the factors except $Prob_G$ are dependent on the circuit structure, inputs to the circuit, operating voltage and technology.

3. Methodology for circuit level analysis of soft errors

For a soft error to occur at a specific node in a circuit, the collected charge Q at that particular node should be greater than $Q_{critical}$. $Q_{critical}$ is be defined as the minimum charge collected due to a particle strike that can cause a change in the output. If the charge generated by a particle strike at a node is more than $Q_{critical}$, the generated pulse is latched, resulting in a bit flip. This concept of critical charge is generally used to estimate the sensitivity of SER. The value of $Q_{critical}$ can be found by measuring the current required to flip a memory cell and derived using Equation 3. The particle strike itself is modeled as a piece wise linear current waveform where the waveform's peak accounts for funneling charge collection and the waveform's tail accounts for diffusion charge collection. By changing the magnitude of the peak of the waveform and appropriately scaling the waveform, we try to find the minimum height for which the wrong value is stored in the memory element. A similar approach has been used in prior work [26].

In a logic circuit, a transient change in the value of the circuit does not affect the results of a computation unless it is captured in a memory element like a flip- flop. A logic error can be masked by logical masking, electrical masking and latching- window masking [16]. This is illustrated in the Figure 1(a). In the Figure 1(a), the error pulse generated at the node *B*, will translate into an error at node *Out1* on the path *B-D-Out1*, but not on the node *Out2* using the path *B-E-Out2*, this is called *logical masking*. Next the magnitude of the electrical pulse reaching the output is usually damped by the electrically properties of the gates, such damping is referred to as *electrical masking*. In addition, such a pulse reaching the input of the register should also satisfy the register's timing constraints namely, the pulse should arrive within the hold and setup time of the register. If the error pulse violates these timing constraints, the pulse is not captured by the register, this is masking is called *latching window masking*. For our study, the circuits were setup such that an error pulse generated by an injected current pulse is always translated to an error at the output. Hence, the circuits were setup to prevent the masking effects. For memory elements, the internal nodes, where the logical value is stored as charge, were chosen for injecting the current pulse.

The actual magnitude of the charge is given by Equation 3.

$$Q_{critical} = \int_0^{T_f} I_d dt \tag{3}$$

Where, I_d is the drain current induced by the charged particle. T_f is the flipping time and in memory circuits it can be defined as the point in time when the feedback mechanism of the back-to-back inverter will take over from the incident ion's current. For logic circuits, T_f is simply the duration of the pulse. We inject a pulse such that it reaches the input of the register within the latching window and repeat this procedure such that we sweep the entire latching window, as shown in Figure 1(b). Among these pulses, we choose the pulse which can be injected closest to the hold time and still cause an error. Next we attempt to change the magnitude of this pulse to determine the minimum value of the pulse that can cause a error. The $Q_{critical}$ of the pulse is determined using the formulation provided by Equation 3. In this work, we focus primarily on $Q_{critical}$ in comparing the SER of our designs, since the other parameters, such as charge collection efficiency are quite similar across designs. In [12, 22] we have characterized SER of different SRAM and flip-flop designs using similar procedure. In our study we use two types of designs; memory elements which include 6T-SRAM, asymmetric SRAMs(ASRAM), flip-flops, and logic elements which include 6-inverter chain, 4-FO4 nand chain, 1-bit transmission gate (TG) based adders. All the circuits are custom designed using 70nm Berkeley predictive technology [1] and the netlists are extracted. The netlists are simulated using Hspice. The normal V_t of these devices is 0.22V, and the supply voltage of 1V is used. V_t is changed using *delvto* option of Hspice [3].

Delvto changes the V_t of the transistors by the amount specified. We analyzed all circuits by changing V_t by 0.1V and 0.2V for both PMOS and NMOS.

4. Power and Soft Errors

Power consumption is a major design constraint towards building more powerful chips and systems. Supply voltages have continued to scale down with feature size with every technology generation. The transistor threshold voltage has been scaled down commensurate with the supply voltage scaling in order to maintain a high drive current to avoid performance penalties. However, the threshold voltage scaling results in the substantial increase of the subthreshold leakage current [8]. Hence, as technology scales, the leakage power consumed in circuits increases. On the other hand, even though the operating voltage is reduced the dynamic power increases due to higher operating frequency of the new generation circuits. Subsequently, there have been several efforts spanning from the circuit level to the architectural level at reducing the energy consumption (both dynamic and leakage). Circuit mechanisms include adaptive substrate biasing, dynamic supply scaling, dynamic frequency scaling, and supply gating [29, 25]. Many of the circuit techniques have been exploited at the architectural level to control leakage at the cache bank and cache line granularities. These optimizations influence the susceptibility of the circuits to soft errors. The subsequent sections will present the effect of two of the widely used power reduction techniques namely, reducing supply voltage using clustered voltage designs and increasing the threshold voltage, on soft error rates. In the Section 4.3, we will also address the issue of delay balancing in the context of high V_t devices.

4.1 Impact of supply voltage scaling in clustered voltage designs on soft error rate

Voltage scaling is a very common technique to reduce dynamic and leakage power consumption. Dynamic power of the circuit is proportional to the square of the supply voltage. Hence, supply voltage is decreased to reduce the power consumption of the circuit. To maximize the gains from this technique, it is a common practice to employ clustered voltage design. In clustered voltage design, parts of a circuit operate at a lower voltage. Figure 2, provides a schematic view of the clustered voltage design. Voltage level converters are used to move from one voltage cluster to another [19]. While no level convertering logic is needed to move from a high voltage cluster to a low voltage cluster, a level convertering circuit is need to when we move from a low voltage cluster to a high voltage cluster. Level converters are needed in the second case because in this case low voltage based devices need to drive high voltage based devices. In clustered voltage designs, the error can be generated in either

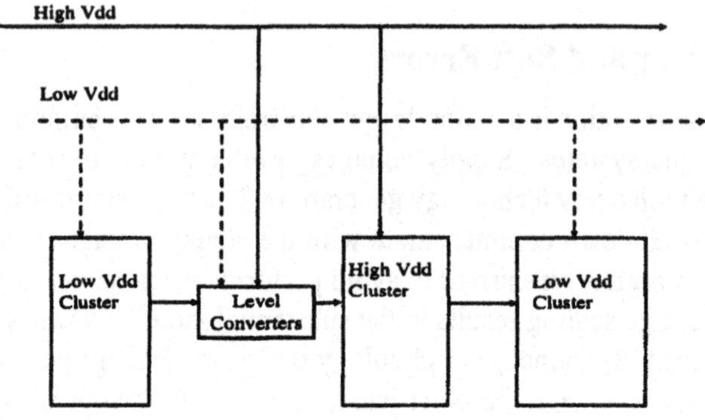

Figure 2. Clustered voltage design

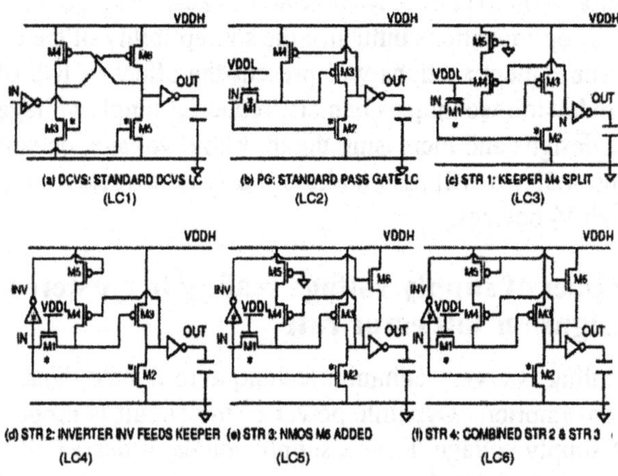

Figure 3. Level converters for clustered voltage design

the clusters or the level converters. Based on Equation 1 we know that SER increases exponential with reduction in $Q_{critical}$. $Q_{critical}$ is proportional to the supply voltage. Hence, the SER is exponentially dependent on supply voltages.

To examine the effect of supply voltage scaling in clustered voltage designs on SER, we analyzed the soft error susceptibly of six level converters. The level converters are shown in Figure 3. The effect of voltage scaling on the soft error susceptibility of six level converters is presented in Figure 4. We find that the $Q_{critical}$ of these level converters is linearly dependent on the supply voltage. Next we analyze the influence of voltage scaling in adders circuits on SER. The adder circuit can be considered as an example of a circuit representing the cluster. We find that, similar to level converters, the adders consume less power at a lower voltage but their $Q_{critical}$ is also reduced. Figure 5, presents the $Q_{critical}$ of a 4-bit Kogge-Stone adder at the output nodes for a transient pulse generated at the carry-in node. In comparison with Brent-Kung adders, Kogge-Stone adders are shown to have lesser $Q_{critical}$ and higher soft error rate [21]. While there are differences due to structural variations, in absolute $Q_{critical}$ values, the $Q_{critical}$ reduces as voltage reduces for both designs.

Supply voltage scaling is also employed to reduce the leakage energy. It is a common practice to reduce the supply voltage of the circuit when the circuit is not active and the overheads do not facilitate turning off the supply. For example in caches, when a cache line is not in use, the supply to the cache line can be reduced while still ensuring that the line retains the values [13]. On examining a custom designed cell, we found that a cell designed to operate at a voltage of 1V, can retain the values when the supply is reduced to 300mV. At 300mV the leakage is reduced by 70% but the $Q_{critical}$ also reduces by 65% [12, 20]. Hence there is a trade off between the power saved and the immunity to soft error. Based on the above results it can be seen that even though voltage scaling reduces the dynamic and static power, there is also a loss of immunity to soft errors.

4.2 Impact of high threshold voltage devices on SER

There are two distinct factors that affect soft error rates due to increase in threshold voltages. First, due to the physical properties of high V_t silicon, higher energy is requires to create electron-hole pairs in the substrate. This effect can potentially reduce SER. Second, higher V_t increases the gain and delay of circuits. This affects attenuation of the transient pulse.

4.2.1 Charge creation under high threshold voltages . This subsection gives a simplified theory of the semiconductors and we use this analysis to explain the phenomenon of charge creation under high V_t. Equation 4 represents the factors on which the threshold voltage depends.

$$V_t = V_{fb} + V_b + V_{ox} \tag{4}$$

where,

V_t is the threshold voltage of the MOS device

V_{fb} is the flat band voltage

V_b is the voltage drop across the depletion region at inversion

V_{ox} stands for potential drop across the gate oxide

When we change the threshold voltage of a device we change the flat band voltage(V_{fb}) of the device. Flat band voltage is the built in voltage offset across the MOS device [11]. It is the workfunction difference θ_{ms} which exists between polygate and silicon. By increasing the threshold voltage, we increase the energy required to push the electrons up the valence band. This is the same reason for which the device slows down. Consequently, when we increase the threshold voltage, the charge creation and collection characteristics change.

4.2.2 Logic attenuation due to high threshold voltage device . The transient pulses attenuate, when they propagate through pass-transistors and transmission gates, due to V_t drop across the devices. However, static CMOS sees different trends. In static CMOS, the gain of the circuit is positive. The gain of an inverter is given by Equation 5

$$GainG = \frac{1+r}{(V_m - V_t - V_{dsat}/2)(\lambda_n - \lambda_p)} \tag{5}$$

where, r is a ratio which compares the relative driving strength of NMOS transistor in comparision with PMOS transistor, V_M is the switching threshold (usually V_m is half of the supply voltage), V_{dsat} is drain saturation current, and λ_n, λ_p are channel length modulation factors for an n-channel and p-channel respectively. We can see that due to higher gain, when V_t is increased, a transient pulse will propagate in a system for a longer time and travels more logic stages.

4.2.3 Effect of V_t on SER of SRAM and Flip-flops. Table 1 presents the $Q_{critical}$ of the SRAM and Asymmetric SRAM cell. From Table 1, we observe that the threshold change does not affect $Q_{critical}$ of the standard 6T SRAM significantly. By increasing V_t by 0.2V, we do not notice any significant change in $Q_{critical}$. Because the threshold voltage of both PMOS and NMOS in the back-to-back inverter configuration was changed, the regenerative property of the circuit ensures that there is no loss of charge and hence relatively no gains in terms of $Q_{critical}$. However, when we analyze an ASRAM [4] optimized for leakage while storing a preferred logic state, we observe a different trend.

	ΔV_t	$Q_{critical}$ /C	*Leakage* /W
ASRAM	0	4.75e-14	2.20e-07
	0.1	6.58e-14	9.10e-09
	0.2	7.58e-14	3.42e-10
SRAM	0	4.75e-14	2.40e-07
	0.1	4.04e-14	9.66e-09
	0.2	3.82e-14	9.46e-10

Table 1. $Q_{critical}$ and leakage power of SRAM and ASRAM with different V_t. Nominal V_t was 0.22V

	ΔV_t	$Q_{critical}$ at input /C	$Q_{critical}$ at most susceptible node /C
SDFF	0	6.06e-21	1.24e-20
	0.1	5.08e-21	1.33e-20
	0.2	-	-
C^2MOS	0	3.69e-20	7.12e-21
	0.1	5.64e-20	7.12e-21
	0.2	1.68e-19	-
TGFF	0	1.99e-20	7.36e-21
	0.1	1.77e-19	7.36e-21
	0.2	3.87e-17	7.36e-21

Table 2. $Q_{critical}$ of different flip-flops. Nominal V_t was 0.22V

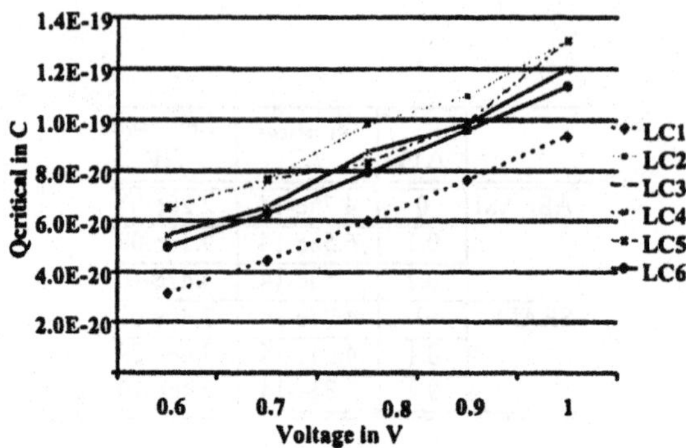

Figure 4. $Q_{critical}$ vs the supply voltage for different level converters

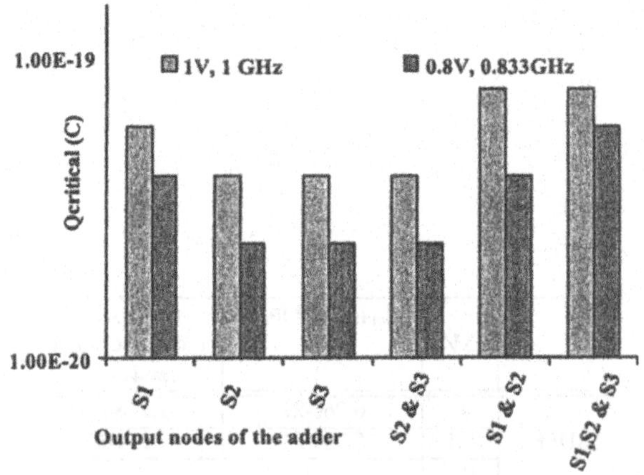

Figure 5. Effect of frequency and voltage scaling on 4-bit Kogge Stone adder

Figure 7 shows a circuit schematic of ASRAM optimized for reducing leakage when storing a 0. In ASRAM, the threshold voltage of transistors in the leaky path of circuit is increased to reduce leakage. For a stored value of 0, the transistors in the leaky path are shown. The V_t of these transistors are increased to reduce the leakage. The $Q_{critical}$ of this SRAM in its preferred state (i.e, when storing a 0) increases significantly, however for the non preferred state it remains the same. Specifically, when V_t is increased by 0.2V, $Q_{critical}$ increases by 59%. This is due to the fact that if we try to charge the node to

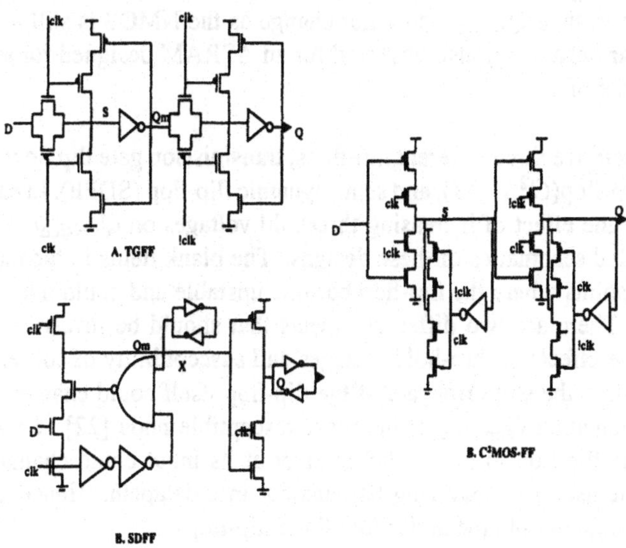

Figure 6. Flip-Flops evaluated for SER

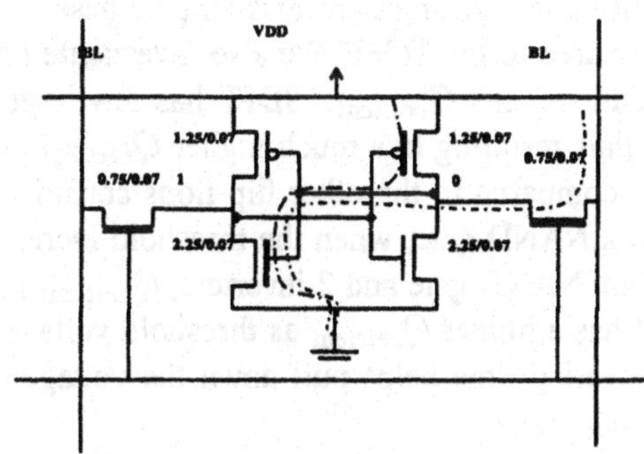

Figure 7. Asymmetric SRAM: Optimized for 0

1, the PMOS due to its high V_t will not be able to provide necessary feedback to quickly change the bit. However, if a value of 1 is stored, and we attempt to discharge it, then $Q_{critical}$ does not change as the NMOS is still at normal V_t. A similar behavior is also observed for an ASRAM designed for storing a preferred state of 1.

We characterize three different flip-flops, transmission gate flip flop(TGFF), C^2MOS flip-flop(C^2MOS), and semi-dynamic flip-flop (SDFF). In each case we estimate the effect of increasing threshold voltages on $Q_{critical}$. Figure 6 shows detailed schematics of these designs. The blank fields in the table represents the points where the flip-flop became unstable and could not latch the input data. There are two different aspects that should be investigated with respect to the effects of threshold voltages and susceptibility of soft errors on flip-flops. First, the soft error rate of the flip-flop itself could change. This is found by evaluating $Q_{critical}$ at the most susceptible node [22]. Second the ability of the flip-flop to latch onto an error at its input could change. This effect will be useful in analyzing its behavior in a datapath. Hence Table 2, lists the $Q_{critical}$ at both the nodes for all the flip-flops.

From Table 2, we can note that, for a TGFF, $Q_{critical}$ at the input node, when V_t is increased, while the $Q_{critical}$ at the node S is same. We ascribe this trend to the presence of to the transmission gate present at the input. On the other hand for the node S, the higher gain of the inverter cancels out the effect of the transmission gate at the slave stage and hence the $Q_{critical}$ remains almost constant. Similar testing was done on a C^2MOS flip-flop which also has master-slave stages similar to that of the transmission gate flip-flop. Since C^2MOS flip-flop does not have any transmission gate based structures it has a lower $Q_{critical}$ compared to the TGFF. We also investigate one of the pulse triggered designs SDFF for it's $Q_{critical}$. SDFF has few large sized devices in its feedback path thus resulting in a much higher $Q_{critical}$ at the most susceptible node (X) as compared to the other flip-flops considered. Since this node feeds back into a NAND gate, when the threshold increases, due to the increase in delay of the NAND gate and 2 inverters, $Q_{critical}$ increases. Thus the flip-flop by itself has a higher $Q_{critical}$ as threshold voltage increases. At the input the larger overlap time helps pull down the voltage at node X and hence reduces the $Q_{critical}$.

4.2.4 Effect of V_t on Combinational Logic . We analyze three kinds of logic circuits: chain of 6-inverters, chain of 4-nand gates and transmission gate based full adders. For all of these circuits we check for an error by latching the transient pulse at the end of the logic chain. A transmission gate flip-flop (TGFF) was used to latch the values. TGFF was chosen as it is one of

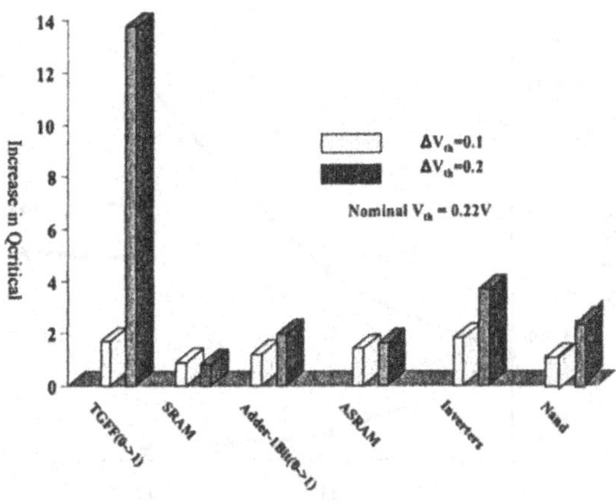

Figure 8. Increase in $Q_{critical}$ of different designs with respect to operating nominal V_t of o.22V

the most commonly used flip-flop. From Table 3, we note that the $Q_{critical}$ of the circuit is increasing for increasing threshold voltages. For TG based adders, the threshold drop across transmission gates accounts for the increase in $Q_{critical}$ as the V_t increases. However, for static logic this is counter intuitive. Based on the pulse propagation characteristics, the $Q_{critical}$ of the circuits should be lower. This can be accounted for the robustness of flip-flops. In Figure 8, we find that the $Q_{critical}$ increase for the flip-flop is many orders of magnitude higher than the others. To confirm our observations we simu-

	ΔV_t	$Q_{critical}$ /C	*Leakage* /W
Nand	0	1.31e-20	22.56e-07
	0.1	2.26e-20	9.92e-09
	0.2	2.83e-20	4.90e-10
Inverters	0	1.28e-20	2.20e-07
	0.1	2.3e-20	4.90e-10
	0.2	4.73e-20	41.99e-11
TG Adder	0	4.60e-20	1.18e-07
	0.1	1.35e-19	3.42e-08
	0.2	5.87e-17	3.40e-08

Table 3. $Q_{critical}$ and leakage power of various designs with different V_t. A high V_t TGFF was used at the output of the logic chain

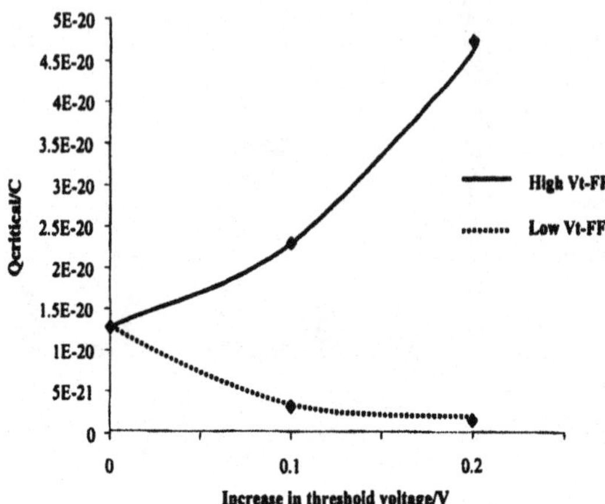

Figure 9. $Q_{critical}$ of typical inverter chain with either high V_t or low V_t flip-flop at the output of the chain. Nominal V_t was 0.22V

lated the 6-inverter chain again, but this time we used normal-V_t flip-flops at the output of the logic chain and we found that as the V_t increased, the $Q_{critical}$ values decreased. The results are presented in Figure 9. In the next section we show how this fact can be leveraged to reduce power and increase robustness of the circuit.

4.3 Effect of delay balance using high V_t devices on soft errors

Figure 10 shows a typical pipeline. The logic between pipeline stages is distributed across slow and fast paths, with the slowest path determining the clock frequency. Thus, slow paths become critical paths and fast paths become non-critical paths. It is an accepted practice to use high V_t devices on non-critical paths. Since these are not delay sensitive, we achieve high leakage power savings with minimal performance penalty. This is some times referred to as *delay balancing*. To examine the effect of delay balancing on $Q_{critical}$, we simulate two circuits, one with a 6-inverter chain which forms the critical path and the other with a 3-inverter chain. Figure 11, shows the $Q_{critical}$ of the 6-inverter chain as compared to the $Q_{critical}$ of the 3-inverter chain with both low and high V_t TGFFs. If we perform delay balancing on this logic with low V_t TGFF, and high V_t 3-inverter chain, we can observe the $Q_{critical}$ of 3-inverter

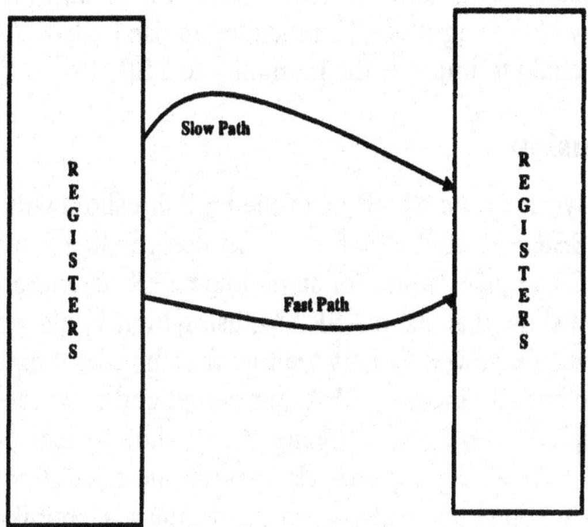

Figure 10. Delay Balancing

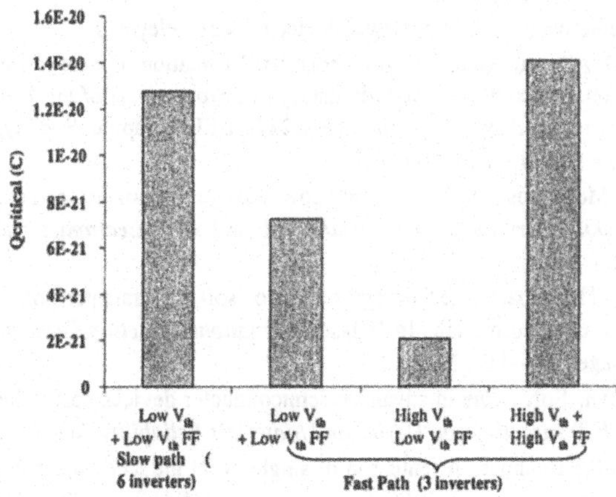

Figure 11. Effect of delay balancing

chain reduces. Thus, we see that this path now becomes more vulnerable to soft errors. Consequently, if a high-V_t flip-flop is used for latching, the $Q_{critical}$ of the 3-inverter chain (relative to the 6-inverter chain) is still high. So, while performing delay balancing it is recommended to use high V_t flip-flops at the end of the logic chain to improve the immunity to SER.

5. Conclusion

In this work, we examine the effect of the high threshold voltages and voltage scaling on SER. We find that for certain designs like transmission gate based designs SER reduces while for static logic SER deteriorates as V_t is increased. Also we show that, as in ASRAM, using high V_t cleverly can reduce both SER and leakage power. Finally we find that the use of high V_t for delay balancing can potentially increase SER, but the reliability can be bought back by the use of high V_t flip-flops. In general, we showed that use of high V_t devices not only reduces leakage but also affects the reliability of circuit. In contrast, voltage scaling almost always increases the susceptibility to SER.

Acknowledgments

This work was supported in part by GSRC and NSF Grants CAREER 0093085 and 0103583.

References

[1] Berkeley predictive model. http://www-device.eecs.berkeley.edu.

[2] L. Anghel, D. Alexandrescu, and M. Nicolaidis. Evaluation of a soft error tolerance technique based on time and/or space redundancy. In *Proceedings of the 13th symposium on Integrated circuits and systems design*, page 237. IEEE Computer Society, 2000.

[3] Avant! *Hspice User Manual*, 2003 edition.

[4] N. Azizi, A. Moshovos, and F. N. Najm. Low-leakage asymmetric-cell sram. In *Proceedings of the 2002 International Symposium on Low Power Electronics and Design*, pages 48–51, 2002.

[5] R. Baumann. The impact of technology scaling on soft error rate performance and limits to the efficacy of error correction. In *Digest. International Electron Devices Meeting, 2002. IEDM '02*, pages 329–332, 2002.

[6] R. C. Baumann. Soft errors in advanced semiconductor devices-part i: the three radiation sources. *IEEE Transactions on Device and Materials Reliability*, 1(1):17–22, 2001.

[7] M. Baze and S.P.Buchner. Attenuation of single event induced pulses in cmos combinational logic. *Nuclear Science, IEEE Transactions on*, 44(1):2217–2223, December 1997.

[8] S. Borkar. Design challenges of technology scaling. *IEEE Micro*, 19(4):23–29, 1999.

[9] S. Borkar, T. Karnik, and V. De. Design and reliability challenges in nanometer technologies. In *Proceedings of the 41st annual conference on Design automation*, pages 75–75, 2004.

[10] S. Buchner, M. Baze, D. Brown, D. McMorrow, and J. Melinger. Comparison of error rates in combinational and sequential logic. *Nuclear Science, IEEE Transactions on*, 44(1):2209–2216, December 1997.

[11] J. Y. Chen. *CMOS Devices and Technology for VLSI*. Prentice-Hall, Englewood Cliffs, NJ, 1990.

[12] V. Degalahal, N. Vijaykrishnan, and M. J. Irwin. Analyzing soft errors in leakage optimized sram designs. In *6th International Conference on VLSI Design*, Jan. 2003.

[13] K. Flautner, N. S. Kim, S. Martin, D. Blaauw, and T. Mudge. Drowsy caches: simple techniques for reducing leakage power. In *Proceedings of the 29th annual international symposium on Computer architecture (ISCA-29)*, pages 148–157, 2002.

[14] S. Hareland, J. Maiz, M. Alavi, K. Mistry, S. Walsta, and C. Dai. Impact of cmos process scaling and soi on the soft error rates of logic processes. In *Digest of Technical Papers. 2001 Symposium on VLSI Technology*, pages 73–74, 2001.

[15] P. Hazucha and C. Svensson. Impact of cmos technology scaling on the atmospheric neutron soft error rate. *IEEE Transactions on Nuclear Science*, 47(6), 2000.

[16] K. Johansson, P. Dyreklev, O. Granbom, M. C. Calver, S. Fourtine, and . Feuillatre. Inflight and ground testing of single event upset sensitivity in static rams. *Nuclear Science, IEEE Transactions on*, 45(3):1628–1632, June 1998.

[17] T. Karnik, B. Bloechel, K. Soumyanath, V. De, and S. Borkar. Scaling trends of cosmic ray induced soft errors in static latches beyond 0.18u. In *Digest of Technical Papers. 2001 Symposium on VLSI Circuits*, pages 61–62, 2001.

[18] T. Karnik, S. Vangal, V. Veeramachaneni, P. Hazucha, V. Erraguntla, and S. Borkar. Selective node engineering for chip-level soft error rate improvement [in cmos]. In *VLSI Circuits Digest of Technical Papers, 2002. Symposium on*, pages 204–205, 2002.

[19] S. H. Kulkarni and D. Sylvester. High performance level conversion for dual v/sub dd/ design. *Very Large Scale Integration (VLSI) Systems, IEEE Transactions on*, 12(9):926–936, September 2004.

[20] L. Li, V. Degalahal, N. Vijaykrishnan, M. Kandemir, and M. J. Irwin. Soft error and energy consumption interactions: a data cache perspective. In *Proceedings of the 2004 international symposium on Low power electronics and design*, pages 132–137, 2004.

[21] R. Rajaraman, N. Vijaykrishnan, Y. Xie, M. J. Irwin, and K. Bernstein. Soft errors in adder circuits. In *MAPLD*, 2004.

[22] R. Ramanarayanan, V. Degalahal, N. Vijaykrishnan, M. J. Irwin, and D. Duarte. Analysis of soft error rate in flip-flops and scannable latches. In *SOC Conference, 2003. Proceedings. IEEE International [Systems-on-Chip]*, pages 231 – 234, September 2003.

[23] J. Ray, J. C. Hoe, and B. Falsafi. Dual use of superscalar datapath for transient-fault detection and recovery. In *Proceedings of the 34th annual ACM/IEEE international symposium on Microarchitecture*, pages 214–224. IEEE Computer Society, 2001.

[24] E. Rotenberg. Ar-smt: A microarchitectural approach to fault tolerance in microprocessors. In *Proceedings of the Twenty-Ninth Annual International Symposium on Fault-Tolerant Computing*, page 84. IEEE Computer Society, 1999.

[25] K. Roy, S. Mukhopadhyay, and H. Mahmoodi-Meimand. Leakage current mechanisms and leakage reduction techniques in deep-submicrometer cmos circuits. *Proceedings of the IEEE*, 91(2):305–327, Feb 2003.

[26] N. Seifert, D. Moyer, N. Leland, and R. Hokinson. Historical trend in alpha-particle induced soft error rates of the $alpha^{TM}$ microprocessor. In *39th Annual IEEE International Reliability Physics Symposium*, pages 259–265, 2001.

[27] P. Shivakumar, M. Kistler, S. W. Keckler, D. Burger, and L. Alvisi. Modeling the effect of technology trends on the soft error rate of combinational logic. In *Proceedings of the 2002 International Conference on Dependable Systems and Networks*, pages 389–398, 2002.

[28] J. Wallmark and S. Marcus. Minimum size and maximum packaging density of non-redundant semiconductor devices. *In Proc. IRE*, 50:286–298, 1962.

[29] L. Wei, K. Roy, and V. K. De. Low voltage low power cmos design techniques for deep submicron ics. In *VLSI Design, 2000. Thirteenth International Conference on*, pages 24–29, 2000.

[30] J. Ziegler. Terrestrial cosmic ray intensities. *IBM Journal of Research and Development*, 40(1):19–39, 1996.

DYNAMIC MODELS FOR SUBSTRATE COUPLING IN MIXED-MODE SYSTEMS

João M. S. Silva
jmss@algos.inesc-id.pt
INESC ID Lisboa
Instituto Superior Técnico
Technical University of Lisbon
R. Alves Redol, 9, Sala 133, 1000-029 Lisboa, Portugal

Luís Miguel Silveira
lms@inesc-id.pt
Cadence Laboratories
INESC ID Lisboa
Instituto Superior Técnico
Technical University of Lisbon
R. Alves Redol, 9, Sala 136, 1000-029 Lisboa, Portugal

Abstract

In modern monolithic integrated circuits, substrate coupling is a major concern in mixed-mode systems design. Noise injected into the common substrate by fast switching digital blocks may affect the correct functioning or performance of the overall system. Verification of such systems implies the availability of accurate and simulation-efficient substrate coupling models. For frequencies up to a few gigahertz pure resistive models are considered sufficient, but increasing frequencies of operation imply that capacitive coupling analysis also becomes mandatory.

In this paper, we motivate the use of dynamic resistive-capacitive (RC) models of substrate coupling as a natural extension to the standard purely resistive models. We propose an extraction methodology that starts from information about the process parameters and the contact layout of the circuit, and leads to a contact-to-contact RC element model. The underlying algorithm is based upon a Finite Difference discretization of the substrate, leading to a large tridimensional mesh which is solved by means of a fast Multigrid algorithm.

The proposed model is trivially incorporated into circuit simulation tools. Comparisons are also made to a model obtained using standard model order reduction algorithms and it is shown to be of similar accuracy. The formulation proposed can accurately model substrate coupling effects for frequencies up to several tens of gigahertz.

Please use the following format when citing this chapter:

Silva, João, M. S., Silveria Luis, M., 2006, in IFIP International Federation for Information Processing, Volume 200, VLSI-SOC: From Systems to Chips, eds. Glesner, M., Reis, R., Indrusiak, L., Mooney, V., Eveking, H., (Boston: Springer), pp. 21-37.

1. Introduction

Substrate behavior in integrated circuits has long ceased to be considered as a perfect insulator [Joardar, 1994; Kup et al., 1991; Gharpurey, 1995]. As MOS process' transistor channel widths decrease to the size of a few nanometers, digital clock frequencies have been steadily increasing, so that current injection into the polysilicon substrate becomes a great concern. Along with technology miniaturization, die area has shrunk on behalf of package count and production yield purposes. Consequently, different cells and blocks are built closer to each other, in a way that facilitates injected substrate currents to migrate among the substrate layers and reach arbitrarily distant parts of the circuit [Gharpurey, 1995; Su et al., 1993; Verghese, 1995].

Current injection into the substrate can occur through active and channel areas, as well as through substrate and well contact ties. Such currents can cause substrate voltage fluctuations leading to changes in the device's bulk-to-substrate voltage. For purely digital circuits, this is still not a major concern since, from a functional perspective, digital logic is somewhat immune to substrate voltage fluctuations. However, performance degradation can still occur as millions of logic gates switching induces significant additional noise and can cause power supply voltage levels to fluctuate. This can affect logic gates delay and circuit overall time performance.

It is however in the context of mixed-signal design that the issue of substrate coupling has received the most attention in recent years. Industry trends aimed at integrating higher levels of circuit functionality, resulting from an emphasis on compactness in consumer electronic products, and a widespread growth and interest in wireless communications, have triggered a proliferation of mixed analog-digital systems. The design of such systems is an increasingly difficult task owing to the various coupling problems that result from the combined requirements for high-speed digital and high-precision analog components.

Analog circuitry relies on accurate levels of currents and voltages, so that analog transistors are correctly biased and projected performance is met. When substrate injected currents migrate through the substrate, substrate voltages fluctuate, causing havoc in sensitive analog transistors and possibly leading to malfunctioning circuitry [Gharpurey, 1995; Su et al., 1993; Johnson et al., 1984; Nauta and Hoogzaad, 1997].

Analyzing the effects of substrate coupling requires that a model of such couplings is obtained and used in a verification framework. Typically such a verification is done at the electrical level by means of a circuit simulator. Therefore, the usual strategy is to generate an electric coupling model and feed it to a circuit simulator together with the remaining circuitry. Since potentially, everything couples to everything else through the common substrate, leading to extremely large models, special care must be taken to make sure that the model is accurate but will not unnecessarily slow down the verification step. This has traditionally been achieved in a variety of ways all aiming at a simplified model.

When generating the model, a common simplifying assumption is to consider that the major coupling mechanism is due to the finite resistivity of the substrate and derive a resistive model. Such an approximation is valid when the dielectric relaxation time of the layers composing the substrate translates into an insignificant susceptance at the frequencies of interest. Thus, such an approximation becomes questionable beyond a few gigahertz, specially since harmonics of significant amplitude, generated by circuit nonlinearities, may fall in the range of frequencies where reactive effects are of importance.

In this paper, a methodology is proposed for generating dynamic resistive-capacitive (RC) models of substrate coupling. The methodology proposed for model extraction is detailed and the model is analyzed in terms of its validity and accuracy. In section 2, the mechanisms for substrate coupling are briefly discussed and background work in this area is reviewed. In Section 3 the proposed model extraction algorithm is presented and its extension to dynamic analysis is detailed. In Section 4 the validity, accuracy and relevance of the obtained model is discussed through some example simulations. The model obtained is also compared to those generated using standard model order reduction techniques and it is shown to be of similar accuracy. Finally in Section 5 some conclusions are drawn.

2. Background

Substrate Coupling Mechanisms

Coupling through the substrate occurs, mainly, due to substrate finite resistivity. Devices built into the same substrate are consequently not perfectly isolated from each other. Considering a typical substrate profile like the one shown in Figure 1, MOS transistors are based on channel formation, so substrate resistivity is not desired to be infinite. However, when a transistor is on, while current flows through the corresponding channel, part of it is injected into the substrate and is free to migrate to arbitrarily distant substrate zones. Another equally important type of noise injection into the substrate occurs through VDD and GND contact ties. When devices switch, currents are drawn from or injected into the power supply which typically also bias the substrate. In this manner, current is also directly injected into the substrate.

At higher frequencies, when active areas are charged and discharged, source-bulk and drain-bulk parasitic capacitances show a lower impedance and current is also directly injected into the substrate by these active areas.

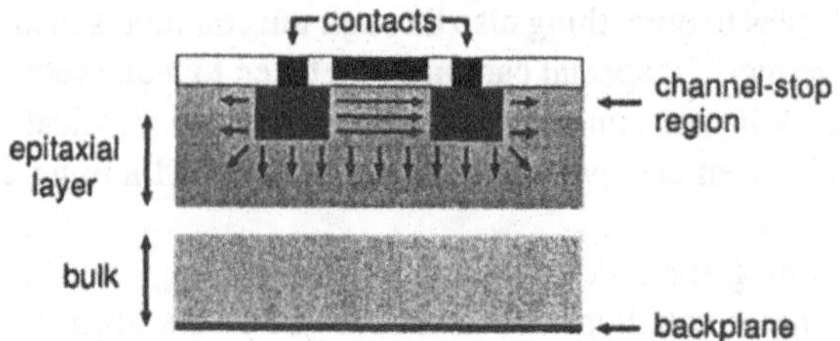

Figure 1. Depiction of typical substrate profile.

The fact that such currents are in a sense free to roam around the substrate and may be captured wherever appropriate conditions are met, makes the verification process much harder. While it is true that most of the coupling may occur locally, designer experience and good design practices lead to designs where such local couplings are explicitly minimized. As a consequence, the assumption of mostly local coupling is not necessarily valid and unexpected long range couplings may appear where least expected. As such, not only is it mandatory that some kind of substrate model be used to account for substrate couplings between different blocks built on the same substrate, but that model must also account for all or at least large portions of the substrate. Substrate coupling mechanism and corresponding electric model examples are shown in Figure 2.

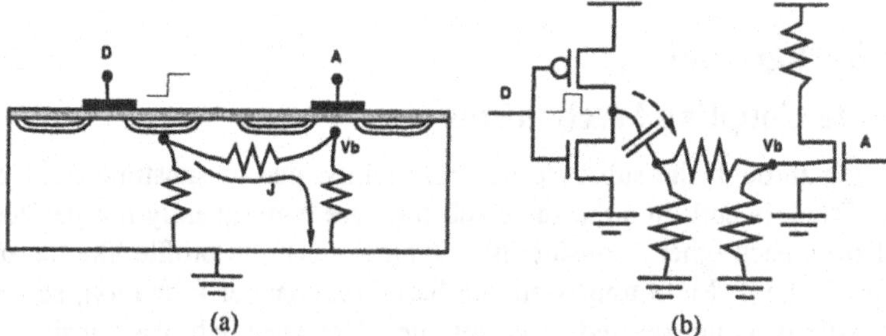

Figure 2. Example of resistive coupling: current injection mechanism into the substrate (a) and corresponding electric model (b).

Previous Work

Previous work in the area of substrate model extraction is profuse. The general trend, however, has been toward the generation of resistive coupling models. Capacitive coupling through the substrate has been generally considered neglectable based on the properties of current technologies which present a silicon relaxation time in the order of picoseconds. This fact is said to typically limit the validity of pure resistive models to be up to the order of gigahertz. As today's frequency of operation increases, the gigahertz frontier has clearly been surpassed, so that mixed systems with aggressive fast digital components may require more accurate modeling.

Several extraction methodologies were studied in the past and, based on them, several extracting tools were developed. The simplest modeling methodologies consist on finding coupling elements based on heuristic rules. Such methods are very attractive since the extraction overhead is minimal and they lead to simple first order models which also have low simulation costs [Su et al., 1993; Johnson et al., 1984; Nauta and Hoogzaad, 1997; van Genderen et al., 1996; Mitra et al., 1995]. These models are, however, generally very imprecise. Furthermore, heuristic models are only really useful to the designer, for they are unable to account for higher order effects and, in fact, rely on designer's experience to prune out the expected relevant couplings [Phillips and Silveira, 2001]. Moreover, once that is accomplished they do not provide any form of verification as to whether the performed approximation enables correct circuit simulation. With the geometric complexity and dimensions of actual circuits, heuristic models are unable to predict the real functioning of the full system substrate interactions and such models are somewhat imprecise and unreliable.

On the other hand, methodologies that avoid a-priori heuristic pruning and work at the electrical level directly are typically based on a full description of the media and all the possible couplings. A problem that arises from model extraction in those cases is the extraction time and the size of the final model. Coupling can occur from any substrate contact to any other, so that a full interaction matrix can be drawn from it.

Several methods have been proposed to generate such a model.One of these families of methods are Boundary Element Methods (BEM). In BEM methods, only the surface of the substrate contacts is discretized which leads to a system of equations that corresponds to small but full matrices. Extraction of such models requires intensive computations which restrains the range of applicability of this method to small and medium sized circuits [Smedes et al., 1995; Gharpurey and Meyer, 1995; Verghese and Allstot, 1995]. Fortunately, significant progress in BEM methods performance has been lately achieved [Costa et al., 1998; Chou and White, 1998; Kanapka et al., 2000].

A different but also efficient family of extraction methods are Finite Difference (FD) or Finite Element Methods (FEM). In these methods, the whole tridimensional volume of the substrate is discretized leading to large but sparse matrices. The sparsity pattern of these matrices can be taken advantage of depending on the system solving methods which are used. Notwithstanding, FEM still face difficulties, mostly related to memory resources, due to the large matrices

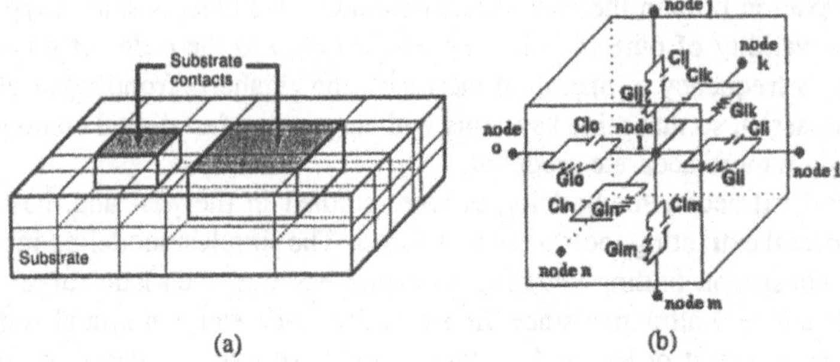

Figure 3. (a) FD method discretization and (b) resulting substrate resistive-capacitive (RC) model.

required. This type of methods have also been recently enhanced with fast solution techniques [Clement et al., 1994; Wemple and Yang, 1995; Stanisic et al., 1994; Kanapka et al., 2000; Silveira and Vargas, 2002; Silva, 2003].

As mentioned previously, RC models of substrate coupling are less common that purely resistive models. Notwithstanding, previous work has been published in this field that considers the needs for capacitive effects in substrate coupling [Clement et al., 1994; Gharpurey and Hosur, 1997]. RC models are also partially used in some commercial tools, typically in an heuristic way, but there is no systematic assessment of their relevance. In this work, a Finite Difference method for the extraction of resistive-capacitive (RC) substrate models is proposed and its usefulness and validity are assessed.

3. Substrate Model Extraction

In the following, our method for the extraction of a model of the couplings through the substrate is presented.

Finite Difference Tridimensional Model

We use a FD method accounting for geometric discretization accuracy. This implies a discretization of the substrate volume into a large number of small cuboid elements. An example of such a discretization is shown in Figure 3-a) where nodes on a tridimensional mesh are immediately visible. This method provides infinite accuracy when discretization spacing tends to zero.

The next step is to model the whole substrate by applying electromagnetic laws to each cuboid element (which we will call node) of the mesh. Starting from Maxwell's equations and neglecting the effect of magnetic fields, we use the identity $\nabla(\nabla \times a) = 0$ and Ampère's Law to write:

$$\nabla J + \nabla \frac{\partial D}{\partial t} = 0 \tag{1}$$

where J is the current density and D is the electric displacement. Equation (1) is the continuity equation and expresses the conservation of electric charge. Recalling that:

$$\begin{cases} J = \sigma E \\ D = \varepsilon E \end{cases} \tag{2}$$

where σ and ε are the conductivity and the permitivity of the medium, respectively, and E the electric field, Equation (1) can also be written as:

$$\sigma \nabla E + \varepsilon \frac{\partial \nabla E}{\partial t} = 0. \tag{3}$$

Assuming an homogeneous medium in each substrate layer, consider a cuboid whose center is node i with neighbor cuboid whose center is j. If E_{ij} denotes the electrical field normal to the cuboid side surface between nodes i and j, the Finite Difference approximation leads to:

$$E_{ij} \approx \frac{V_i - V_j}{l_{ij}} \tag{4}$$

where l_{ij} is the distance between adjacent nodes i and j, and V_i and V_j the scalar potentials at those nodes. At this point, a simple box integration technique can be used to solve Equation (3) since the substrate is spatially discretized. Applying Gauss' law:

$$\nabla E = \frac{r}{\varepsilon} \tag{5}$$

where r is the charge density of the medium (time dependence omitted but implicit). From the divergence theorem, we know that:

$$\int_{S_i} E \, dS_i = \int_{\mathcal{V}_i} \nabla E d\mathcal{V}_i = \int_{\mathcal{V}_i} \frac{r}{\varepsilon} \, d\mathcal{V}_i \tag{6}$$

where $\mathcal{V}_i$ is the volume of the i-th cuboid and S_i its surface. The left hand side of Equation (6) can be approximated as:

$$\int_{S_i} E \, dS_i \approx \sum_j E_{ij} S_{ij} = \frac{r}{\varepsilon} \, \mathcal{V}_i \tag{7}$$

where the summation is performed on cuboids that are neighbors of cuboid i, and S_{ij} is the common surface between cuboids i and j. Now using Equation (7) in (5) leads to:

$$\nabla E = \frac{1}{\mathcal{V}_i} \sum_j E_{ij} S_{ij}. \tag{8}$$

The above derivation assumed a common resistivity, i.e. a single layer. The extension to multiple layers is trivially handled by assuring that the layer interface is delimited with mesh nodes.

Finally, replacing Equations (4) and (8) into (3) results in:

$$\sum_j \left[G_{ij}(V_i - V_j) + C_{ij} \left(\frac{\partial V_i}{\partial t} - \frac{\partial V_j}{\partial t} \right) \right] = 0 \tag{9}$$

where:

$$\begin{cases} G_{ij} = \sigma \frac{S_{ij}}{l_{ij}} \\ C_{ij} = \varepsilon \frac{S_{ij}}{l_{ij}} \end{cases} . \tag{10}$$

Equation (9) can readily be interpreted in terms of the electrical model depicted in Figure 3-b). In fact, applying Nodal Analysis (NA) to the 3D mesh model (9) leads to the following system of equations:

$$(sC + G)V = I \tag{11}$$

where C and G are, respectively, the capacitance and conductance matrices of the system, V is the voltage on all nodes of the discretization mesh and I the corresponding injected currents. From (9), entries of G and C in (11) can be approximated with Equation (10) here applied to each element in the model.

The size of the model in (11) is directly determined by the chosen discretization. For very fine discretizations, required for accuracy considerations, this implies that the model in (11) could be very large indeed. For a discretization of dx, dy, dz in each direction, the matrix size will be $N = dx \times dy \times dz$. On the other hand, one should note that the model is very sparse, since matrices C and G correspond to the 3D discretization pencil and only have at most 7 non-zero entries in each row or column (corresponding, for each node, to the diagonal entry and to entries for each of the 6 neighbor nodes in a 3D mesh).

Circuit-level Model Extraction

Using the 3D mesh model from (9) in any electrical simulator is prohibitive. As such, a reduced model must be sought. A possible solution to this problem

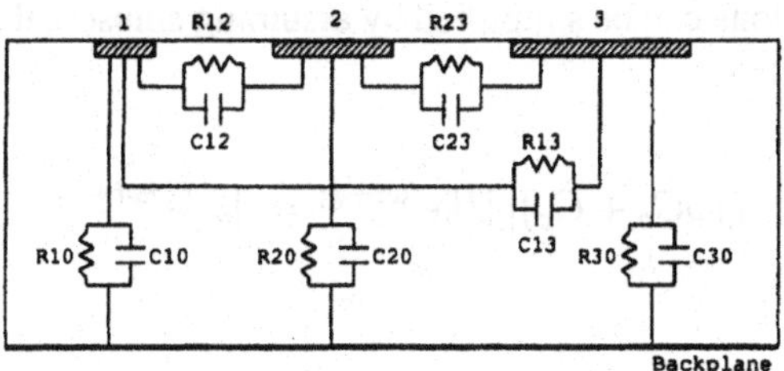

Figure 4. RC model for a three contact configuration.

is to apply standard model order reduction techniques to the problem and obtain a reduced model [Feldmann and Freund, 1995; Silveira et al., 1995; Odabasioglu et al., 1998]. For such methods, the size of the resulting model is directly proportional to the product of the approximation order by the number of ports (inputs/outputs or contacts in our case). This causes two potential problems. First, an appropriate reduction order must be devised. Second, for systems with large numbers of contacts, small increases in the approximation order lead to large increases in model size and potentially to overly large models. We will come back to this discussion in the next section, but for the time being we propose a constructive methodology and seek to obtain a simple model whose size is uniquely determined by the number of substrate contacts and thus independent from the chosen discretization or any other parameter. We will illustrate this methodology by means of a simple example such as the one depicted in Figure 4 for a three-contact setup. Furthermore, we note this model is an obvious extension of the typical resistive models whereby a coupling resistance is computed between pairs of contacts. Here that resistance is replaced with the parallel impedance of a resistor and a capacitor.

Assuming a generic model similar to that of Figure 4, and using NA, the corresponding system of equations is given by:

$$Y_c(s)U = (sC_c + G_c)U = J \tag{12}$$

where Y_c is the admittance of the contact's system, C_c and G_c are, respectively, the capacitive and resistive coupling elements between contacts, and U and J are the vectors of contact voltages and injected currents. This system is naturally analog to (11) but much smaller.

The substrate model in (12) can readily be obtained from the 3D model in (11) by means of simple computations. An algorithm to perform this task is presented in Algorithm 1. Again, not surprisingly, this is the obvious extension of the standard FD procedure used nowadays to obtain resistive substrate models [Clement et al., 1994; Silveira and Vargas, 2002; Silva, 2003].

The computations can be simplified by assuming sinusoidal steady state and rewriting (12) as:

$$(j\omega C_c + G_c)\|U\|e^{j\omega t+\phi_1} = \|J\|e^{j\omega t+\phi_2} \qquad (13)$$

leading to:

$$(j\omega C_c + G_c)\|U\|e^{\phi_1} = \|J\|e^{\phi_2} \Leftrightarrow (j\omega C_c + G_c)\overline{U} = \overline{J}. \qquad (14)$$

If the imposed voltage in contact k is $U_k = 1\ sin(j\omega t + 0)$ then only the k-th component of $\overline{U}$ in (14) is at 1 volt and thus $\overline{J}$ is equal to the k-th column of $G_c + j\omega C_c$.

This process can be repeated as many times as the number of contacts so that the full admittance matrix $Y_c = G_c + j\omega C_c$ is formed, one column at a time. This may, in general, require complex numbers arithmetic. The cost of computing the contact model, $Y_c(s)$, for a system of m contacts is thus equal to m times the cost of solving the 3D mesh to determine the node voltages. This can be performed very efficiently by means of a fast Multigrid algorithm with a cost of $\mathcal{O}(N)$ per solve [Silveira and Vargas, 2002; Silva, 2003], albeit using complex arithmetic.

Algorithm 1 Admittance model extraction algorithm.

For contact k = 1 ... number of contacts:

1 Put contact k nodes at a pre-determined voltage (e.g. 1 V);

2 Using Norton's equivalent, obtain the currents injected into adjacent nodes;

3 Solve the 3D system (11) obtaining all nodes voltages, V;

4 From V and 3D model admittances use Ohm's law to compute current injected in all contact nodes;

5 Use Gauss' law and sum injected node currents to obtain contact injected currents J;

6 By Equation (12) and as only nodes on contact k had a fixed voltage, k-th column of $G_c + C_c$ equals J.

Table 1. Resistance and capacitance values for dynamic model for a three contact example.

Contact 1	Contact 2	Resistance	Capacitance
1	backplane	23 kΩ	687 aF
1	2	243 kΩ	65.0 aF
1	3	1.6 MΩ	9.91 aF
2	backplane	16.3 kΩ	970 aF
2	3	113 kΩ	139 aF
3	backplane	11.4 kΩ	1.38 fF

4. Validity Spectrum of Dynamic Models

In this section the significance of RC models is shown and its limitations evaluated.

RC Model Significance

In order to evaluate the importance and need to account for capacitive coupling through the substrate, the 3D model of Figure 3-b) is taken as example. In particular, looking at any branch in the 3D model and assuming that the capacitive part becomes relevant when the susceptance reaches 10% of the conductance, then:

$$\omega C_{ij} \geq 0.1 G_{ij} \Leftrightarrow \omega \varepsilon \frac{S_{ij}}{l_{ij}} \geq 0.1 \sigma \frac{S_{ij}}{l_{ij}} \Leftrightarrow \omega \geq 0.1 \frac{\sigma}{\epsilon}. \tag{15}$$

Applying this result to a technology of a single layer substrate with a resistivity $\rho = \sigma^{-1} = 15\ \Omega\, cm$ and $\varepsilon_r = \frac{\varepsilon}{\varepsilon_0} = 11.9$, the previous equation leads to $\omega = 6.33\ Grad/s$ which corresponds to approximately 1 gigahertz. This confirms the usual assumption about the validity of resistive models for frequencies up to a few gigahertz, depending on the technology.

Table 1 lists values extracted using the method proposed for a simple configuration such as shown in Figure 4. Using, for instance, computed values of R_{10} and C_{10}, and assuming the same error factor of 10%, leads to:

$$\omega C_{10} \geq 0.1\, G_{10} \Leftrightarrow \omega \geq 6.33\, Grad/s \Leftrightarrow f \approx 1\, GHz. \tag{16}$$

This result, at the contact-level, using extracted data, is compatible with the result at the 3D model mesh level, as expected. It serves as additional validation for the extracted model values. Clearly for frequencies upwards of a few gigahertz, a purely resistive model will be inaccurate as it will not take into account the increase in admittance due to the susceptance term.

RC Model Accuracy

As seen in the previous section, for frequencies greater than a few gigahertz, it becomes necessary to use dynamic coupling models. The model proposed in this paper attempts to fulfill that need and it is necessary to verify its accuracy and limitations.

To simplify the description, and without loss of generality, consider the three contact system from Figure 4 for which we want to compute the admittance description using the method described previously. The input of this system is a vector with the voltages imposed at the contacts, $[U_1, U_2, U_3]$. In the extraction methodology proposed, after discretization, a system such as (11) is obtained. Setting a contact's voltage to some value is equivalent to setting the voltages of all nodes in the mesh that fall within the contact to that value. In our case this can be written as $V = M[U_1, U_2, U_3]^T$, with $M \in \Re^{n\times 3}$ an appropriate contact incidence matrix. As nodal analysis is used, the inputs to (11) should be currents, applied to nodes adjacent to the contact nodes. The values of such currents can easily be obtained from the corresponding Norton equivalent circuits seen by those nodes. The complete transformation can be written as:

$$I = Y_{adj} \begin{bmatrix} U_1 \\ U_2 \\ U_3 \end{bmatrix} \tag{17}$$

where I is the vector of injected currents on all nodes of the 3D mesh and $Y_{adj} \in \mathbb{C}^{n\times 3}$ is a matrix, combining the incidence matrix M mentioned above and the Norton equivalent admittances seen by the nodes in the mesh. Clearly, most of the entries in Y_{adj} are zero, with the exception of lines related to the nodes adjacent to the contacts.

On the other hand, the output of the system is given by the current on the destination contact, $[J_1, J_2, J_3]$. Combining (11) with (17), it is easy to see that these can be obtained as:

$$\begin{bmatrix} J_1 \\ J_2 \\ J_3 \end{bmatrix} = Y_{adj}^T V = \underbrace{Y_{adj}^T (G + sC)^{-1} Y_{adj}}_{Y_c(s)} \begin{bmatrix} U_1 \\ U_2 \\ U_3 \end{bmatrix} \tag{18}$$

which exposes the admittance of our simplified three contact system. Obviously this derivation extends trivially to the generic m contacts case.

Several experiments have been elaborated using typical typical substrate profiles, like the ones presented in Figure 5. Properties of the system (18), like pole and zero location, pole residues, Bode plots, etc., were studied. We have come to the conclusion that in single layer isotropic substrates the system behaves approximately like having a single admittance zero. This is due

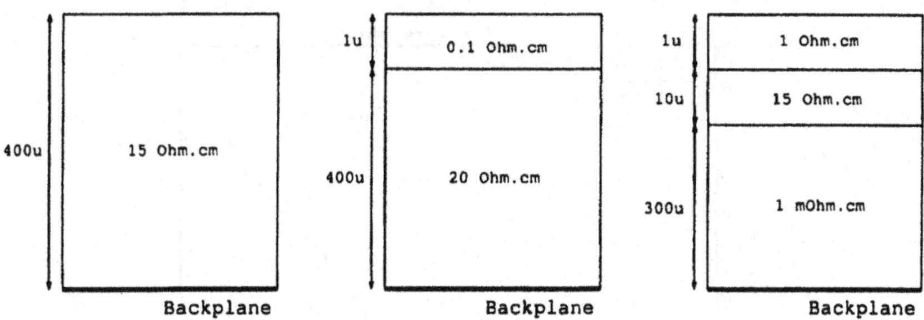

Figure 5. Typical substrate profiles.

to the 3D system having all poles and zeros clustered around a specific frequency, corresponding to the single intrinsic time-constant of the system, given by σ/ε. In multiple layer substrates, each layer possesses a different intrinsic time-constant. However, it turns out that a very similar behavior still occurs. For higher frequencies, more dynamic features are exhibited, but for lower frequencies one can see the effect of a dominant admittance "corner" frequency which is now determined by the properties of the top layers where the contacts are contained.

When the frequency ω of the least conductive of those layers, layer k, is such that $\omega > \sigma_k/\epsilon$, its intrinsic admittance starts to increase, turning into a very low impedance path between contacts, and eventually dominating the overall admittance.

This does not mean system's admittance immediately increases, for the conductance of that particular layer might still be smaller than that of other layers from where contact currents can flow, but eventually it starts to dominate as the path impedance decreases.

Since we now know that there is a dominant pole/zero behavior, we also computed a first-order PRIMA [Odabasioglu et al., 1998] approximation to the system's behavior. In Figure 6 the Bode diagrams of the full 3D model, the reduced order model and the PRIMA approximation are presented. These plots correspond to the admittance between two contacts for the three-layer substrate profile.

Here, the reduced model parameters were obtained by solving Equation (11) twice: once for $\omega = 0\ rad/s$ (DC) in order to obtain the resistive component of the coupling, G_c, and the second time for $\omega = \sigma_2/\varepsilon = 9.491 \times 10^{11}\ rad/s$ (corresponding to the intrinsic cutoff frequency of the middle layer) in order to obtain the capacitive component of the coupling, C_c (contacts were assumed to have a depth of 4 μm). Using just one solve would lead to large coupling estimation errors either at DC or high frequencies. The complete model is obtained by adding G_c with C_c like in Equation (12).

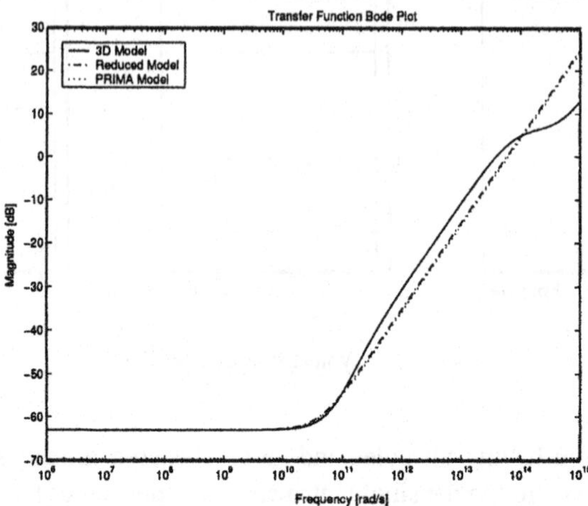

Figure 6. Magnitude Bode diagram of 3D transfer function, proposed model and PRIMA model.

As can be seen from the plots, the proposed reduced model and the PRIMA approximation are indistinguishable and have in fact quite similar accuracy. Both of them present a good approximation to the 3D model, accurately capturing the dynamics around the dominant pole/zero and boosting an error smaller than 5 dB for frequencies up to a few hundred gigahertz. Furthermore, the plot also shows quite effectively the limits of using a purely resistive model for substrate coupling.

RC Model Simulation

In order to assert for the significance of RC substrate models in circuit simulation, a simple experimental configuration was designed (cf. Figure 7) and simulated. Three CMOS inverters were implanted next to each other and an analog NMOS transistor (m_7) built near them. A substrate coupling model between all contacts has been extracted. In the simulation phase, the chain of inverters was driven by a 10 GHz sinusoidal wave and the noise injected through the inverters' NMOS diffusion and channel areas was coupled to the sensitive m_7 bulk.

The sensitive transistor has been biased in a way that its drain voltage is constant and equal to 2.33 V in perfect isolation conditions. Figure 8 shows the analog transistor drain voltage in three different situations: when using no substrate coupling model, when using purely resistive coupling models between noise generators and the sensitive transistor, and when using RC coupling models. From the figure, it becomes immediately apparent that the injection of

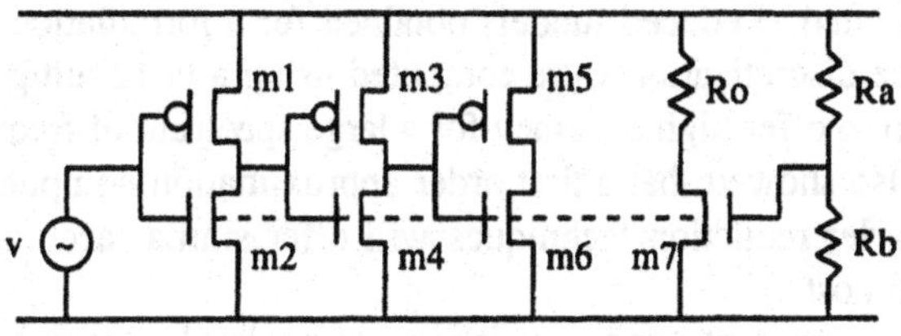

Figure 7. Simulation test circuit.

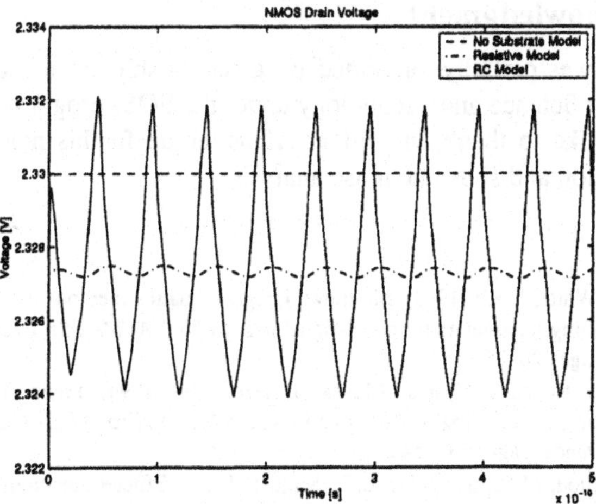

Figure 8. Substrate coupling simulation results.

noise into the substrate by the inverters induces substrate voltage fluctuation. Through the body effect of transistor m_7 its drain voltage also bounces instead of being steady as expected with substrate coupling. The difference from resistive to RC models is that resistive models do not account for substrate intrinsic capacitance properties, which at higher frequencies enhance coupling effects. Resistive models are therefore unable to predict correct functioning of the analog transistor in this case.

Clearly, this example demonstrates the need for substrate RC dynamic models for frequencies higher that a few gigahertz and it also validates the accuracy of the proposed method for frequencies up to several tens of gigahertz.

5. Conclusions

A methodology for the extraction of dynamic RC substrate coupling models, that naturally extends the traditional resistive-only modeling techniques, has been presented. Reduced models obtained for a formulation based on Finite Difference discretization were computed using a fast Multigrid algorithm and are shown to offer high accuracy for a large spectrum of frequencies. Further studies also showed that a first order approximation computed with standard model order reduction techniques will offer similar accuracy at similar computational cost.

Extensive experiments and simulations of a simple example circuit performed using the proposed model demonstrate both its relevance and accuracy for frequencies up to several tens of gigahertz.

6. Acknowledgment

This work was partially supported by a scholarship from the Portuguese Foundation for Science and Technology under the POSI program.

We would like to thank Dr. Edgar Albuquerque for his helpful hints on model simulation and substrate noise analysis.

References

Chou, Mike and White, Jacob (1998). Multilevel integral equation methods for the extraction of substrate coupling parameters in mixed-signal ic's. In 35th *ACM/IEEE Design Automation Conference*, pages 20–25.

Clement, F. J. R., Kayal, E. Zysman M., and Declercq, M. (1994). Layin: Toward a global solution for parasitic coupling modeling and visualization. In *Proc. IEEE Custom Integrated Circuit Conference*, pages 537–540.

Costa, Joao P., Chou, Mike, and Silveira, L. Miguel (1998). Efficient techniques for accurate modeling and simulation of substrate coupling in mixed-signal IC's. In *DATE'98 - Design, Automation and Test in Europe, Exhibition and Conference*, pages 892–898, Paris, France.

Feldmann, P. and Freund, R. W. (1995). Efficient linear circuit analysis by Pade approximation via the Lanczos process. *IEEE Transactions on Computer-Aided Design of Integrated Circuits and Systems*, 14(5):639–649.

Gharpurey, R. and Meyer, R. G. (1995). Modeling and analysis of substrate coupling in integrated circuits. In *IEEE 1995 Custom Integrated Circuits Conference*, pages 125–128.

Gharpurey, Ranjit (1995). *Modeling and Analysis of Substrate Coupling in Integrated Circuits.* PhD thesis, Department of Electrical Engineering and Computer Science, University of California at Berkeley, Berkeley, CA.

Gharpurey, Ranjit and Hosur, Srinath (1997). Transform domain techniques for efficient extraction of substrate parasitics. In *Proceedings of the Int. Conf. on Computer-Aided Design*, pages 461–467.

Joardar, Kuntal (1994). A simple approach to modeling cross-talk in integrated circuits. *IEEE Journal of Solid-State Circuits*, 29(10):1212–1219.

Johnson, T. A., Knepper, R.W., Marcellu, V., and Wang, W. (1984). Chip substrate resistance modeling technique for integrated circuit design. *IEEE Transactions on Computer-Aided Design of Integrated Circuits*, CAD-3(2):126–134.

Kanapka, J., Phillips, J., and White, J. (2000). Fast methods for extraction and sparsification of substrate coupling. In *Proc. 37th Design Automation Conference*.

Kup, B.M.J., Dijkmans, E.C., Naus, P.J.A., and Sneep, J. (1991). A bit-stream digital-to-analog converter with 18-b resolution. *IEEE Journal of Solid-State Circuits*, 26(12):1757–1763.

Mitra, Sujoy, Rutenbar, R. A., Carley, L. R., and Allstot, D. J. (1995). A methodology for rapid estimation of substrate-coupled switching noise. In *IEEE 1995 Custom Integrated Circuits Conference*, pages 129–132.

Nauta, Bram and Hoogzaad, Gian (1997). How to deal with substrate noise in analog CMOS circuits. In *European Conference on Circuit Theory and Design*, pages Late 12:1–6, Budapest, Hungary.

Odabasioglu, A., Celik, M., and Pileggi, L. T. (1998). PRIMA: passive reduced-order interconnect macromodeling algorithm. *IEEE Trans. Computer-Aided Design*, 17(8):645–654.

Phillips, Joel R and Silveira, L. Miguel (2001). Simulation approaches for strongly coupled interconnect systems. In *International Conference on Computer Aided-Design*.

Silva, Joao M. S. (2003). Modeling substrate coupling in mixed analog-digital circuits (in portuguese). Master's thesis, Instituto Superior Tecnico, Technical University of Lisbon, Lisboa, Portugal.

Silveira, L. Miguel, Kamon, Mattan, and White, Jacob K. (1995). Algorithms for Coupled Transient Simulation of Circuits and Complicated 3-D Packaging. *IEEE Transactions on Components, Packaging, and Manufacturing Technology, Part B: Advanced Packaging*, 18(1):92–98.

Silveira, L. Miguel and Vargas, Nuno (2002). Characterizing substrate coupling in deep submicron designs. *IEEE Design & Test of Computers*, 19(2):4–15.

Smedes, T., van der Meijs, N. P., and van Genderen, A. J. (1995). Extraction of circuit models for substrate cross-talk. In *International Conference on Computer Aided-Design*, San Jose, CA.

Stanisic, Balsha, Verghese, Nishath K., Rutenbar, Rob A., Carley, L. Richard, and Allstot, David J. (1994). Addressing substrate coupling in mixed-mode IC's: Simulation and power distribution systems. *IEEE Journal of Solid-State Circuits*, 29(3):226–237.

Su, David K., Loinaz, Marc J., Masui, Shoichi, and Wooley, Bruce A. (1993). Experimental results and modeling techniques for substrate noise in mixed-signal integrated circuits. *IEEE Journal of Solid-State Circuits*, 28(4):420–430.

van Genderen, A. J., van der Meijs, N. P., and Smedes, T. (1996). Fast computation of substrate resistances in large circuit. In *European Design and Test Conference*, Paris.

Verghese, Nishath (1995). *Extraction and Simulation Techniques for Substrate-Coupled Noise in Mixed-Signal Integrated Circuits*. PhD thesis, Department of Electrical and Computer Engineering, Carnegie Mellon University, Pittsburgh, PA.

Verghese, Nishath K. and Allstot, David J. (1995). Subtract: A program for the efficient evaluation of substrate parasitics in integrated circuits. In *International Conference on Computer Aided-Design*, San Jose, CA.

Wemple, Ivan L. and Yang, Andrew T. (1995). Mixed-signal switching noise analysis using voronoi-tesselation substrate macromodels. In 32nd *ACM/IEEE Design Automation Conference*, pages 439–444, San Francisco, CA.

HINOC: A HIERARCHICAL GENERIC APPROACH FOR ON-CHIP COMMUNICATION, TESTING AND DEBUGGING OF SOCS

Thomas Hollstein, Ralf Ludewig, Heiko Zimmer, Christoph Mager, Simon Hohenstern, Manfred Glesner

Darmstadt University of Technology
Institute of Microelectronic Systems
Karlstrasse 15, D-64283 Darmstadt, Germany
{hollstein|ludewig|zimmer|glesner}@mes.tu-darmstadt.de

Abstract This paper presents a new generic system architecture and design methodology for the design, debugging and testing of complex systems-on-chip (SoC). Starting from a hierarchical generic system architecture, platforms for dedicated application scenarios will be customized. In order to be able to handle very complex submicron designs, the system is based on a globally asynchronous and locally synchronous (GALS) concept. The problem of the increasing functionality versus outer access capabilities ratio is faced by novel embedded and combined debugging and test structures. The integration of debugging possibilities is essential for an efficient co-design of SoC integrated hardware and software, especially for systems with integrated reconfigurable hardware parts.

Keywords:
Networks-on-Chip, Silicon Debug, Built-In Self Test

1. Introduction

Increasing system complexities, driven by ongoing technology improvements and application requirements, create a demand for highly efficient system design methods. Furthermore time-to-market becomes the primary factor for economical success ([Jager, 2002]). In order to gain the required design productivity, design reuse and efficient behavioral synthesis methods are indispensable. Advanced methods for low power design will be essential for an increasing number of battery-driven applications.

Currently complexity is mainly handled by *intellectual property (IP)* based approaches. Existing design components are used in several projects and frequently designs are enhanced and completed by buying IP components from

Please use the following format when citing this chapter:
Hollstein, Thomas, Ludewig, Ralf, Zimmer, Heiko, Mager, Christoph, Hohenstern, Simon, Glesner, Manfred, 2006, in IFIP International Federation for Information Processing, Volume 200, VLSI-SOC: From Systems to Chips, eds. Glesner, M., Reis, R., Indrusiak, L., Mooney, V., Eveking, H., (Boston: Springer), pp. 39-54.

external suppliers. At present the effort for the integration of external IPs (IP components) into SoC designs can be high, especially if wrappers have to be developed in order to adapt the component's ports to the system's communication architecture and protocols. A common way to cope with the IP block integration problem are platform-based design approaches, where a fixed or limited generic communication architecture is presumed. These platforms are adapted and scaled in every design generation to fulfill the currently required system constraints (incremental adaption). The advantage of state-of-the-art platform-based design approaches is that the migration effort between different design generations is limited. Major changes of the platform architecture are in most cases complex and very costly. The integration of IP components may be difficult if the components are available on register transfer level or on layout level, since the clocking scheme of the IP block may differ considerably from the timing of the blocks designed inhouse.

On an long term view, efficient system-level synthesis methods will be needed as well as IP-based design. In the recent decade synthesis methods have been presented on different levels of abstraction. Behavioral synthesis tools, mapping behavioral specifications on register-transfer level are in the meantime commercially available. Since the target space of high-level-synthesis is extremely huge, improved methods are currently still under construction. On system level several codesign methods for combined hardware-/software systems have been presented. Compared to software synthesis (compilation), automatic hardware transformations are much more complex, since the number of design parameters is comparatively high (area, performance, power) as well as the extent of implementation possibilities. Nevertheless in future synthesis methods will be a helpful complement to reuse approaches.

Concerning future SoC design challenges the following aspects have to be considered (SIA roadmap 2001 [SIA, 2001]): for large chip die sizes it won't be possible to transmit signals over long on-chip distances within only one clock cycle. Additionally it cannot be assumed any more, that all SoC components are running with the same clock frequency and phase. This results in a demand of communication-centric design. Communication structures with asynchronous communication of synchronously operated SoC components (Globally Asynchronous, Locally synchronous (GALS)) have to be developed in order to face the requirements of huge SoC systems, comprising a large number of components([Muttersbach et al., 2000, Sgroi et al., 2001]). Furthermore the functional density (amount of functionality related to the port width for outer access (e.g. number of pins)) of SoCs is permanently increasing. Therefore an outer access (controllability and observability) for manufacturing tests is becoming more difficult. Furthermore the debugging of heterogeneous SoCs, especially in the case of malfunction, is difficult.

The aspect of power minimisation is not focused in this paper. For an description of our investigations concerning SoC low power communication we'd like to refer to [Murgan et al., 2003].

This contribution is organized as follows: in the next section future SoC design challenges based on ongoing research of different groups are analysed in detail. Based on this analysis in section 3 a novel network-on-chip (NoC) communication architecture for future SoCs is presented. Section 4 describes a new concept of embedding combined on-chip debugging and testing structures. Further on in section 5 simulation results of the SystemC NoC-model are presented and discussed with respect to packet-oriented extensions. A new design flow for NoC-based HW/SW Codesign is presented in the last section. Finally we end up with some conclusions and an outlook on future investigations.

2. Design Challenges

SoC/NoC Design

Due to permanently increasing technology capabilities future on-chip integrated systems will be composed of a large number of subcomponents. As stated in the introduction, a central clock supply won't be feasible any more and a lot of heterogenous components have to be interconnected. Therefore global intermodule connections will become asynchronous, connecting synchronously operating subdomains (GALS structure). New methods are required to ensure the testability of such systems: locally embedded and independently operating testing structures for each synchronous subdomain with a central control and evaluation of the results.

Generally, future SoC design methodologies have to face the following aspects:

- system specification/modeling
- system simulation
- design methods for components (IP-based design, synthesis approaches)
- overall system architecture and communication structure
- design verification
- debugging capabilities
- testability concepts

The design of SoC communication architectures will be a central task which has to face the following demands:

- general approach, which is generic and scalable
- customizable to any kind of application topology
- adaptability to different application classes
- data transmission between different clock domains (synchronisation)
- flexible use and configurability of communication resources
- coverage of quality of service (QoS) aspects
- provision of required communication services

In future the on-chip communication will become more vulnerable to transmission faults. Decreasing feature sizes and voltages let single event upsets occur more frequently. Therefore networks-on-chip (NoC) have to be designed fault tolerant [Dally and Towles, 2001, Zimmer, 2002].

Generally for the design of NoCs, existing know-how from the field of computer networks can be evaluated. But in difference the topology for NoCs is fixed, the parameters of interconnections are known, and the amount of buffer memory in network nodes is limited. Therefore it seems to be reasonable to have several local merges in adjacent OSI protocol layers.

In the context of current research investigations several approaches for the structuring of SoC communication networks have been proposed. In [Wielage and Goossens, 2002],[Sgroi et al., 2001],[Benini and Micheli, 2002] and [Dally and Towles, 2001] basic NoC issues are discussed. Several proposals for SoC architectures have been presented: NOSTRUM [Kumar et al., 2002], SOCBUS [Wiklung and Liu, 2003], PROTEO [Saastamoimen et al., 2002], SPIN [Benini and Micheli, 2002], PROPHID [Guerrier and Greiner, 2000], MESCAL [Sgroi et al., 2001], a RAW (MIT) [Taylor et al., 2002], torus architecture (NTNU) [Natvig, 1997] and HiNoC [Hollstein et al., 2003] (among others).

In the following section we present a new NoC approach, which combines a GALS concept with a new hierarchical multilayer network structure. In contrast to other approaches, it allows the generation of heterogeneous architectures and the integration of any kind of existing design platforms based on a constant grid size communication approach.

Testing and Debugging

To ensure the correct function and in order to detect manufacturing errors it is indispensable to test an ASIC after fabrication. This is done by loading test patterns in the internal registers, performing one clock cycle and comparing the resulting pattern with the simulation results (Fig: 1). Therefore the internal registers are replaced by special scan registers that are linked together as a shift register in a so called scan path. In the traditional approach the sink and the source of the test patterns are realized off-chip. This requires the use of external Automatic Test Equipment (ATE). Special attention has to be paid to the generation of the test pattern necessary achieve a good fault coverage.

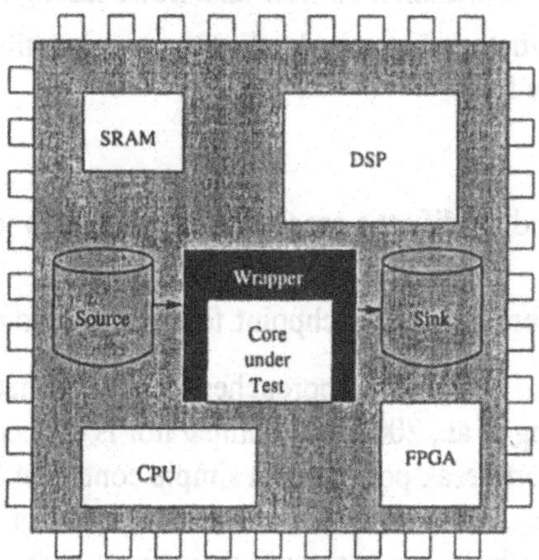

Figure 1. SoC Testing

Due to the shrinking feature size that allows for higher complexity and higher clock frequencies the requirements for the ATE speed and memory size are continuously increasing. Therefore the ATE solution can be unacceptable expensive and inaccurate. Thus on-chip test solutions, often referred to as build-in self-test (BIST), are becoming more and more popular.

Three additional components have to be implemented on the chip in order to perform BIST: a test pattern generator, a test response analyzer and a BIST control unit. The classical way to implement a test pattern generator is to use linear feedback shift registers to generate pseudo-random test patterns. In the test response analyzer the results from the applied test patterns are analyzed and compared to a pre-stored table of valid results. Due to some drawbacks like low fault coverage, long test application time and high area overheads, several proposals have been made to combine these pseudo-random tests with deterministic patterns [Jervan et al., 2002].

BIST allows at-speed tests and eliminates the need for an external tester. It cannot only be used for manufacturing test, but also for periodical field maintenance tests.

For complex systems it is also necessary to add debugging capabilities to the programmable components. For embedded system debugging the IEEE-ISTO Nexus 5001 Forum Standard defines a "Global Embedded Processor Debug Interface" [Nexus 5001 Forum, 1999]. The following basic needs when debugging an embedded processor are defined in the standard:

- Logic Analysis:
  - Access instruction trace information
  - Retrieve information on how data flows through the system
  - Assess whether the embedded software is meeting the required performance level
- Run Control:
  - Query and modify the processor state/registers when the processor is halted
  - Support breakpoint/watchpoint features in hardware and software

Prior to this, some JTAG-based approaches were published (e.g. [de la Torre et al., 2002] and [Jung et al., 2002]). A comparator is connected to the address and data bus to support break points and a simple controller is added for single step operation mode. The data of the internal registers of the processor can be shifted out and in by using the scan chain from testing. This technique requires only a small amount of additional hardware, but does not fulfill all requirements (like trace informations) listed above.

For reconfigurable components the same requirements are valid as for embedded processors. For FPGA-prototyping a commercial solution is available [Synplicity, 2003], which uses the resources of the reconfigurable unit to provide the debugging capabilities. It is also possible to use hardware-based debugging and facilitate only the routing resources to get access the internal signals.

For future SoCs it will be necessary to provide debugging units for all reprogrammable/reconfigurable units. Linking this units to each other and thereby enable the units to communicate with each other will greatly improve the debugging performance of the development engineers.

3. HiNoC: A Novel Network-on-Chip Architecture

In recent times simple communication structures have been used within SoCs, since the number of connected components has been small [Sgroi et al., 2001]. Future applications will comprise of a large number of heterogeneous processing units, which have to be interconnected efficiently [Kumar et al., 2002].

We have developed a new hierarchical NoC topology which is capable to implement GALS systems in a very efficient way. The system is, in difference to [Kumar et al., 2002], structured in two levels (Fig. 2).

The top level is an asynchronous mesh structure. This topology is a special case of a general torus structure, which can be implemented more efficiently (smaller and similar delays between routing units).

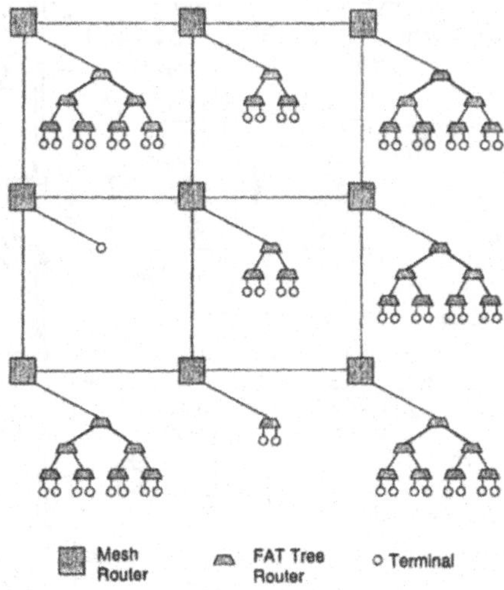

Figure 2. HiNoC: Hierarchical Topology of Communication Network

The general HiNoC concept is to have a basis for an automatic generation of application platforms. Based on a system-level application simulation, a customized mesh will be generated by leaving out several of the mesh nodes. The resulting structure will allow to interconnect applications comprising components with heterogeneous granularity (Fig. 3).

Adjacent routers in the HiNoC (Hierarchical NoC) mesh are connected by two unidirectional connections (one in each direction) with generic bitwidth. Furthermore the asynchronous communication is handled by a handshake protocol. Each network router (as far as it is not an network edge router) is connected to four adjacent routers and to a second layer node. This second level of the hierarchy is formed by FAT trees with generic depth. Each FAT tree is in a general case a binary tree which is capable to provide a full connect of all its leaf terminals (connection to processing nodes). In our implementation the connectivity of of the FAT tree can be thinned out in order to have a trade-off between full connectivity and the amount of required wires and multiplexers. FAT trees are a very cost efficient topology for implementing VLSI systems ([Leiserson, 1985]). The communication within the FAT tree is completely synchronous and it has an asynchronous port towards the level one network router. In the special case that the FAT tree consists of one leaf node only, this node is directly attached to the mesh router. All FAT trees connected to mesh routers may have different sizes and different thin out factors. In difference to the definitions in [Tanenbaum, 1996], the HiNoC terminal concept basically in-

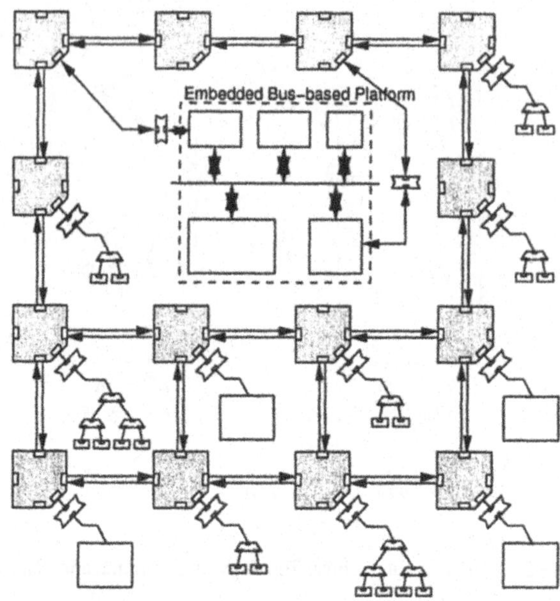

Figure 3. Customized Mesh (FAT trees are not depicted for the sake of simplicity)

cludes the possibility of multiple terminal connections of processing elements to different FAT tree leaf terminals (Fig. 3).

Of course, the FAT tree can be replaced by any kind of proprietary communication architecture (e.g. crossbar or buses) or complex components. Existing design platforms can be integrated completely as NoC attached components. If the size of an embedded platform exceeds the area defined by the router distances, one or several routers can be left out in order to integrate the platform component.

HiNoC offers two data transmission modes, which can be interleaved in order to gain best use of available communication resources:

Packet-Switched Communications: Packets are routed by a local routing algorithm (local connections required only) and transmitted by application of a virtual cut-through / wormhole switching scheme. Stucked packets are resolved and backtraced to the packet source. Packet transmission uses the available network links efficiently at the cost of a non-deterministic latency.

QoS-based Communications: if circuit-switched data transmission is applied, then dedicated virtual communication channels are set up with the full link bandwidth or any binary divided portion of it. This mode provides

Quality-of-Service based communications, which are required for multimedia applications (e.g. video streaming). The setup of the connection is done by a control packet which is sent towards the stream target router, reservating the requested bandwidth on each link (if possible). Having reached its sink, the packet is back-traced to its origin and activating the channel on each passed mesh link. If a connection has been set up, the bandwidth is guaranteed for the full transmission of a data stream (Quality-of-Service, QoS).

The routing is done based a combined static and dynamic metric. The static information states, which target node of the mesh can be reached via a specific mesh router output link at which metric (distance). If parts of the SoC/Mesh are powered down (power reduction), the static routing tables are updated in order to adapt the routing behaviour to the new NoC configuration. Additionally adjacent mesh routers exchange information about the current traffic situation by sending control packets in free time slots (dynamic metric).

Fig. 4 depicts the basic functional schematic of the first version of a level one mesh router.

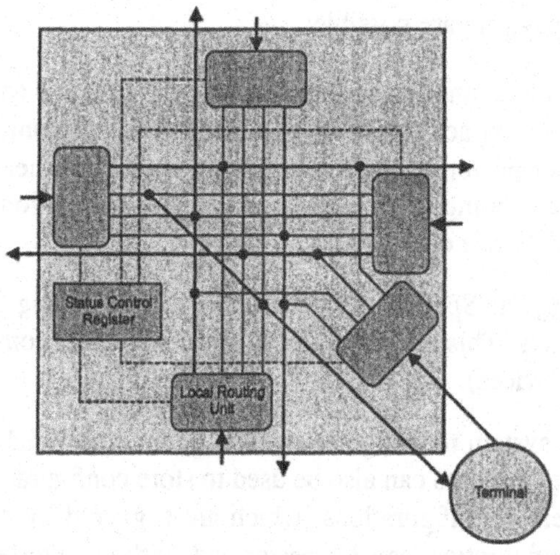

Figure 4. HiNoC Mesh Router Structure

The input FIFO can be enhanced in order to implement several priority levels. In an advanced implementation this input buffer design has been migrated to a middle buffer design with virtual channels, which are groupwise assigned to one of the potential four output connections.

For the routing of streams and packets an efficient local routing algorithm is applied, which also considers the possibility of routing around high density spots and which is avoiding cycles.

The mesh and FAT tree models are implemented in SystemC and the simulator is writing a log file on all network events. The log file can be processed by a JAVA based graphical viewer and stepped through offline.

4. Integrated Test and Debugging Concept for SoCs

Comparing the requirements for build-in self-test and debugging support for SoC components, it becomes obvious that the main requirement, the access to the internal registers, is the same. Furthermore running a self test is possible only if the tested component is not in an operational mode. For running the test procedure it has to be in a special system maintenance mode. On the other hand debugging only make sense when the component is in normal operational mode. Thus we have developed an integrated hierarchical concept, which benefits from resource sharing for both tasks. The method provides controllability and observability over several levels of hierarchy of the SoC topology.

Having a test/debug unit connected to internal registers and external input/output signals of SoC components, additional applications besides build-in self-test and debugging are possible:

- On-chip signal analysis: The interface signals of a component can be analysed to extract statistical data from the communication. This can be used for power estimation and optimization by measuring the activity and calculating the temporal and spatial correlation of the interface signals [Ludewig et al., 2002].

- Event logging: Special events could be logged during the whole period of operation. This can be used for mission critical components (e.g. in medical devices).

- Dynamic system reconfiguration: additional embedded memory for test and debug purposes can also be used to store configuration data for (partial) system reconfigurations, which are triggered by dedicated events. The reconfiguration can be performed without interfering with other components of the SoC.

If the embedded test and debug units are interconnected by global network-on-chip communication resources, new collaborative functionalities become possible. For instance an embedded processor could be stopped if a breakpoint condition in a reconfigurable component is reached.

For SoCs with more than one clock domain a hierarchical structuring of the interconnection of the test and debug units is proposed, so that all of these units which are located in one clock domain are connected to a communication controller that implements the communication with controllers in other (potentially asynchronously running) clock domains. This also reflects the architecture of the proposed HiNoC architecture in which the components connected to one FAT-tree are clocked synchronous, while the components connected via the switched network could belong to other clock domains. Therefore one communication controller has to be integrated in every FAT tree/synchronous domain.

5. Results

The goal of a first SystemC implementation of the HiNoC communication architecture had been the analysis of the setup and the achieved bandwidth of connection-oriented data transmissions (build up connection, transmit data, cancel connection). Therefore first we instantiate a 4x4 mesh within our generic SystemC model and set the height of the FAT tree to zero. In order to stress the net, all network terminals permanently try to send 500 data words to a randomly selected destination address. The following parameters have been analysed: the average relative bandwidth (active data transmission time) of all terminal's sending processes:

$$t_{av,rel} = \frac{1}{n_t}\sum_{i=1}^{n_t}\frac{t_{send,i}}{t_{measure,i}} = \frac{1}{n_t}\sum_{i=1}^{n_t} t_i \qquad (1)$$

and the minimum relative bandwidth $t_{min,rel}$ over all i nodes. n_t represents the number of terminals (in our case 16). Furthermore we have been tracking the average waiting time $t_{wait,av}$, which nodes had to wait (because of blocked resources) after finishing a transmission until they could start the next data send cycle. $t_{wait,max}$ stands for the maximum waiting time which occurred during the whole simulation. Fig. 5 depicts $t_{av,rel}$ under the assumption, that the transmission of 500 data words is clustered to 5 or 25 blocks of equal size, which are independently transmitted. The partitioning of the transfer reduces the $t_{wait,max}$ values by 40% to 50%. Of course $t_{av,rel}$ is decreased, since much more connection setups have to be processed (overhead). If the mesh size is varied, the time parameters show an expected quadratic dependency (Tab. 1).

The modification of the maximum waiting time is increasing correlated to these results. Of course the experiment is very pessimistic, since a reasonable mapping of intensively communicating terminals on adjacent positions on the grid can reduce the problem dramatically. Further improvement can be achieved by enhancing the concept to a two layer NoC architecture with synchronous subdomains with synchronous intra module communication using the FAT tree structure. For heavily communicating pairs of terminals an assignment to the same synchronous domain generates a huge gain in speed, since

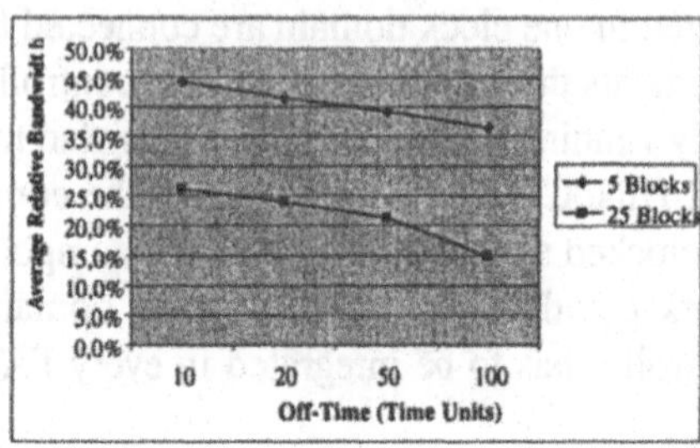

Figure 5. Measured Average Bandwidth for different Terminal Parameters

Mesh size	tav, rel	$t_{min,rel}$
4x4	41,6%	35,0%
6x6	33,1%	21,4%
8x8	24,7%	7,2%
10x10	18,2%	1,0%

Table 1. Influence of the Mesh size

synchronous communication is much faster than asynchronous data transmission (requires a certain effort to avoid metastability of registers). Table 2 compares the simulation results of two different 64 terminal network-on-chip configurations. The first configuration is a 4x4 mesh with four terminal FAT trees

Mesh size	$t_{av,rel}$	$t_{min,rel}$
4x4x4	50,0%	21,9%
8x8x0	24,7%	7,2%

Mesh size	Average max. wait time	$\max(t_{wait,max})$
4x4x4	1500	7230
8x8x0	5686	17657

Table 2. Simulation results: pure mesh versus two-level HiNoC architecture

attached to each mesh router. Additionally the assumption is applied, that 75%

of the communication can be routed within a FAT tree (simulation of intelligent mapping). The second topology is a 8x8 mesh without FAT trees. For the given assumption the resulting average bandwidth in the 4x4x4 topology is 50% and the maximum waiting time is comparatively low.

6. The proposed Dynamic Co-Design Flow

Fig. 6 depicts a new design flow based on communication-centric hyper-platforms. In the static design phase (the SoC hardware design phase) a set of applications to be run on the target system is profiled by simulation. Based on the profiling information, a design constraint library and an existing component implementation library, a set of required execution components is determined (Mapping Step). In a second Placement Step the required IP components are placed on the hyper-platform and the NoC is customized, considering the communication closeness of the IP blocks. After manufacturing, a dynamic recon-

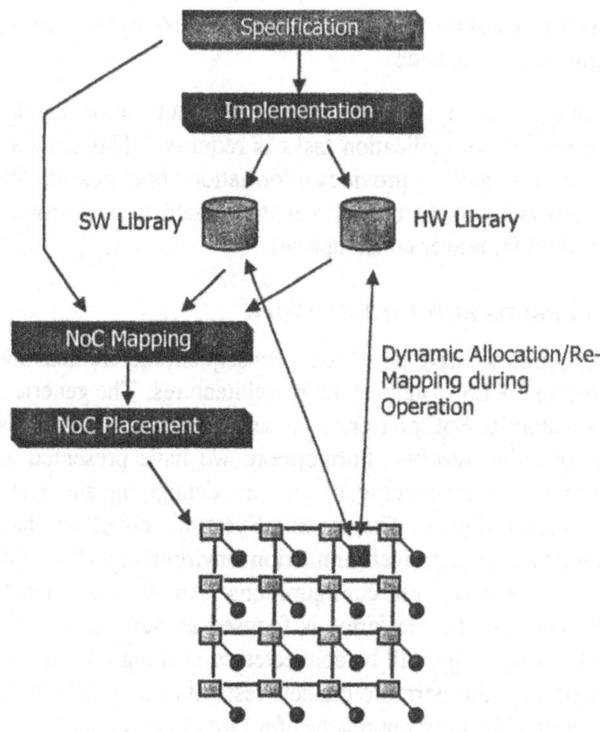

Figure 6. DesignFlow: Static and Dynamic Hardware/Software Co-Design

figuration of the system can be done based on the flexibility provided by the use of processor (software) and reconfigurable hardware IP components. Possible reasons for the need of system reconfiguration during operation can be:

- availability of new implementations optimized to power and performance
- reaction on changed constraints in the system's environment
- improved power efficiency due to low battery status in mobile devices
- additional services to be executed on the system, implementation requires a dynamic re-assignment of resources

The dynamic hardware/software reconfiguration of a hyper-platform is defined as "Dynamic Hardware/Software Co-Design". Two modes of dynamic reconfiguration can be distinguished:

1 minor reconfiguration for inclusion of additional services. An on-chip task scheduler will assign the new tasks to available resources or organize a minor rearrangement of task assignments.

2 major reconfiguration of the whole system: has to be done by a CAD environment from outside

For the dynamic system reconfiguration an implementation database (HW, SW, FPGA) for concrete application tasks is required. This database, which will typically not be complete, provides information about possible implementations of system tasks on IP blocks and the related design properties (e.g. throughput, resolution, power consumption).

7. Conclusions and Future Work

In this contribution we have presented a consequent hierarchical and generic topology for future GALS system-on-chip architectures. The generic approach can be used to customize SoC platforms for dedicated application classes based on simulation profiling results. Furthermore we have presented an hierarchical approach for an embedding of on-chip debugging and testing functionality in GALS topologies. The existing SystemC model/simulator is currently enhanced to a mixed-level simulation environment (SystemC/VHDL co-simulation)for irregular mesh configurations. An automatic generation of application domain specific platforms is focused as next step. The build-in debugging and test topology will be connected to existing probing techniques for power monitoring. Furthermore the new testability concept will combined with classical hierarchical test approaches for circuit components.

References

[Benini and Micheli, 2002] Benini, L. and Micheli, G. D. (2002). Networks on Chips: A New SoC Paradigm. *IEEE Computer*, Vol. 35:70–78.

[Dally and Towles, 2001] Dally, W. J. and Towles, B. (2001). Route Packets, Not Wires: On-Chip Interconnection Networks. In *Proc. of the Design Automation Conference (DAC 2001)*, pages 684–689.

[de la Torre et al., 2002] de la Torre, E., Garcia, M., Riesgo, T., Torroja, Y., and Uceda, J. (2002). Nonintrusive debugging using the JTAG interface of FPGA-based prototypes. In *Proceedings of the 2002 IEEE International Symposium on Industrial Electronics (ISIE 2002)*, volume Vol. 2, pages 666–671.

[Guerrier and Greiner, 2000] Guerrier, P. and Greiner, A. (2000). A Generic Architecture for On-Chip Packet-Switched Interconnections. In *Proc. of Automation and Test in Europe Conference and Exhibition 2000*, pages 250–256.

[Hollstein et al., 2003] Hollstein, T., Ludewig, R., Mager, C., Zipf, P., and Glesner, M. (2003). A hierarchical generic approach for on-chip communication, testing and debugging of SoCs. In *Proc. of the VLSI-SoC 2003*, pages 44–49.

[Jager, 2002] Jager, S. (2002). "System on Chip"-Design: Technische und Økonomische Herausforderungen. Studienarbeit, Darmstadt University of Technology.

[Jervan et al., 2002] Jervan, G., Peng, Z., Ubar, R., and Kruus, H. (2002). A Hybrid BIST Architecture and its Opimization for SoC Testing. In *Proc. of the Int. Symposium on Quality Electronic Devices*. IEEE.

[Jung et al., 2002] Jung, D.-Y., Kwak, S.-H., and Lee, M.-K. (2002). Reusable embedded debugger for 32 bit RISC processor using the JTAG boundary scan architecture. In *Proc. of IEEE Asia-Pacific Conference on ASIC 2002*, pages 209–212.

[Kumar et al., 2002] Kumar, S., Jantsch, A., Soininen, J.-P., Forsell, M., Millberg, M., Øberg, J., Tiensyrja, K., and Hemani, A. (2002). A Network on Chip Architecture and Design Methodology. In *Proc. of VLSI Annual Symposium (ISVLSI 2002)*, pages 105–112.

[Leiserson, 1985] Leiserson, C. (1985). Fat Trees: Universal Networks for Hardware-Efficient Supercomputing. *IEEE Transaction on Computers*, Vol. C-34(No. 10):892–901.

[Ludewig et al., 2002] Ludewig, R., Garcia, A., Murgan, T., and Glesner, M. (2002). Power Estimation based on Transition Activity Analysis with Architecture Precise Rapid Prototyping System. In *Proc. of the 13th IEEE Int. Workshop on Rapid System Prototyping (RSP)*, pages 138–143.

[Murgan et al., 2003] Murgan, T., Petrov, M., Ortiz, A. G., Ludewig, R., Zipf, P., Hollstein, T., Glesner, M., Θlkrug, B., and Brakensiek, J. (2003). Evaluation and Run-Time Optimization of On-Chip Communication Structures in Reconfigurable Architectures. In *Proc. of 13th International Conference on Field Programmable Logic and Applications (FPL2003)*, pages 1111–1114.

[Muttersbach et al., 2000] Muttersbach, J., Villinger, T., and Fichtner, W. (2000). Practical Design of Globally-Asynchronous Locally-Synchronous Systems. In *Proc. of the Sixth Int. Symposium on Advanced Research in Asynchronous Circuits and Systems (ASYNC 2000)*, pages 52–59.

[Natvig, 1997] Natvig, L. (1997). High-level Architectural Simulation of the Torus Routing Chip. In *Proc. of International Verilog HDL Conference, California.*

[Nexus 5001 Forum, 1999] Nexus 5001 Forum (1999). IEEE-ISTO Nexus 5001 Forum Standard for a Global Embedded Processor Debug Interface. http://www.nexus5001.org.

[Saastamoimen et al., 2002] Saastamoimen, I., SigÃijenza-Tortosa, D., and Nurmi, J. (2002). Interconnect IP Node for Future System-on-Chip Designs. In *Proc. of the First IEEE Int. Workshop on Electronic Design, Test and Applications (DELTA 02).*

[Sgroi et al., 2001] Sgroi, M., Sheets, M., Mihal, A., Keutzer, K., Malik, S., Rabaey, J., and Sangiovanni-Vincentelli, A. (2001). Addressing the System-on-a-Chip Interconnect Woes Through Communication-Based Design. In *Proc. of the Design Automation Conference (DAC 2001)*, pages 667–672.

[SIA, 2001] SIA (2001). SIA roadmap 2001. http://public.itrs.net.

[Synplicity, 2003] Synplicity (2003). Synplicity. http://www.synplicity.com.

[Tanenbaum, 1996] Tanenbaum, A. (1996). *Computer Networks*. Prentice Hall.

[Taylor et al., 2002] Taylor, M. B., Tim, J., Miller, J., and et. al., D. W. (2002). A Computational Fabric for Software Circuits and General-Purpose Programs. In *IEEE Micro*.

[Wielage and Goossens, 2002] Wielage, P. and Goossens, K. (2002). Networks on Silicon: Blessing or Nightmare? In *Euromicro Symposium On Digital System Deisign (DSD 2002)*, pages 423–425.

[Wiklung and Liu, 2003] Wiklung, D. and Liu, D. (2003). SOCBUS: Switched Network on Chip for Hard Real Time Embedded Systems . In *Proc. of the Int. Parallel and Distributed Processing Symposium.*

[Zimmer, 2002] Zimmer, H. (2002). Fault Modelling and Error-Control Coding in a Network-on-Chip. Studienarbeit, Darmstadt University of Technology.

AUTOMATED CONVERSION OF SYSTEMC FIXED-POINT DATA TYPES

Axel G. Braun, Djones V. Lettnin, Joachim Gerlach, Wolfgang Rosenstiel

University of Tuebingen,
Wilhelm-Schickard-Institute for Computer Science,
Department of Computer Engineering,
Sand 13
72076 Tuebingen, Germany
{abraun|lettnin|gerlach|rosenstiel}@informatik.uni-tuebingen.de

Abstract This article describes a methodology for the automated conversion of SystemC fixed-point data types and arithmetics to an integer-based format for simulation acceleration and hardware synthesis. In most design flows the direct synthesis of fixed-point data types and their related arithmetics is not supported. Thus all fixed-point arithmetics have to be converted manually in a very time-consuming and error-prone procedure. Therefore a conversion methodology has been developed and a tool enabling an automated conversion of SystemC fixed-point data types as well as fixed-point arithmetics has been implemented. The article describes the theory and transformation rules of the conversion methodology, their implementation into a tool solution, and its application in terms of an experimental case study.

Keywords: Fixed-Point; Data Type Conversion; SystemC; System Design; Hardware Synthesis.

1. Introduction

The increasing complexity of today's and future electronic systems is one of the central problems that must be solved by modern design methodologies. A broad range of functionality which should be adaptable to market requirements in a very fast and flexible way, and thereby a decreasing time-to-market phase are only some of the most important issues. There are several promising approaches like system-level modelling, platform-based design, and IP re-use, to face these problems. The economic success of a product highly depends on the power and flexibility of the design flow and the design methodology.

A very popular approach for system-level modelling is SystemC [14, 15]. SystemC is a C++-based system-level specification language that covers a

---

Please use the following format when citing this chapter:
Braun, Axel, G, Lettnin, Djones, V., Gerlach, Joachim, Rosenstiel, Wolfgang, 2006, in IFIP International Federation for Information Processing, Volume 200, VLSI-SOC: From Systems to Chips, eds. Glesner, M., Reis, R., Indrusiak, L., Mooney, V., Eveking, H., (Boston: Springer), pp. 55-72.

broad range of abstraction levels. The language is already a de-facto standard for system-level design. Several commercial design and synthesis products from Synopsys, Forte, Coware, Summit and others will be a foundation for system-level design flows. An important property of a design flow is that it must be free of gaps for all paths in a design cycle.

Algorithm design is usually based on floating-point data types. In a following step word lengths and accuracies are evaluated and fixed-point data types are introduced (e.g. for signal processing applications, etc.). The introduction and the evaluation process of fixed-point data types are well-supported by tools like FRIDGE (Fixed-Point Programming and Design Environment) [3, 6] or Synopsys CoCentric Fixed-Point Designer [11]. SystemC itself offers a very powerful concept of fixed-point data types [8]. In nearly all cases these types are not directly synthesizable to hardware. In common hardware design flows an additional manual conversion has to be applied before the synthesis process. Fixed-point data types are transformed to an integer-based data format containing virtual decimal points (which correspond to binary points in the digital world). The manual translation is very time-consuming and error-prone. To close this gap, it is necessary to develop an appropriate methodology for the conversion of those data types and arithmetics to a synthesizable integer format. In the following sections we present a solution for this step, which is based on SystemC fixed-point data types.

2. Fixed-Point to Integer Conversion

The basis of our methodology is a set of conversion rules. These conversion rules have been developed by analyzing traditional transformations of fixed-point arithmetics and operations, manually done by a designer. We considered variable declarations, assignments, basic arithmetic operations like summation, subtraction, multiplication, and division, as well as comparisons. In the description of the conversion rules of our methodology below, we will use the following definitions. A fixed-point number in binary representation can be described as follows:

$$fix(n,m) := \underbrace{d_{n-1}d_{n-2}\ldots d_0}_{IntegerPart} . \underbrace{d_{-1}\ldots d_{-(m-1)}d_{-m}}_{FractionalPart}$$

$$fix(n,m) := \sum_{i=-m}^{n-1} d_i \cdot 2^i$$

The corresponding data type in SystemC notation is `sc_fixed<n+m,n>`. A fixed-point number can be described using an integer data type without loss of information, if the decimal point will be shifted m positions to the right. All fractional bit positions are now located in the integer part. The resulting integer number has the same word length as the fixed-point representation before. In the following we will also use the definitions below:

- Least Common Data Type (LCDT): The LCDT of two fixed-point variables a and b is a fixed-point data type, which has minimal width that can contain all possible values of a and b without a loss of accuracy.

  $$LCDT(fix(n,m), fix(o,p)) := fix(max(n,o), max(m,p))$$

- Result Data Type (RDT): The RDT is the fixed-point data type, which can contain the result of an operation without loss of accuracy.

Basically, only conversions of those operations where both operands are fixed-point data types will be made (basic arithmetic operations). Constants or other operands must be signed by an explicit type cast to a fixed-point data type. The first step is the calculation of the LCDT of a (binary) operation (see Figure 1). The integer and the fractional part equal to the maximum width of the two operand's integer and fractional parts. Secondly, the word lengths and type of the result (i.e. RDT) are determined. For the preservation of accuracy, both operands are adapted to the word length of the result type before the intrinsic operation takes place. The result will then be reduced to the LCDT. This reduction is important to avoid a steady extension of the word length throughout the further conversion procedure (e.g. for multiplications).

Declarations, Assignments, and Comparisons

Variables or attribute declarations of a fixed-point data type (`sc_fixed` and `sc_ufixed`) are directly converted to their appropriate SystemC integer data type (`sc_int` and `sc_uint`) of the same overall word lengths. A major prerequisite for assignments and comparisons is that both operands are exactly of the same data type. To ensure this, both operands are converted to the LCDT and then the comparison takes place. All C++ comparison operators can be converted directly. In case of a simple assignment (not combined with other operators like `+=`), the conversion has to adapt the data type on the right side to the type of the left side. We assume that combined assignments (e.g. `+=`) are split up in a basic three-address-format (e.g. `a+=b` is expanded to `a=a+b`). Return values of functions or methods are treated in the same manner as assignments: the returned value will be adapted to the type of the function respectively method.

Summation and Subtraction

Summation and subtraction operations are relatively easy to convert. Both operators are treated identically. The integer type of the result's data type

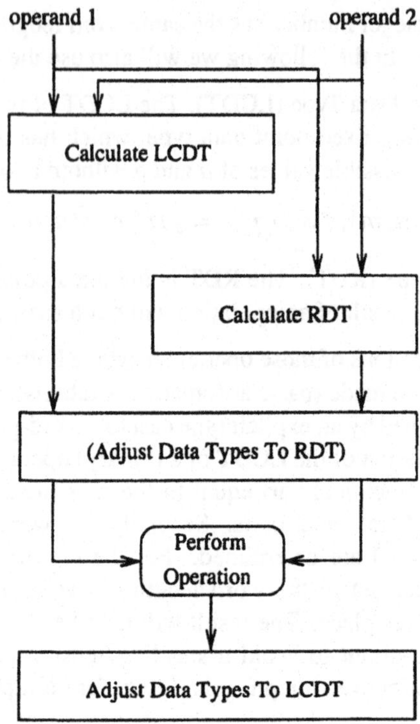

Figure 1. Conversion procedure

equals to the maximum of the integer parts of both operands. The part on the right side of the decimal point is converted in the same manner. In case of a summation or a subtraction an overflow is possible. Therefore the result type word length is enlarged by one bit:

$$fix(s,t) \pm fix(u,v) \rightarrow fix(max(s,u)+1, max(t,v))$$

For the conversion of a summation or subtraction of two fixed-point variables with different bit layouts, i.e. different widths of integer part and different widths on the right side of the decimal point, an adaptation has to be performed. The position of both variables has to be adjusted. If one of the operands has less bits on the right side, a shift to the left is necessary.

Figure 2 (left side) gives a short example for the summation of 2 numbers. Variable x has entirely 5 bits and contains 3 integer bits and 2 fractional bits, variable y has 6 bits and contains 1 fractional bit. The width of 7 bits is determined by the LCDT. In case of summation and subtraction here, the adaptation to the RDT is not needed and therefore kept. The conversion performs one left

```
//Fixed-point                        //Integer base
sc_fixed<5,3> x;                     sc_int<5> x;
sc_fixed<6,5> y;                     sc_int<6> y;
...                                  ...
x + y;                               (sc_int<7>)x + ((sc_int<7>y) << 1);
x - y;                               (sc_int<7>)x - ((sc_int<7>y) << 1);
...                                  ...
```

Figure 2. Summation and subtraction example

shift of variable y to adjust the (virtual) decimal points (right side of Figure 2). Now the summation can be performed correctly.

Multiplication and Division

During the conversion of a multiplication the word lengths of the integer parts of both operands are added. The same holds for the fractional part on the right sides of the decimal points. If there are differences between the word lengths of the fractional parts of the LCDT and RDT, the RDT must be adapted. As described above, both operands will be adapted to the width of the RDT before the multiplication will be performed.

$$fix(s,t) \cdot fix(u,v) \rightarrow fix(s+u, t+v)$$

For the conversion of a division operation the RDT will be calculated similarly to the multiplication, except of the word lengths of the fractional part. These word lengths will be subtracted instead of added. The resulting lower number of bits will be balanced with the integer parts. Instead of a data type adaptation after the operation, the dividend will be scaled. Due to this scaling, the dividend gets a new type, which must not under-run the word lengths of the LCDT. Consequently, the data type (word lengths) of the result also changes and must be adapted.

$$\frac{fix(s,t)}{fix(u,v)} \rightarrow fix(s+v, t-v)$$

In principle the number of bits in the fractional part is inherently defined by the layout of the fixed-point data type of the result. And this type is determined by the algorithm respectively hardware designer or even by a fixed-point evaluation tool in the previous design stage. This accuracy is exactly preserved during the conversions process.

The goal was to perform all these conversions automatically without any manual assistance. Therefore a tool, which will be described more detailed in the following section, has been developed.

3. Automated Fixed-Point to Integer Conversion Tool

The basic criteria for the development of the conversion tool, called **FixTool** [4], were an arithmetical correct application of the conversion rules described above, the preservation of parts of the code, which contain no fixed-point data types and arithmetics, including the code formatting, and finally the automated generation of directly synthesizable SystemC code.

The conversion is structured into four different stages: 1) parsing of the SystemC [7] code; 2) scope analysis and data type extraction; 3) analysis of all expressions in the code; 4) generation of the converted and adapted integer-based code. Figure 3 depicts the basic flow within the FixTool application.

At first the SystemC source code is analyzed syntactically and a complete parse tree is generated. The analysis within this stage cannot be restricted to SystemC keywords only or to keywords related to the fixed-point data types. Besides SystemC-specific syntax, the analysis must cover also C++ syntax [1]. In the code generation stage, the tool inserts e.g. brackets into the SystemC code. This procedure must preserve the evaluation order of expressions given in the original code. Therefore it is a fundamental prerequisite to be able to recognize SystemC (C++) expressions correctly. The parse unit can handle context-free grammars. The grammar itself is combined of C++ and SystemC, but is specified separately, so it can be modified and extended very easily. The parse tree will be the major repository of all information collected in the following analysis steps.

The second stage of the tool analyzes variable declarations and attributes. A very important issue is the scope of variables respectively classes. The FixTool annotates these scopes in detail. Data types (numerical data types) and the related scope information of variables or classes are the basis for the correct conversion. Both, data types and scope information are integrated into the nodes of the parse tree for the following processing stages.

The third phase analyzes all expressions found in the parse tree by evaluating the data types and the scopes of the partial expressions. The final type of a certain expression is evaluated by applying the related grammar rules and the data type information to the expression.

The fourth stage of the tool replaces the fixed-point arithmetics code parts and inserts appropriate integer-based code. These insertions are mainly type casts and shift operations as described in Section 2. Fixed-point variable declarations will be mapped to the appropriate SystemC integer data types. As mentioned above, one of the criteria for the development of the tool was that the original structure and layout of the code will be preserved if possible. It is very important for a designer to be as familiar with the converted code as with the original fixed-point-based model. Therefore the fourth stage does not generate independently new code, it rather uses a replacement strategy for pre-

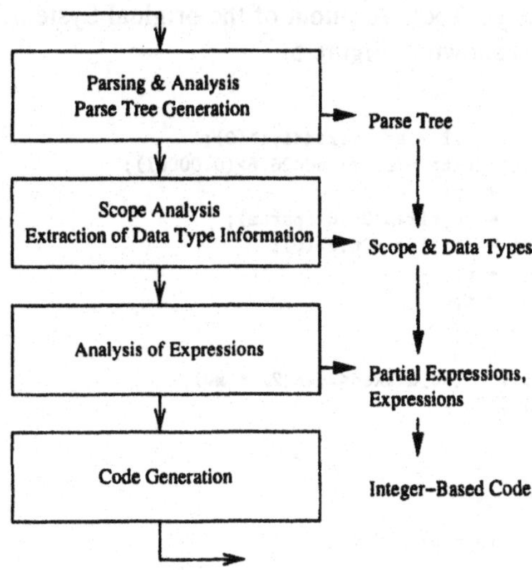

Figure 3. Conversion stages

cisely located code fragments. To be able to handle these code fragments, the parse tree also contains dedicated layout information of the source code file.

The conversion of entire projects consisting of several files requires the usage of a special project file. The project file specifies all source files and their corresponding header files. Similarly to the `make` tool, FixTool only converts files if they have changed since the last conversion run. Figure 4 shows a sample project description, which specifies the dependencies of different source modules (`source1.cpp`) with their related header files (`header1_1...`).

```
source1.cpp {header1_1.h ...
             header1_n.h source1.cpp}
source2.cpp {header2_1.h ...
             header1_m.h source2.cpp}
...
```

Figure 4. Project file

4. Simple Example

In this section we demonstrate the conversion methodology by applying it to a SystemC model that calculates square roots according to Newton's algorithm. The specification is using SystemC fixed-point data types and has been converted using the FixTool. A cutout of the original SystemC code, the calculation function, is shown in Figure 5.

```
void newton() {
  const sc_ufixed<1,1> null = sc_ufixed<1,1>(0);
  const sc_fixed<26,6> delta = sc_fixed<26,6>(0.00001);
  sc_uint<5> param    = in;
  sc_fixed<26,6> c    = sc_fixed<26,6>(param);
  sc_fixed<26,6> diff = sc_fixed<26,6>(0);
  sc_fixed<30,12> x0  = c;
  sc_fixed<30,12> x1  = c;

  do {
    sc_fixed<26,6> div = (sc_ufixed<2,2>(2) * x0);
    if (div == null) {
      x0 = null;
      break;
    }
    x0   = x1;
    x1   = x0 - (x0 * x0 - c) / div;
    diff = x0 - x1;
    if (diff<null) diff = sc_fixed<2,2>(-1) * diff;
  } while(diff>delta);
  out = x0;
}
```

Figure 5. Fixed-point-based Newton example

The resulting converted code (see Figure 6) shows the mapping of the SystemC fixed-point data types to SystemC integer data types. E.g. the constant `delta`, which is of type `sc_fixed<26,6>`, a fixed-point type with 6 bits integer width and 20 bits fractional part will be inserted. This constant is converted in a `sc_int<26>` type, the decimal point is only virtual. The underlying Newton's algorithm is quite simple, but it uses subtraction operations as well as multiplications and divisions and therefore gives a good example for demonstrating the methodology.

A comparison of this very simple example shows the changes introduced during the conversion process. The changes made in a manual translation are similar. It is evidently clear that the manual conversion is time-consuming and error-prone. The manual conversion of more complex algorithms is therefore very difficult and often messing. Optimized fixed-point data type word lengths are often softened during the debugging after a manual conversion process. Especially algorithms containing calculations with very different data type layouts (very different integer and fractional part word lengths) are critical for manual conversion. If certain variables are not local to some expressions, but rather used in the entire module in different expressions, it is also very difficult to do a manual conversion step by step. This requires additional type casting and data conversion effort during a stepwise conversion of the code.

```
void newton() {
  const sc_uint<1> null  = sc_uint<1>(0*1);
  const sc_int<26> delta = sc_int<26>(0.00001*1048576);
  sc_uint<5> param   = in;
  sc_int<26> c    = sc_int<26>(param*1048576);
  sc_int<26> diff = sc_int<26>(0*1048576);
  sc_int<30> x0  = (((sc_int<30>)(c))>>2);
  sc_int<30> x1  = (((sc_int<30>)(c))>>2);

  while(true) {
  do {
    sc_int<26> div = ((sc_int<26>)(((((sc_int<30>)
          (((((sc_uint<32>)(sc_uint<2>(2*1))))*
          (((sc_int<32>)(x0))))))))<<2));
    if (div == (((sc_int<26>)(null))<<20)) {
      x0 = (((sc_int<30>)(null))<<18);
      break;
    }
    x0 = x1;
    x1 = ((sc_int<30>)(((((sc_int<32>)(x0))<<2) -
         ((sc_int<32>)(((((sc_int<52>)
         (((((sc_int<32>)(((sc_int<30>)
         (((((sc_int<60>)(x0)))*(((sc_int<60>)
         (x0))))>>18))))<<2)-(((sc_int<32>)
         (c))))))<<20) /(((sc_int<52>)
         ( div)))))))>>2));
    diff = ((sc_int<26>)((x0 - x1)<<2));

    if (diff<(((sc_int<26>)(null))<<20))
      diff = ((sc_int<26>)(((((sc_int<28>)
              (sc_int<2>(-1*1)))) *
              (((sc_int<28>)(diff))))));
  } while(diff>delta);
  out = ((sc_int<26>)((x0)<<2));
}
```

Figure 6. Converted Newton example

These models can only be handled efficiently using an automated fixed-point to integer conversion. The automated conversion process of the Newton's algorithm example shown in Figure 4 takes 0.96 seconds. In opposition to that, the manual conversion of this code may take several hours of work, including the correction of conversion errors. The simulation of both, the original fixed-point-based model and the converted integer-based model, leads to the same output values. Figure 7 shows the example output for the fixed-point-based calculation of square roots starting from 0 up to 9.

The corresponding output of the converted model using integer arithmetics is shown in Figure 8. The integer numbers in the output protocol are representing bit patterns containing a virtual decimal point with 6 bits on the left and 20 bits on the right, e.g. $3145728_d = 1100000000000000000000_b$. The virtual

```
        SystemC 2.0 --- Dec 11 2001 15:28:26
  Copyright (c) 1996-2001 by all Contributors
                   ALL RIGHTS RESERVED
sqrt(0) = 0
sqrt(1) = 1
sqrt(2) = 1.4142112731933359375
sqrt(3) = 1.732051849365234375
sqrt(4) = 2
sqrt(5) = 2.2360687255859375
sqrt(6) = 2.449493408203125
sqrt(7) = 2.645748138427734375
sqrt(8) = 2.828426361083984375
sqrt(9) = 3
```

Figure 7. Fixed-Point model output

```
        SystemC 2.0 --- Dec 11 2001 15:28:26
  Copyright (c) 1996-2001 by all Contributors
                   ALL RIGHTS RESERVED

sqrt(0) = 0
sqrt(1) = 1048576
sqrt(2) = 1482912
sqrt(3) = 1816192
sqrt(4) = 2097152
sqrt(5) = 2344688
sqrt(6) = 2568484
sqrt(7) = 2774272
sqrt(8) = 2965820
sqrt(9) = 3145728
```

Figure 8. Integer-based model output

Table 1. Results

	Fixed Point	Integer
Simulation Speed	0.53 sec.	0.03 sec.
Synthesizable	no	yes
Conversion Effort (Manual.)	> 2 hours	
Conversion Effort (FixTool)	0.96 sec.	

decimal point is located at position 20 (seen from the least significant bit) and therefore the represented number is $11.00000000000000000000_b = 3.0_d$. A simulation run for the fixed-point-based version takes 0.53 seconds, whereas the integer version takes 0.03 seconds for 100 square root computations[1]. The conversion results for the Newton's example are summarized in Table 1.

Table 2. Synthesis results.

Combinational area:	17677.474609
Noncombinational area:	1644.525757
Net Interconnect area:	61395.625000
Total cell area:	19321.390625
Total area:	80717.625000

This automatically generated integer-based code has been taken directly into the CoCentric SystemC Compiler flow without any additional modification for hardware synthesis. Parts of the result are shown in Table 2.

In addition to the simple Newton's algorithm example above, we also applied the tool to a more complex design of a hand prosthesis control unit [5]. The control unit is based on artificial neural networks, which contain subunits for the signal-processing of nerve signals and classification algorithms based on Kohonen's self-organizing maps (SOM) [9]. All computations in the nerve signal recognition stage and the signal classification module are originally based on floating-point arithmetics. In the refinement process of our SystemC design flow, the appropriate accuracies have been determined and fixed-point data types have been introduced. Finally, this fixed-point-based model was automatically converted. A comparison with the fixed-point version of the control unit shows that all output values of both simulations are exactly identical.

For comparison purposes an integer-based version of the prosthesis control unit has also been implemented manually. This implementation took several days, whereas the automated conversion was done in minutes for the entire design. This shows that the automated conversion will drastically save time. Especially against the background of the refinement philosophy of SystemC and the practical work in a design flow, this becomes important: All changes and modifications concerning the algorithm or the word lengths in a fixed-point-based model have to be propagated to the integer-based modelling level.

The FixTool ensures that the (synthesizable) integer version of a design is basically up-to-date at any time of the refinement process. It allows to evaluate and modify the fixed-point version of a model instead of the integer version (see Figure 9). In the fixed-point version of a model, the designer can easily exercise full control over the bit-layout and parameters of fixed-point data types.

Similar to the Newton's algorithm example described above, parts of the prosthesis control unit have already successfully been taken through the same hardware synthesis design flow to hardware.

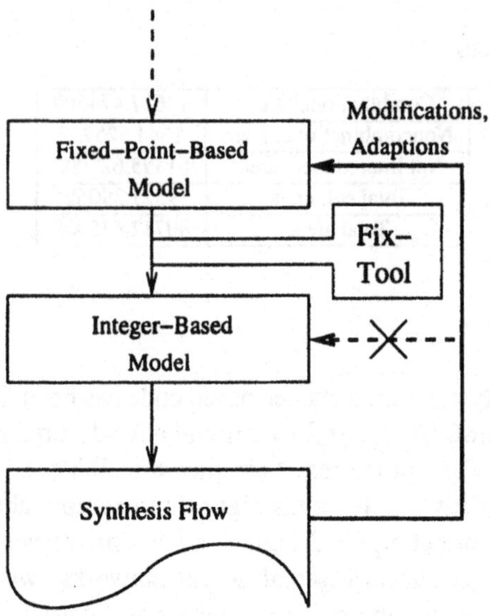

Figure 9. FixTool design flow

5. FixTool Application in a Real Case Study: Electrocardiogram Classification System

In addition to the last section, where a simple example of conversion of the Newton's algorithm was presented, we also have the objective to exemplify the application of the FixTool with a more complex design corresponding to the classification of electrocardiogram (ECG) signals. This system aims to present a new proposal in order to integrate the signal acquisition and the automatic ECG data classification in a single chip (e.g., SoC), close to the hospital patient [2]. Therefore, an IP (Intellectual Property component) was designed using the connectionist model (e.g., Artificial Neural Network - ANN) of the Artificial Intelligence area, which is capable to solve complex non-linear problems in many different areas.

ANNs are usually non-linear adaptive systems that can learn how to map input/output functions based on an input vector set. The adaptation process happens through well-defined rules that compose the training phase of the neural network. These training rules adapt the free parameters (e.g., synaptic weights) until the moment that the neural network achieves the pre-established stop criteria (for instance, minimum error or maximum number of epochs). After the training occurs, the free parameters should be fixed on the neural network in order to verify it in test phase [9].

This electrocardiogram classification system consists of a multilayer perceptron (MLP) neural network, which is composed of parallel units (i.e., neurons) responsible to process and to classify the electrocardiogram based on the well known backpropagation algorithm [9]. The system is initially described on floating point arithmetic to capture the requirements of the system. In Figure 10, we present the design flow to convert floating-point arithmetic to integer-based arithmetic in the function AF-derivate(), which is responsible to derive the error signal to each neuron of the network. The system is initially simulated to achieve the specifications requirements. After that, the floating-point to fixed-point translation is applied for the determination of the word length and its precision. The FRIDGE and CoCentric Fixed-Point Designer are examples of tools for automating this process. Also, it can be done manually, in which it is necessary to explore the system in several stages of simulation to extract variable value ranges and their precision in number of bits.

Based on fixed-point system representation, the translation to integer base can be started. This process using FixTool can be done in two modes: a) using a GUI (Graphical User Interface), where some parts of the code can be copied to the application and the translation will occur automatically; b) using a project file, where all the source files and their corresponding headers are specified. This project file is executed in the shell console and then all the system will be translated from fixed-point to integer base.

In Figure 10 we can see also that not only the variables and the arithmetic expressions can be translated, but also the function declarations with their arguments. It is possible to figure out also how the result data types are determined. The explicit type cast is inserted to adapt the result of the expression to the capacity range of the variables.

Having the integer base description of the system, it is necessary to perform the simulation to verify if the system attends the same characteristics as before. Another fact that should be pointed out is the possibility to improve the simulation speed with the integer-based representations. This process was observed based on the number of epochs[2] that the neural network is trained. As is presented in Table 3, the simulation speed based on integer arithmetic can be around 2 times faster, compared to fixed-point representation. This slower simulation speed occurs due to the higher complexity transformations used to simulate the fixed-point representation.

The last phase of the design flow is the synthesis of the system. The hardware neural network design specified in SystemC was validated on a Xilinx SPYDER-VIRTEX-X2E board. This board contains one XCV2000E FPGA of the VirtexE family (i.e., equivalent to 2.000.000 logics ports) and one CPLD responsible for controlling the 32 address/date bits of the local bus. In addition, it contains SRAM memory blocks that operate synchronously. The frequency of the system can be selected up to 33MHz.

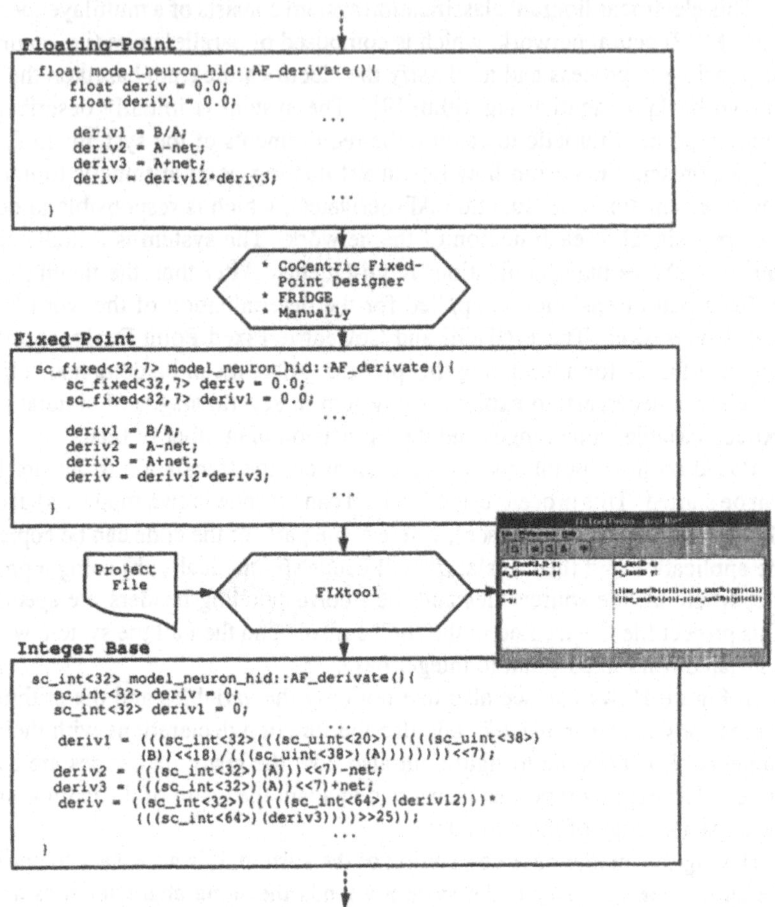

Figure 10. From floating point to integer base

Figure 11 shows the tool chain to perform the synthesis process of the system originally described in SystemC. Initially, the specified system could be analyzed and verified by CoCentric System Studio [13, 10] and DAVIS visualization environments. After its functional verification, the floating-point to fixed-point data type conversion was performed using SystemC fixed-point types. The refinement of the fixed point date types to integer base was performed by the FixTool.

Table 3. Simulation speed improvement[3].

Epochs	Floating(sec.)	Fixed(sec.)	Integer(sec.)
1	2.56	6.15	3.06
10	26.33	62.56	30.25
100	257.80	629.02	304.09

The *CoCentric SystemC Compiler*, is capable to synthesize integrated RTL and behavioral modules in logic gates level (EDIF) or HDL RTL descriptions. In this work, an EDIF description was obtained and applied to the synthesis tools of the Xilinx platform. This process is composed by mapping to the desired prototyping board. The automatic synthesis was completed after the generation of FPGA configuration file.

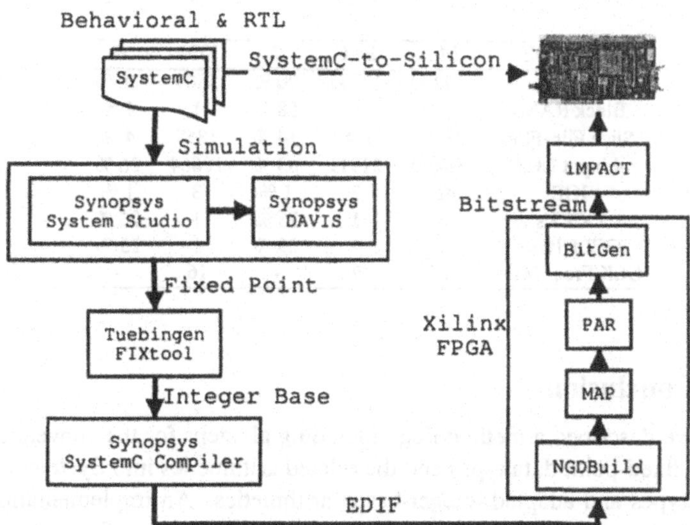

Figure 11. Synthesis process toolchain: from SystemC to Silicon.

Besides the synthesis of the electrocardiogram classification system, the classic XOR problem[4] was also synthesized so as to compare synthesis of a complex system (e.g., ECG) with a simpler one (e.g., XOR). These two examples are non-linear complex systems, which use neural network architecture with at least two layers. Thus, two ANNs were synthesized having the following architectures:

- ECG - 7 inputs / 5 neurons in hidden layer / 1 neuron in output layer;
- XOR - 3 inputs / 2 neurons in hidden layer / 1 neuron in output layer;

The synthesis results on the Xilinx platform are presented in Table 4. The first two columns present the elements and the maximum amount of them available in this FPGA. The last four columns present the synthesis results of the main system integrated with its testbench. The presented results also consider the memory block synthesis of both systems (ECG and XOR). One can notice that the ECG classification system occupied 90% of the present slices[5]. The XOR system is simpler because it has a smaller number of: neurons, interconnections and amount of local registers to store the synaptic weights. In both systems three I/O blocks (IOB) are used and also an input for reset signal and two outputs to indicate neural network training process.

Table 4. Synthesis process summary.

XCV200E	Max	ECG	%	XOR	%
Slices	19200	17391	90 %	5509	28 %
Block RAMs	160	93	58 %	1	1 %
Slice Flip-Flop	38400	5441	14 %	1858	4 %
4 input LUT	38400	31911	83 %	10064	26 %
IOB	404	3	1 %	3	1 %
GCLKs	4	1	25 %	1	25 %
GCLKIOBs	4	1	25 %	1	25 %
CLK Freq. MHz	33	7	-	16	-

6. Conclusion

We have described a methodology including all steps for the conversion of SystemC fixed-point data types and the related arithmetics into SystemC integer data types and adapted integer-based arithmetics. An implementation of the conversion tool FixTool allows the automated generation of integer-based designs out of fixed-point-based models. The tool covers all basic arithmetics like summation, subtraction, multiplication, and division. During the conversion process the structure of the original model code is preserved as far as possible in order to allow the designer to stay familiar with the design code at any time. The conversion results can be passed directly to a hardware synthesis tool, demonstrated by two examples.

The FixTool closes the gap in a system-level design flow e.g. between design tools, their floating-point-to-fixed-point features [12, 13], and hardware synthesis tools. It can avoid labour-intensive and error-prone manual conversion procedures within the SystemC refinement process.

Currently, further adaptations for a tight integration into the design flow are made. This also includes optimizations of the generated code in order to get an efficient adaptation to the synthesis tool, e.g. an optimized splitting of complex arithmetical expressions containing several divisions. These optimizations are mainly not related to the core conversion methodology, but to special requirements of the synthesis tools and the design flow beneath.

Notes

1. SystemC 2.0 on a SunBlade 100 at 500 MHz.
2. Epoch is the presentation of the whole N vectors of the training set to artificial neural network input layer.
3. The simulations were executed in a Sun-Blade-100, with 1536 Mbytes of RAM memory and with 2201 Mbytes of Swap memory.
4. The XOR problem has four input vectors (0,0), (0,1), (1,0), (1,1). The first and the fourth belong to the class 0 and consequently the second and the third belong to the class 1.
5. One CLB is composed by two *slices*.

References

[1] B.Stroustrup. *The C++ Programming Language (Special Edition). 2000.* Addison Wesley. Reading Mass. USA., 2000.

[2] D.Lettnin/A.Braun/M.Bodgan/J.Gerlach/W.Rosenstiel. Synthesis of embedded systemc design: A case study of digital neural networks. In *Design, Automation and Test in Europe Conference and Exhibition (DATE04)*, 2004.

[3] H.Keding/M.Coors/O.Luetje/H.Meyr. Fast bit-true simulation. In *38. DesignAutomation Conference (DAC)*, 2001.

[4] J.Freuer. *Entwurfsprozess einer Handprothesen-Steuerung in SystemC.* Diplomarbeit, Universitaet Tuebingen, 2002.

[5] M.Bogdan. *Signalverarbeitung biologischer Nervensignale zur Steuerung einer Prothese mit Hilfe kuenstlicher neuronaler Netzwerke.* Dissertation, Universitaet Tuebingen, 1998.

[6] M.Coors/H.Keding/O.Luetje/H.Meyr. Integer code generation for the ti tms320c62x. In *International Conference on Acoustics, Speech and Signal Processing (ICASSP)*, 2001.

[7] Open SystemC Initiative. *Functional Specification for SystemC 2.0*, version 2.0-q edition, March 2002.

[8] Open SystemC Initiative. *SystemC User's Guide*, version 2.0 edition, 2002.

[9] S.Haykin. *Neural Networks: A Comprehensive Foundation.* New Jersey:Prentice-Hall, USA., 1999.

[10] Synopsys, Inc. *CoCentric SystemC Compiler Behavioral User Guide*, version 2000.11 edition, March 2001.

[11] Synopsys, Inc. *CoCentric Fixed-Point Designer User Guide*, version 2002.05 edition, 2002.

[12] Synopsys, Inc. *CoCentric System Studio Reference Manual*, version 2002.05 edition, June 2002.

[13] Synopsys, Inc. *CoCentric System Studio User Guide*, version 2002.05 edition, June 2002.

[14] T.Groetker/S.Liao/G.Martin/S.Swan. *System Design with SystemC*. Kluwer Academic Publishers, Boston/Dodrecht/London, 2002.

[15] W.Mueller/W.Rosenstiel/J.Ruf. *SystemC Methodologies and Applications*. Kluwer Academic Publishers, Boston/Dordrecht/London, 2003.

EXPLORATION OF SEQUENTIAL DEPTH BY EVOLUTIONARY ALGORITHMS

Nicole Drechsler
Institute of Computer Science
University of Bremen
28359 Bremen, Germany
nd@informatik.uni-bremen.de

Rolf Drechsler
Institute of Computer Science
University of Bremen
28359 Bremen, Germany
drechsle@informatik.uni-bremen.de

Abstract Verification has become one of the major bottlenecks in today's circuit and system design. Up to 80% of the overall design costs are due to checking the correctness. Formal verification based on Bounded Model Checking (BMC) is a very powerful method that allows to prove the correctness of a device. In BMC the circuits behavior is considered over a finite time interval, but for the user it is often difficult to determine this interval for a given *Device Under Verification* (DUV).

In this paper we present a simulation based approach to automatically determine the sequential depth of a *Finite State Machine* (FSM) corresponding to the DUV. An *Evolutionary Algorithm* (EA) is applied to get high quality results. Experiments are given to demonstrate the efficiency of the approach.

Keywords: Verification of sequential circuits, evolutionary algorithms, simulation based approach

1. Introduction

Modern circuits contain up to several hundred million transistors. In the meantime it has been observed that verification becomes the major bottleneck in circuit and system design, i.e. up to 80% of the overall design costs are due to verification. This is one of the reasons why recently several methods have been proposed as alternatives to classical simulation, since it cannot guarantee sufficient coverage of the design. E.g. in [Bentley, 2001] it has been reported

Please use the following format when citing this chapter:

Drechsler, Nicole, Drechsler, Rolf, 2006, in IFIP International Federation for Information Processing, Volume 200, VLSI-SOC: From Systems to Chips, eds. Glesner, M., Reis, R., Indrusiak, L., Mooney, V., Eveking, H., (Boston: Springer), pp. 73-83.

that for the verification of the Pentium IV more than 200 billion cycles have been simulated, but this only corresponds to 2 CPU minutes, if the chip is run with 1 GHz.

As alternatives, formal verification or symbolic simulation have been proposed and in the meantime these techniques have been successfully applied in many industrial projects. To allow for an early detection of design errors, model checking has been used. While "classical" CTL-based model checking [Burch et al., 1990] can only be applied to medium sized designs, approaches based on *Bounded Model Checking* (BMC) as discussed in [Biere et al., 1999] give very good results when used for complete blocks with up to 100k gates.

But there is one inherent problem when applying BMC: The circuit is considered over a fixed time interval and to give complete proofs it is important to determine the sequential depth of the circuit. Recently in [Yen et al., 2002] an approach based on simulation in combination with a toggle-heuristic has been proposed, but experiments have shown that the method might result in over- or under-approximations and often gives sub-optimal results. This makes the technique hard to use for a designer or a verification engineer. An exact solution to this problem based on a problem formulation as quantified Boolean functions has been proposed in [Mneimneh and Sakallah, 2003]. A SAT-solver is applied to compute the optimal result, but due to the complexity of real-world circuits this technique cannot be applied to larger problem instances.

In this paper we present a simulation based algorithm for computation of the sequential depth of FSMs. The quality of the simulated vectors is evaluated using techniques from *Evolutionary Algorithms* (EAs). It has been observed that EAs work very well in testing applications [Corno et al., 1996b; Corno et al., 1996a; Rudnick et al., 1997; Keim et al., 2001; Drechsler and Drechsler, 2002] and here the underlying problem is very similar. Experiments show that the same quality can be obtained as the exact approach but using simulation techniques only. By this, the EA technique combines the best of the two approaches from [Yen et al., 2002] and [Mneimneh and Sakallah, 2003], i.e. we get the optimal results but for the evaluation no time consuming proof techniques, like BDD or SAT, are used, but only simulation that can be carried out in linear time in the circuit size.

The paper is structured as follows: First, basic definitions of sequential circuits and sequential depth computation are outlined. Then the proposed EA for depth approximation is presented. Experimental results show the quality of the presented approach and finally, the paper is summarized.

2. Preliminaries

A synchronous sequential circuit can be described using a *Finite State Machine* (FSM). An FSM is a 5-tuple $M = (I, O, S, \delta, \lambda)$, where I is the input

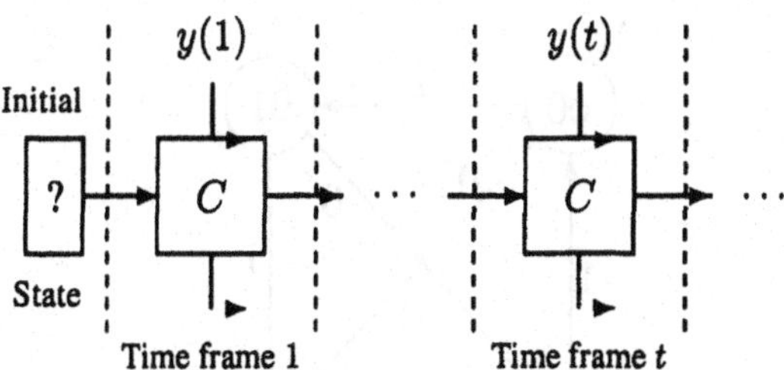

Figure 1. Iterative description of a sequential circuit

set, O is the output set and S is the set of states. $\delta : I \times S \rightarrow S$ is the next-state function and $\lambda : I \times S \rightarrow O$ is the output function. Since we consider a gate level realization of the FSM, we have $I = \mathbf{B}^k$, $O = \mathbf{B}^l$, and $S = \mathbf{B}^m$ with $\mathbf{B} = \{0, 1\}$. k denotes the number of primary inputs, l denotes the number of primary outputs, and m denotes the number of memory elements. The functions δ and λ are computed by a combinational circuit C. The inputs (outputs) of the combinational circuit, which are connected to the outputs (inputs) of the memory elements, are called secondary inputs (outputs). Sometimes the secondary inputs are called *present state variables* and the secondary outputs are called *next state variables.*

For the description of our algorithms we use the following notations: $X = x(1), \ldots, x(n)$ denotes the input sequence of depth n. s_i denotes the next state defined by $x(i)$ and s_{i-1}, $1 \leq i \leq n$.

Using these notations the next state is given by

$$s(s_0, t) = \begin{cases} s_0 & \text{if } t = 0 \\ \delta(x(t), s(s_0, t-1)) & \text{otherwise} \end{cases}$$

In doing so, we consider a synchronous sequential circuit as an iterative network (see Figure 1).

The state transition graph of an FSM is a labeled directed graph $T = (V, E)$ where each node $v \in V$ corresponds to a state s_i, $0 \leq i \leq |S| - 1$, of M, and each edge $e = (v, w)$, $v, w \in V$, corresponds to a transition from state s_i to state s_j. The edge is labeled with $y \in I^k$ which is the input vector that affects the transition from s_i to s_j, i.e. $\delta(y, s_i) = s_j$, $0 \leq i, j \leq |S| - 1$.

A *path* is a sequence of nodes v of T where all nodes are different. Using the definitions above, the *sequential depth* of an FSM is given as follows:

> Consider an FSM M and its corresponding state transition graph T with a single initial state s_0. Find a path of maximum length starting in s_0 such that each node along the path is visited only once and additionally, the path has maximum length.

EXAMPLE 1 *In Figure 2 a state transition graph with four states is illustrated. If the initial state is 00, only path 00-10-11-01 with length 3 exists. All other*

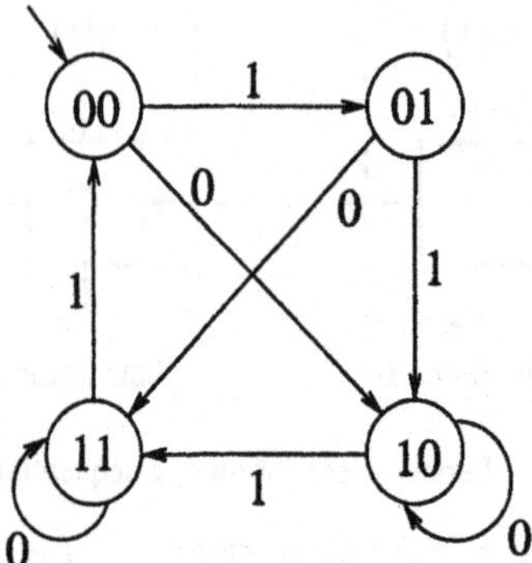

Figure 2. State transition graph

paths have a shorter length, i.e. paths 00-01-11 and 00-10-11 have length 2. The resulting sequential depth of the given example is 3.

In the next section we present a simulation based optimization technique for determining the sequential depth of an FSM.

3. Evolutionary Algorithm

In this section the different components of the EA are described. Instead of a single solution, EAs consider a whole set - also called a population. First, the encoding of these elements and their representation is presented in the following sections. The "critical part" of the EA is to measure the quality of simulation sequences. This is done in several steps using multi-objective optimization. Then, the evolutionary operators used are described and finally, the overall algorithmic flow - including the detailed choices for the parameter settings - is discussed.

Representation

Each individual in the population represents a set of m input vectors $\tilde{Y}$. An upper limit on the size of the vector set is given by the user and the length of one vector is given by the number of input variables k. An individual is a vector set represented by a binary string of length $k \cdot |\tilde{Y}|$.

During the initialization phase, these strings are randomly chosen.

Objective Function

Simulation. Each individual is evaluated by the objective function to determine its quality. For the evaluation of the objective function the set of vectors represented by an individual is simulated starting from the initial state s_0.

- Starting from the initial state s_0 the set of next states is calculated:

$$S = \bigcup_{i=1}^{|\tilde{Y}|} \delta(\tilde{y}_i, s_0),$$

  where $\tilde{y}_i \in \tilde{Y}$.

- Then for each *new* state in S and $\tilde{Y}$ the set of next states is calculated. I.e.:

$$S = \bigcup_{i=1}^{|\tilde{Y}|} \bigcup_{j=1}^{|S_{new}|} \delta(\tilde{y}_i, s_j),$$

  where $y_i \in \tilde{Y}$ and $s_j \in S_{new}$.

- This is repeated, until no *new* state is found.

A sketch of the algorithm is given in Figure 3. The sets S, S_{new} and $S_{present}$ are initialized with the initial state. In set S all states reached during the exploration are included. S_{new} describes only the set of *new* states reached in the present exploration step and $S_{present}$ is set S one time step before. Then for each vector in $\tilde{Y}$ and each state in S_{new} the next states are calculated.

If no new state is found the algorithm terminates and the present value of *depth* is calculated by the input set $\tilde{Y}$.

Multi-objective Optimization. For EAs it has been observed that often a single objective function is not sufficient to allow for high quality results. Using only the computed depth as optimization criterion would prefer input vectors that calculate a *maximum* (instead of the sequential) depth of the given FSM. Thus, several specialized techniques have been developed following the paradigm of *Multi-Objective Optimization* (MOO) [Deb, 2001].

In MOO several criteria are considered in parallel during optimization. The *classical* approach of combining multiple criteria is the *weighted-sum*, i.e. the fitness values of the objectives are combined using linear (or quadratic) combination. One drawback of the method is e.g. that distinct solutions can compute the same fitness. Thus the algorithm is not able to distinguish between these solutions. EAs are very well suited to deal with multi-objective problems, because several solutions are considered in parallel in a population. For this, MOO in EAs has been studied very well in both, theorie and practice. In our application we make use of the MOO technique proposed in [Drechsler et al., 1999] that has been integrated in the software library GAME [Goeckel et al., 1997].

The advantage of GAME's method is that the handling of priorities of selected objectives is supported. For each criterion a priority has to be deter-

```
compute_depth (individual) {
    S := {s_0} ;
    S_new := {s_0} ;
    S_present := {s_0} ;
    depth := 0 ;
    do {
      for i := 1 to |Ỹ| do {
        for j := 1 to |S_new| do {
          s := δ(ỹ_i, s_j) ;
          S := S ∪ {s} ;
        }
      }
      depth := depth +1 ;
      S_new := S \ S_present ;
      S_present := S ;
    } while (S_new ≠ ∅) ;
    return depth ;
}
```

Figure 3. Objective function

priority	objective
1	maximize total number of visited states
1	minimize depth
$k, k = 2, \ldots, depth$	maximize number of visited states in depth k

Figure 4. Optimization objectives and their priorities

mined, that ranks "how important" this objective is. The choices for our application are given in Figure 4. As can be seen, two optimization objectives have the highest priority: the total number of reached states has to be maximized and the computed depth has to be minimized. Thus, the input sets are optimized such that a maximum number of states with a minimum depth is reached. Furthermore, objectives with descending priorities maximize the number of states reached in level k, $2 \leq k \leq depth$. Then the input sets where the states are visited "as fast as possible" are preferred during the optimization process.

Operators

Now the evolutionary operators that are the "core operators" of EA applications are described. First, we distinguish between "standard" crossover operators (well-known for EAs [Davis, 1991]) and problem specific operators [Corno et al., 1996b; Corno et al., 1996a; Keim et al., 2001]. In our framework we only make use of the standard operators and one problem specific "meta operator", that is a generalization of all the others. Additionally, we make use of "classical" mutation operators to explore the local region of proposed solutions.

First, the standard EA operators are briefly reviewed: All operators are directly applied to binary strings of length l that represent elements in the population. The parent(s) for each operation is (are) determined by *Tournament-selection*. For the selection of each parent element two individuals are randomly chosen from the population. Then the better individual - with respect to its ranking in the population - is selected.

Crossover: Construct two new elements c_1 and c_2 from two parents p_1 and p_2, where p_1 and p_2 are split in two parts at a cut position i. The first (second) part of c_1 (c_2) is taken from p_1 and the second (first) part is taken from p_2. (Notice, that a special case of this operator is the *horizontal crossover* from [Corno et al., 1996a], where the cut position is chosen only between two test vectors, i.e. test vectors are not split up.)

2-time Crossover: Construct two new elements c_1 and c_2 from two parents p_1 and p_2, where p_1 and p_2 are split in three parts at cut positions i and j. The first (second) part of c_1 (c_2) is taken from p_1 (p_2), the second part is taken from p_2 (p_1) and the last part is again taken from p_1 (p_2).

Uniform Crossover: Construct two new elements c_1 and c_2 from two parents p_1 and p_2, where at each position the value is taken with a certain probability from p_1 and p_2, respectively.

Next, the problem specific operator is presented. The string representation of a sequence of vectors is interpreted as a two-dimensional matrix, where the

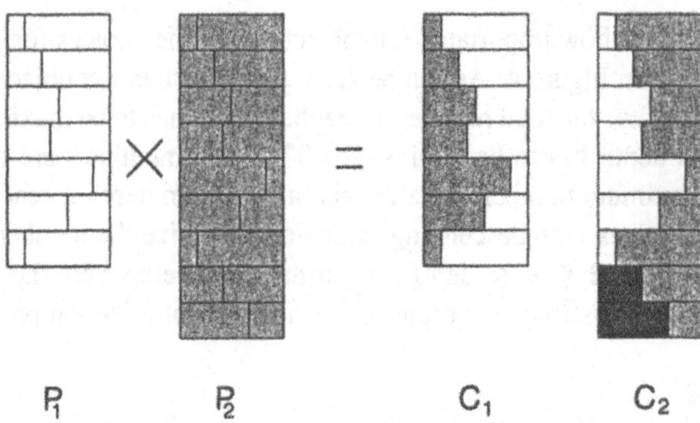

Figure 5. Example for Free-Vertical Crossover

x-dimension represents the number of inputs and the y-dimension represents the number of vectors. The operator works as follows [Keim et al., 2001]:

Free Vertical Crossover: Construct two new elements c_1 and c_2 from two parents p_1 and p_2. Determine for each test vector t a cut position i_t. Divide each test vector t of p_1 and p_2 in two parts at cut position i_t. The first (second) part of each test vector of c_1 (c_2) is taken from p_1 and the second (first) part is taken from p_2. (Notice, that the *vertical crossover* from [Corno et al., 1996b] is a special case of this operator, if i_t is equal for all test vectors t.)

EXAMPLE 2 *The behavior of the* free vertical crossover *is illustrated in Figure 5. The black filled areas result, if vector sets of different size are considered; then, the offsprings are filled with randomly generated values. (But, in our application all individuals have the same length.)*

Moreover, three (standard) mutation operators are applied which are based on bit-flipping at a random position.

Mutation (MUT): Construct one new element c from a parent p by copying the whole element and changing a value at a randomly chosen position i.

2-time Mutation: Perform MUT two times on the same element.

Mutation with neighbour: Perform MUT at two adjacent positions on the same element.

Obviously, all evolutionary operators generate only valid solutions, if they are applied to binary strings.

```
approximate_sequential_depth (circuit) {
    generate_random_population () ;
    evaluate_population () ;
    do {
        apply_evolutionary_operators () ;
        evaluate_offsprings () ;
        update_population () ;
    } while (not terminal case) ;
    return (best_element) ;
}
```

Figure 6. Sketch of basic algorithm

Algorithm

We now introduce the basic EA which describes the overall flow. (A sketch is given in Figure 6.)

- The initial population of size $|\mathcal{P}|$ is generated, i.e. the binary strings of length l are initialized using random values.
- Two parent elements are determined by *Tournament*-selection.
- Two new individuals are created using the evolutionary operators with given probabilities.
- These new individuals are then mutated by one of the mutation operators with a fixed mutation rate.
- The quality of the elements is determined by simulation and MOO ranking.
- The elements which lost the tournament selection in the present parent population are deleted and the offsprings are inserted in the population.
- The algorithm stops if the best element has not changed for 100 generations.

For the experiments the following parameters have been used: The population size is set to $|\mathcal{P}| = 24$. The vertical crossover is carried out with a probability of 80% and one out of the standard crossover operators is carried out with a probability of 20%, respectively. The offsprings are mutated with a probability of 15% by one of the mutation operators.

Table 1. Experiments for ISCAS circuits

name	Sim	SAT	EA
s298	18	18	18
s208	255	n.a.	255
s349	6	n.a.	6
s386	n.a.	7	7
s499	n.a.	21	21
s510	46	n.a.	46
s526	150	n.a.	150
s641	n.a.	6	6
s713	10	6	6
s820	n.a.	10	10
s953	n.a.	10	10
s1196	5	2	2
s1488	21	21	21

4. Experimental Results

The techniques described in the previous section have been implemented using the software library GAME [Goeckel et al., 1997]. All algorithms are written in $C/C++$ and the experiments were all run on a SUN Ultra with 256 MByte main memory. As a simulator for evaluation of the objective function we used a simple functional approach based on the ideas of [Ashar and Malik, 1995]. Here, the underlying BDD package is CUDD from [Somenzi, 2001]. For the experiments a sample of the benchmarks from ISCAS were taken.

The experimental results are given in Table 1. The name of the benchmark is given in the first column. The columns *Sim* and *SAT* give the results from [Yen et al., 2002] and [Mneimneh and Sakallah, 2003], respectively. It is important to notice that *Sim* obtains estimations only, while *SAT* give the exact numbers. As can be seen, compared to *SAT* the other technique gives over- as well as under-approximations, what makes them hard to use in real-world scenarios.

The results of our EA approach are given in the last column. It can be observed that in all cases the exact results (where this is known) is computed. But since the EA is based on simulation, it can also be applied to larger circuits.

In this way, the presented technique combines the best of [Yen et al., 2002] and [Mneimneh and Sakallah, 2003], i.e. very high-quality results are computed but for the evaluation no time consuming proof techniques, like BDD or SAT, are used. Instead, simulation that can be carried out in linear time in the circuit size is successfully applied.

5. Conclusions

In this paper a simulation-based approach for the computation of the sequential depth of a FSM has been presented. Due to the choice of the objective function results of high quality can be obtained. This finds direct application in BMC, since the depth of the FSM corresponding to the DUV gives the results for the maximal time interval that has to be considered.

The run time of the algorithm is dominated by the simulation time. For this, it is a focus of current work to integrate a more efficient parallel simulator in the GAME software library.

References

Ashar, P. and Malik, S. (1995). Fast functional simulation using branching programs. In *Int'l Conf. on CAD*, pages 408–412.

Bentley, B. (2001). Validating the Intel Pentium 4 microprocessor. In *Design Automation Conf.*, pages 244–248.

Biere, A., Cimatti, A., Clarke, E.M., Fujita, M., and Zhu, Y. (1999). Symbolic model checking using SAT procedures instead of BDDs. In *Design Automation Conf.*, pages 317–320.

Burch, J.R., Clarke, E.M., McMillan, K.L., and Dill, D.L. (1990). Sequential circuit verification using symbolic model checking. In *Design Automation Conf.*, pages 46–51.

Corno, F., Prinetto, P., Rebaudengo, M., and Reorda, M.S. (1996a). GATTO: A genetic algorithm for automatic test pattern generation for large synchronous sequential circuits. *IEEE Trans. on CAD*, 15(8):991–1000.

Corno, F., Prinetto, P., Rebaudengo, M., Reorda, M.S., and Mosca, R. (1996b). Advanced techniques for GA-based sequential ATPG. In *European Design & Test Conf.*, pages 375–379.

Davis, L. (1991). *Handbook of Genetic Algorithms*. van Nostrand Reinhold, New York.

Deb, K. (2001). *Multi-objective Optimization using Evolutionary Algorithms*. John Wiley and Sons, New York.

Drechsler, N., Drechsler, R., and Becker, B. (1999). A new model for multi-objective optimization in evolutionary algorithms. In *Int'l Conference on Computational Intelligence (Fuzzy Days)*, volume 1625 of *LNCS*, pages 108–117. Springer Verlag.

Drechsler, R. and Drechsler, N. (2002). *Evolutionary Algorithms for Embedded System Design*. Kluwer Academic Publisher.

Goeckel, N., Drechsler, R., and Becker, B. (1997). GAME: A software environment for using genetic algorithms in circuit design. In *Applications of Computer Systems*, pages 240–247.

Keim, M., Drechsler, N., Drechsler, R., and Becker, B. (2001). Combining GAs and symbolic methods for high quality test of sequential circuits. *Jour. of Electronic Testing: Theory and Applications*, 17:141–142.

Mneimneh, M. and Sakallah, K. (2003). SAT-based sequential depth computation. In *ASP Design Automation Conf.*

Rudnick, E.M., Patel, J.H., Greenstein, G.S., and Niermann, T.M. (1997). Genetic algorithm framework for test generation. *IEEE Trans. on CAD*, 16(9):1034–1044.

Somenzi, F. (2001). Efficient manipulation of decision diagrams. *Software Tools for Technology Transfer*, 3(2):171–181.

Yen, C.-C., Chen, K.-C., and Jou, J.-Y. (2002). A practical approach to cycle bound estimation for bounded model checking. In *Int'l Workshop on Logic Synth.*, pages 149–154.

VALIDATION OF ASYNCHRONOUS CIRCUIT SPECIFICATIONS USING IF/CADP

Dominique Borrione[1], Menouer Boubekeur[1], Laurent Mounier[2], Marc Renaudin[1] and Antoine Siriani[1].
[1]TIMA, 46 avenue Félix Viallet, 38031 Grenoble Cedex, France; [2] VERIMAG, Centre Equation, 2 avenue de Vignate, 38610 Gières, France

Abstract: This work addresses the analysis and validation of modular CHP specifications for asynchronous circuits, using formalisms and tools coming from the field of distributed software. CHP specifications are translated into an intermediate format (IF) based on communicating extended finite state machines. They are then validated using the IF environment, which provides model checking and bi-simulation tools.

1. INTRODUCTION

Asynchronous circuits show interesting potentials in several fields such as the design of numeric operators, smart cards and low power circuits [1]. An asynchronous circuit can be seen as a set of communicating processes, which read data on input ports, perform some computation, and finally write on output ports. In our work, asynchronous circuit specifications are written in CHP, an enriched version of the CSP-based language initially developed by Alain Martin [2].

Even medium size asynchronous circuits may display a complex behavior, due to the combinational explosion in the chronology of events that may happen. It is thus essential to apply rigorous design and validation methods. This paper describes the automatic validation of asynchronous specifications written in CHP, prior to their synthesis with the TAST design flow [12, 14].

Please use the following format when citing this chapter:
Borrione, Dominique, Boubekeur, Menouer, Mounier, Laurent, Renaudin, Marc, Siriani, Antoine, 2006, in IFIP International Federation for Information Processing, Volume 200, VLSI-SOC: From Systems to Chips, eds. Glesner, M., Reis, R., Indrusiak, L., Mooney, V., Eveking, H., (Boston: Springer), pp. 85-100.

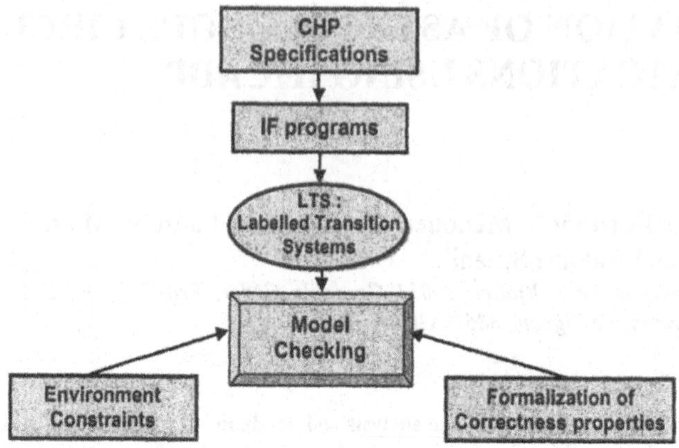

Figure 1. Formal verification flow for CHP

To this aim, we use formalisms and tools coming from the field of software validation, in particular distributed systems, whose execution model is similar to the asynchronous circuits one. We start from an asynchronous specification written in CHP and compile it into the IF format. Resulting IF programs are compiled towards a LTS and eventually submitted to the CADP toolset for verification (Figure 1).

This paper is organized as follows. Section 2 reviews the TAST design flow from CHP. Section 3 describes the validation of concurrent systems with the IF environment. Section 4 discusses the translation of CHP specifications into the IF format, with an emphasis on the CHP concepts that have no direct correspondence in IF. As a case study, section 5 presents the application of our method to an asynchronous FIR filter. Finally, we review related works and present our conclusions.

2. THE TAST DESIGN FLOW

In the TAST asynchronous design flow [14], the compiler translates CHP programs into Petri Nets (PN) and Data Flow Graphs (DFG) (Figure 2). The PN model of a CHP specification is translated to behavioral VHDL for simulation purposes. The synthesizer performs process decomposition and refinements on the PN formalization, depending on the selected architectural target: micro-pipeline, quasi delay insensitive circuit (QDI), or synchronous circuit. A dedicated compiler produces a structural gate network, in source

VHDL, for simulation and back-end processing using commercial CAD tools.

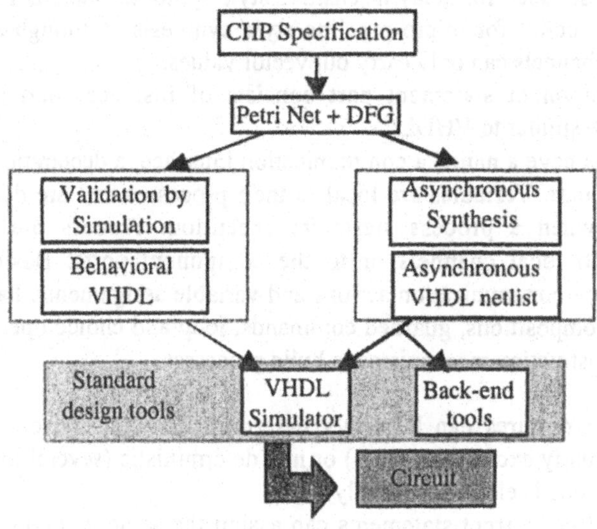

Figure 2. The TAST design flow

2.1 CHP: the TAST Specification Language

CHP Specifications are organized as lists of components. Each component has a name, a communication interface, a declaration part and a statement part. Hierarchy is managed through local component declarations and instance specifications like in VHDL.

Data types are of three kinds: unsigned multi-rail (MR), signed multi-rail (SMR) and single-rail (SR) with base, length and dimension attributes standing for the range of digits, the number of digits in a vector and the number of vectors in an array, respectively.

This allows supporting arbitrary precision bi-dimensional arrays of numbers, thus modeling memories, registers as well as pure protocol signaling (using SR type) independently of the precision of the machine.

Constants are declared at the component level. They are visible throughout the component and cannot be masked by process variables.

Processes communicate via point-to-point, asymmetric, memory-less message passing, and compatible protocol links. Channels have a name, an encoding and a type. They connect ports of opposite directions (in/out) and opposite protocols (active/passive). Data can be encoded in two ways: DI (one of n code) used for delay insensitive style synthesis and BD for bundled data (binary code) for micro-pipeline style synthesis. Although arrays are supported, channels can only carry out vector values.

The component statement part consists of instances and processes. Instances are similar to VHDL.

Processes have a name, a communication interface, a declaration part and a statement part. Variables are local to their processes and are dynamically initialized when a process starts its execution. Process ports can be connected to local channels or to the component ports. Basic process statements are communication actions and variable assignments. Parallel and sequential compositions, guarded commands, loop and choice operations are the main construction mechanisms to build processes.

Control structures can be either deterministic (the environment must provide mutually exclusive guards) or non deterministic (several guards may be true, only one is elected randomly).

Note that concurrent statements can assign the same variable, or access the same channel, within a process. It may be desirable to check the absence of such behaviors in a given design context.

3. THE IF VALIDATION ENVIRONMENT

IF[10] is a software environment developed in Verimag for the formal specification and validation of asynchronous systems. It provides both a description language (the so-called IF intermediate format), and a set of integrated validation tools. This environment is motivated by two main objectives:

Gather complementary formal validation techniques (interactive simulation, model-checking, test case generation, etc.) into a single open environment, independent of any high-level description language;

Support several representation levels of a system behavior: a syntactic level, expressed by the IF intermediate format, and a semantic level in terms of labeled transition systems.

This environment has already been used for software validation in various application domains: safety critical systems, telecommunication protocols, etc. [18].

3.1 The IF intermediate format

A IF system description consists of a set of communicating processes. Each process can access (private) local data and (shared) global data. Inter-process communication can be performed by message passing (through asynchronous buffers), by *rendez-vous* (through gates), or via shared data.

Several predefined data types are proposed (Boolean, integer, enumerated types, arrays, records, etc.) together with abstract data types definition facilities (their concrete implementation being provided in C).

From a syntactic point of view, each process is described as an extended automaton with explicit control states and transitions. Each transition is labelled by an atomic execution step and may include a Boolean guard, an inter-process communication, and assignments to local or global data. The execution model is asynchronous: internal process steps are performed independently from each other, and their parallel composition is expressed by interleaving. However a notion of *unstable* control states allows tuning the atomicity level of process transitions: sequences of transitions between unstable states are considered atomic.

Rather than a "user oriented" description language, IF is a general intermediate format for asynchronous systems. Its main advantages are its expressiveness, flexibility, formal operational semantics, and its well-accepted underlying automaton model. Several high-level specification languages are already automatically translated into IF, such as SDL [10] and UML profiles [19].

3.2 The IF validation toolbox

The IF toolbox is partitioned into two layers:

The *syntactic layer* provides source level operating tools based on static analysis. They allow performing static optimizations (dead code elimination, live variable analysis) and program slicing.

The *semantic layer* relies on the so-called model-based validation technique. A generic simulation engine builds the labeled transition system (LTS) expressing the exhaustive behaviour of an IF description. Various validation tools can then operate on this LTS to check whether it satisfies a given property. Particular available tools are the ones offered by the CADP toolbox [7], namely a general temporal logic verifier, and a bi-simulation checker.

In model-based validation of asynchronous systems one of the major concern is to avoid (or at least limit) the state explosion problem (the size of the underlying LTS). In the IF toolbox this problem is tackled at several levels:

- static optimization and program slicing are very efficient since performed at the source level;
- the IF simulation engine is able to produce "reduced" LTS with respect to a preorder relation between execution sequences (still preserving some trace properties);
- large descriptions can be verified in a compositional way (by generating and reducing in turn the LTS associated to each part of the specification).

4. TRANSLATION OF CHP DESCRIPTIONS INTO IF

In this section, the semantics of the CHP language and its translation into the IF format are briefly presented.

4.1 Concepts in direct correspondence

The CHP notions of *component, port, variable* and *process* are in direct correspondence with the notions of *system, gate, var* and *process* in IF.

All ports and declared local variables must be typed in both formalisms, but many CHP types give implementation information (e.g. one-hot coding) and their translation must be decided individually.

The three simple statements found in a process body are identical (syntax and semantics) in CHP and IF:

- var_Id := expression   variable assignment
- In_port ? var_Id   read an input port into a variable
- Out_port ! var_Id write a variable to an output port

4.2 Concepts of CHP that need a transformation step

4.2.1 From channels to synchronization expressions

In IF the communication architecture is given by a LOTOS-like synchronization expression. To construct this expression, the headings of all processes must be analyzed to identify the synchronization ports between each pair of processes. To lose no information in the translation phase, comments are added (e.g. Input/Output to distinguish channel direction in the IF system).

4.2.2 Process ports

All the component and process *ports* must be declared as *gates* of the IF *system,* even the process ports which are connected to local channels. Thus, ports declared as CHP process ports must be added to the system list of gates if they are not already present (as a component port, or as a port of a previous process).

4.2.3 Process body

The block of statements describing the behavior of the process is translated into a set of states and labeled transitions.

4.3 Statements and their composition

In the following, let S1, ...Si denote general statements,
A1...Aj denote simple assignments, and RW1,...RWi simple read or write communication actions.

4.3.1 Sequential composition

In CHP, the sequential composition of S1 and S2, written S1 ; S2 means that S2 starts its execution after the end of S1. To efficiently translate this composition (and reducing the numbers of IF states) we consider three cases, depending on S1 and S2 structure:

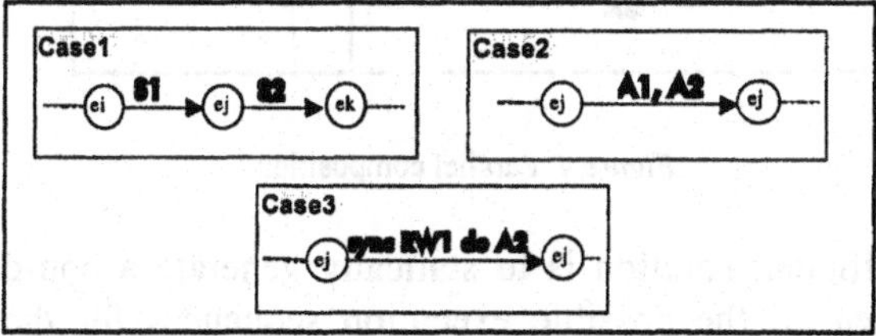

Figure 3. Sequential composition

CASE 1: S1 and S2 are possibly complex, or involve more than one communication (general case): intermediate states must be introduced. In IF syntax: from ei do S1 to ej ;

from ej do S2 to ek;

CASE 2: A1 and A2 are simple assignments: they can all be sequentially executed during a single state transition. In IF syntax: from ei do A1, A2 to ej;

CASE 3: The first statement RW1 is a communication, considered to synchronize one or more following simple assignments. In IF syntax: ei sync RW1 do A2 to ej;

4.3.2 Parallel composition

In CHP, statements separated by commas are concurrent. In IF, no concurrent statements may exist inside a process, parallelism exists only between processes.

To translate the CHP concurrent composition [S1 , S2] the general solution involves the creation of a sub-process *comp* for statement *S2*, and the explicit synchronization of its execution start and completion, as shown on Figure 4. This solution however involves a significant overhead in model size, as it generates all inter-leavings of *comp* with all the model processes.

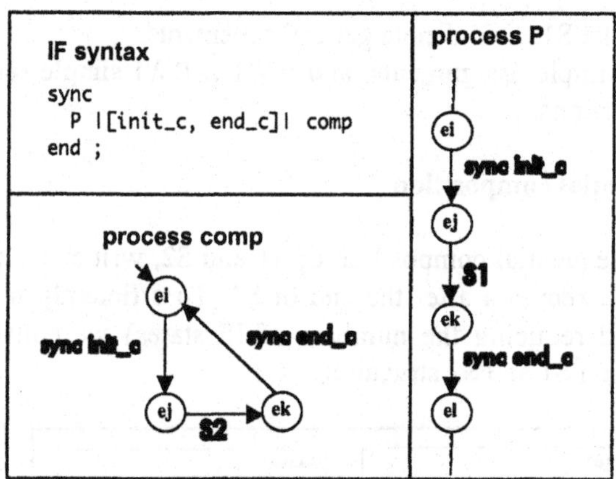

Figure 4. Parallel composition

A more efficient solution is to statically generate a non-deterministic choice between all the possible execution sequences for the concurrent statements, provided these statements initially label a single transition. This keeps the inter-leavings local, and prevents the proliferation of states for the overall model (Figure 5).

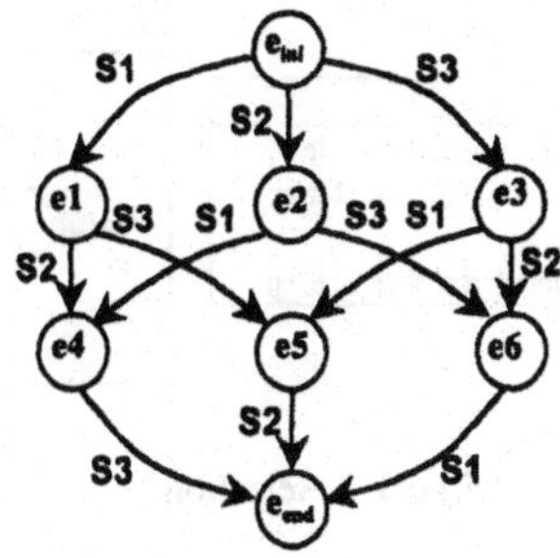

Figure 5. Interleaving generation for [S1, S2, S3]

4.3.3 Repetition

The loop statement of CHP allows repeating the execution of a simple or composed statement S. A repeated simple statement labels a transition from a state to the same state (case 1). If S is complex, one or more intermediate states may be generated for its translation, but the first and final states are the same (case 2).

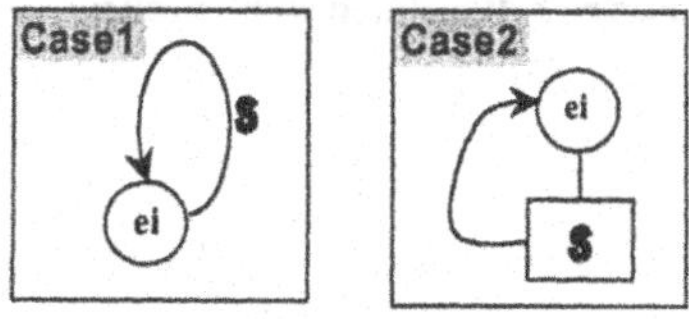

Figure 6. Repetition

4.3.4 Deterministic / non deterministic selection

Let Ci be Boolean expressions and Si be simple or composed statements, guarded by Ci. If Ci is false, Si is stalled; if Ci is true, Si is executable.

CHP offers two selection operators. The non-deterministic selection @@ encloses one or more guarded statements, and makes no assumption on the number of guards than can be true. If more than one Ci is true, one among the corresponding Si is randomly selected and executed. The deterministic selection @ imposes on its environment that only one of the Ci be true, a property to be verified. Both are translated as a set of guarded statements that label a set of transitions between the same two states (see Figure 7).

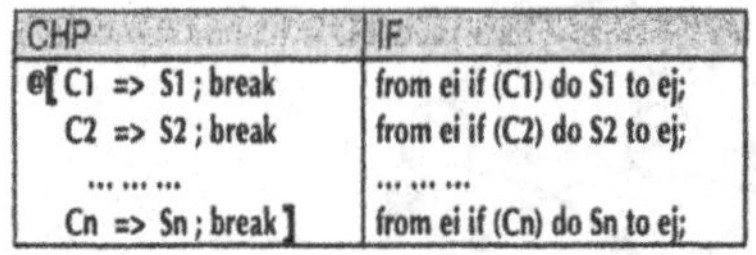

CHP	IF
@[C1 => S1 ; break	from ei if (C1) do S1 to ej;
C2 => S2 ; break	from ei if (C2) do S2 to ej;
... ...	... ...
Cn => Sn ; break]	from ei if (Cn) do Sn to ej;

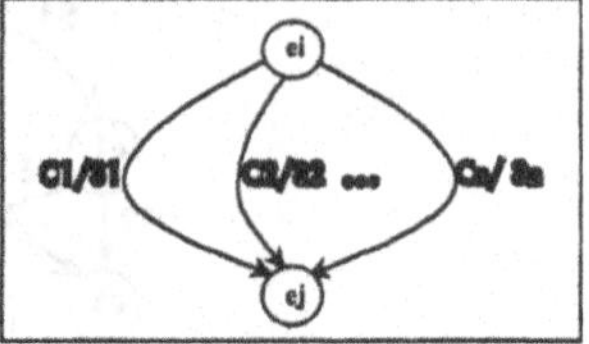

Figure 7. Selection

Repetition and selection may be combined: if one ore more guarded statements end with *loop* instead of *break*, the whole selection block is re-entered upon execution of those statements (Figure 8).

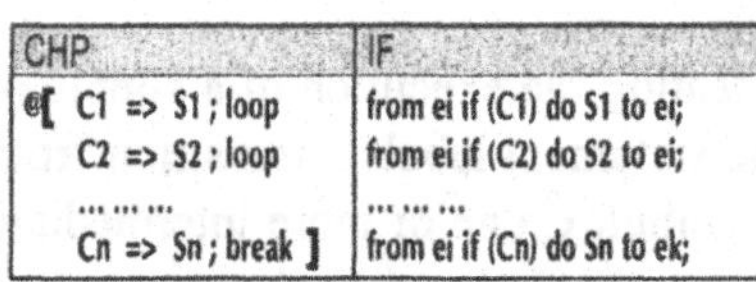

CHP	IF
@[C1 => S1 ; loop	from ei if (C1) do S1 to ei;
C2 => S2 ; loop	from ei if (C2) do S2 to ei;
... ...	... ...
Cn => Sn ; break]	from ei if (Cn) do Sn to ek;

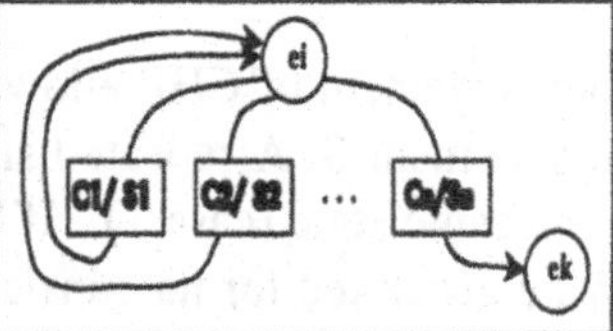

Figure 8. Repetition with selection

4.4 Example

To illustrate the translation, we consider one typical process: *Mux_3L*, taken from the case study of section 5.

In this process, the control channel (*Ctrl_Round1_L*) is typed MR[3][1], i.e. one-of-three data encoding. Control channel is read in the local variable "Ctrl". According to the value of "Ctrl", one of the two channels (*L0*, *Li_buf2*) is read and its value is written on channel *L16* or *Li_1_buf1*.

CHP code of Multiplexer "Mux_3L"

```
process Mux_3L
PORT ( L0, Li_buf2        : IN DI passive DR[32];
    L16,  Li_1_buf1, Crl_round1_l : IN DI passive MR[3]; )
Variable ctrl  : MR[3];
Variable in1  : DR[32];
begin
[Ctrl_round1_l ? ctrl;
@[ Ctrl = "0"[3]  => L0 ? in1 ; Li_1_buf1 ! in1 ; break
```

```
        Ctrl = "1"[3] => Li_buf2 ? in1 ; Li_1_buf1 !
in1 ; break
    Ctrl = "2"[3] => Li_buf2 ? in1 ; L16 ! in1 ; break
  ]; loop ];
end ;
```

The graphical representation of the corresponding IF program is shown on Figure 9.

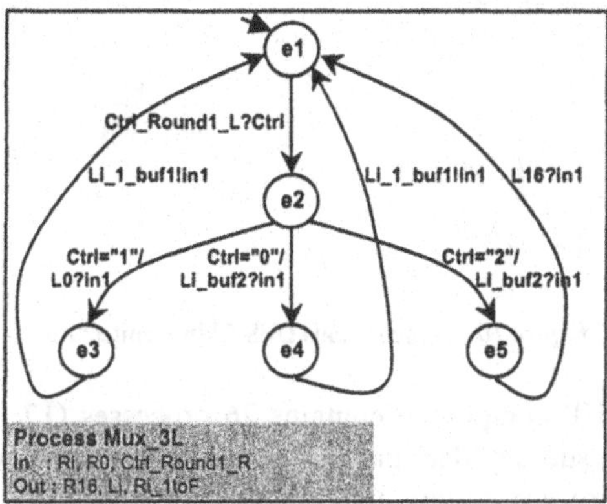

Figure 9. IF representation of Mux_3L

5. CASE STUDY : DES CHIP

The global architecture of a fully asynchronous DES (Data Encryption Standard) chip is described in Figure10. It is basically an iterative structure, based on three self-timed loops synchronized through communicating channels. Channel Sub-Key synchronises the ciphering data-path with the sub-key computation data-path. CTRL is a set of channels generated by the Controller bloc (a finite state machine) which controls the data-paths along sixteen iterations as specified by the DES algorithm [20].

The 1-bit input channel CRYPT/DECRYPT is used by the Controller to configure the chip and trigger the ciphering. The 64-bit channels DATA and KEY are used to respectively enter the plain text and the key. The ciphered text is output through the 64-bit channel OUTPUT.

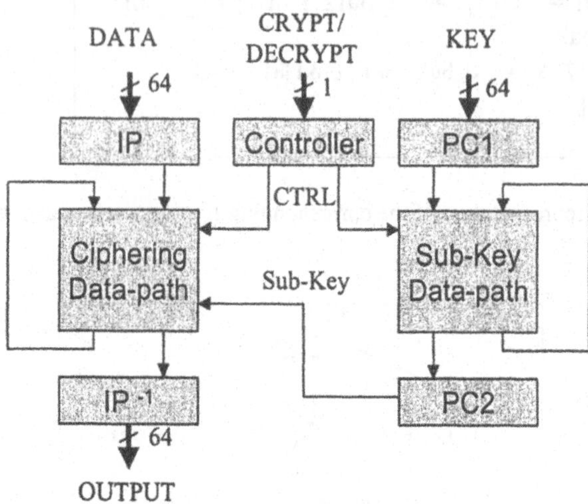

Figure 10. Asynchronous DES Chip architecture

The overall CHP component contains 26 processes (17 in the ciphering data-path, 4 in the sub-key data-path). The translation produces an IF system where each CHP process is represented by an IF process.

5.1 Some verified properties

Characteristic properties of the DES behavior have been expressed in mu-calculus, and automatically verified using CADP, on a SUN Ultra 250 with 1.6 GB memory. The model LTS generation time is 1:05:27.02 and its size is: (3 e+7 transitions, 5.3 e+6 states). The meaning and performances (verification time in h:min:sec; Memory) of typical properties are listed below :

P1: Freedom from deadlock (27:41.23 ; 884 MB)

P2: After reception of the 3 inputs (Key, Data, Decrypt), Output is always produced (27:31.93 ; 865 MB)

P3: The counter of the controller counts correctly. (26:25.64 ; 885 MB)

P4: Each iteration of the ciphering and sub-key data-paths synchronizes correctly. (26:25.65 ; 879 MB)

5.2 Verification by behavior reduction

To verify properties P3 and P4, which relate to the synchronizing channels only, an alternative technique is available. The model behavior is reduced by hiding all the labels which do not relate to CTRL and Sub-Key. This can be obtained by applying property-preserving equivalence reduction techniques which for this model take (11:46.31; 1.71 GB). The resulting LTS is depicted on Figure 11, which shows the cyclic behavior and exhibits the synchronization on channel *Sub-Key*.

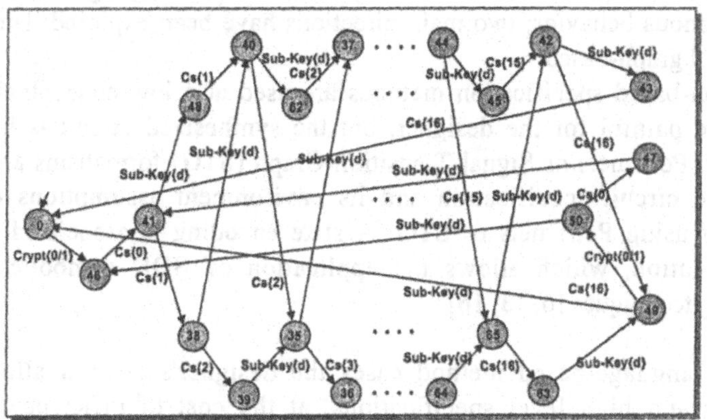

Figure 11. Reduced LTS for property P3

5.3 Handling state explosion

The following techniques have been used during the verification experiments:

- Data abstraction: this is a well known strategy, by which data are reduced to one bit, when their value do not influence the property at hand.
- Explicit generation of interleaving for CHP intra-process concurrent statements (instead of generating synchronized concurrent IF processes, see section 4.3). Without this feature, the model generation faces LTS size explosion.

6. RELATED WORKS

The verification of asynchronous circuits heavily depends on the design approach: timed or un-timed (delay insensitive) circuits. In the first approach, ATACS is a set of tools that supports the synthesis, analysis, and verification of timed circuits [3]; KRONOS is dedicated to timed automata verification [11]. However, only very small examples have been published using these techniques: the verification of timed systems faces serious complexity problems. In contrast, our work stays at a higher abstraction level, focuses on un-timed specifications and leaves open the choice of the target implementation model.

Concurrent processes are widely used for specifying an un-timed asynchronous behavior; two main directions have been explored: language-based and graph-based.

Graph-based specification methods are used at a low conceptual level. They are painful for the designer, but the synthesized circuit is fast and efficient. Petri nets or Signal Transition Graph (STG) formalisms are used. Both the circuit specification and its environment assumptions can be modeled using Petri nets or STG. A state encoding is associated to this representation, which allows the application of BDD symbolic model checking techniques [6, 13, 16].

The language-based method eases the designer's task; it allows for modular and high level specifications, at the cost of efficiency in the synthesized result.

Early validation works include experiments with CIRCAL to model micro-pipe lines: the proof is performed on the parallel composition of the implementation and the properties modeled as processes [4]. The use of LOTOS to specify asynchronous circuits has also been suggested, making available the CADP verification tool box [17]. We question the adequacy of LOTOS for circuit specification as no circuit synthesis flow has been built from it. We do keep however the idea of using CADP, but taking as input a hardware design language.

In a previous feasibility study [5], we used an industrial symbolic model checker intended for property checking on RTL designs. We translated the Petri Net produced by TAST as a pseudo synchronous VHDL description, where the pseudo clock thus introduced was only to make each computation cycle a visible state. In essence, our translation performed a static pre-order reduction. This approach gives good results after the communication expansion, but this occurs too late in the design process to validate the initial specifications.

7. CONCLUSION

We have implemented a first prototype translator that automatically produces the IF model for a CHP specification, along the principles

explained in this paper. Preliminary experiments have shown that the IF/CADP toolbox offers a convenient abstraction level for the formal validation of initial CHP specifications.
Essential properties can be proven on the specification, before synthesis decisions are made visible in the design description. This provides a new service to the TAST user. During the architecture exploration process, the designer may use transformations that have not been formally proven correct, and wishes to check that essential properties are retained on the refined architecture; our automatic link to IF/CADP is a possible answer to this requirement.

The perspectives of this work include some improvements to the current version of the translator (e.g. negative numbers are currently not recognized) and its application to many more benchmarks. The scalability of the approach to large circuits, of the size of a 32-bit microprocessor will be measured. It will certainly involve elaborate model reduction strategies, some of which are still not automated. We also intend to work on the combination of symbolic simulation and model checking, applied to asynchronous circuit verification. Finally, replacing the mu-calculus by a more widely accepted property specification language such as the Accelera PSL would ease the designer's access to the verification toolbox.

8. REFERENCES

[1] M. Renaudin, "Asynchronous Circuits and Systems: a promising design alternative", Microelectronics-Engineering Journal, Elsevier Science, Vol. 54, N° 1-2, Dec 2000, pp. 133-149.

[2] A.J. Martin, "Programming in VLSI: from communicating processes to delay-insensitive circuits", in C.A.R. Hoare, editor, Developments in Concurrency and Communication, UT Year of Programming Series, 1990, Addison-Wesley, p. 1-64.

[3] H. Zheng, E. Mercer, and C. Myers, "Automatic abstraction for verification of timed circuits and systems", Proc. CAV'01, LNCS 2102, Springer, pp. 182-193, July, 2001.

[4] A. Cerone, G. Milne: "A Methodology for the Formal Analysis of Asynchronous Micropipelines", Proc. FMCAD 2000, LNCS N° 1954, Springer Verlag, pp.246-262

[5] D. Borrione et al. "An Approach to the Introduction of Formal Validation in an Asynchronous Circuit Design Flow". Proc. 36th Hawai Int. Conf. on System Sciences (HICSS'03). Jan. 2003

[6] G. Delzanno and A. Podelski,"Model Checking in CLP". Proc. 5th Int. Conf. TACAS'99. R. Cleaveland, ed., Springer Verlag LNCS N°1579, pp.223-239,1999.

[7] http://www.inrialpes.fr/vasy/cadp/

[8] J. Cortadella et al. "Petrify: a tool for manipulating concurrent specifications and synthesis of asynchronous controllers." IEICE Trans. on Information and Systems, E80-D(3): 315-325, Mar. 97.

[9] K. Van Berkel, "Handshake Circuits – An Asynchronous Architecture for VLSI Programming", Cambridge University Press, 1993, ISBN : 0-521-45254-6

[10] M. Bozga, J.-C. Fernandez, et al. "IF : An Intermediate Representation and Validation Environment for Timed Asynchronous Systems". Proc. FM'99, Toulouse, LNCS, 1999.

[11]M. Bozga, H. Jianmin , O. Maler, S.Yovine, "Verification of Asynchronous Circuits using Timed Automata",Proc. TPTS'02 Workshop, Elsevier Science Pub. April 2002.

[12] A Dinh Duc, L. Fesquet, M. Renaudin, "Synthesis of QDI Asynchronous Circuits from DTL-style Petri-Net", IEEE/ACM Int. Workshop on Logic & Synthesis, New Orleans, June 4-7, 02.

[13] A. Kondratyev, J. Cortadella, M. Kishinevsky, et al.: "Checking signal transition graph implementability by symbolic bdd traversal". Proc. EDTC'95, pp 325-332, Paris, March 95.

[14] M. Renaudin, J.B. Rigaud, A. Dinhduc, A. Rezzag, A. Sirianni, J. Fragoso : "TAST CAD Tools", ASYNC'02 TUTORIAL, ISRN: TIMA--RR-02/04/01—FR, 2002

[15] R. Manohar, T.K. Lee, A.J. Martin. "Projection: a synthesis technique for concurrent systems". 5th Int. Symp. Advanced. Research in Asynchronous Circuits and Systems, Apr. 99.

[16] V. Khomenko and M. Koutny: "Towards An Efficient Algorithm for Unfolding Petri Nets". Proc. CONCUR'01. Springer-Verlag, LNCS N° 2154 (2001) 366-380.

[17] M. Yoeli and A. Ginzburg, "LOTOS-based Verification of Asynchronous Circuits", Tech. Report, Dept. of Computer Science, Technion, Haifa, 2001.

[18] M. Bozga et al. "Automated validation of distributed software using the IF environment", Proc. Workshop on Software Model-checking, El. Notes in TCS vol. 55 Elsevier Science Pub. July 00.

[19] http://www.agedis.de and http://www-omega.imag.fr

[20] NIST, Data Encryption Standard (DES), FIPS PUB 46-3, National Institute of Standards and Technology, Reaffirmed 1999 October 25. http://csrc.nist.gov/csrc/fedstandards.html

ON-CHIP PROPERTY VERIFICATION USING ASSERTION PROCESSORS

José Augusto M. Nacif,[1] Claudionor Nunes Coelho Jr.,[1] Harry Foster,[2] Flávio Miana de Paula,[3] Edjard Mota,[4] Márcia Roberta Falcão Mota,[5] and Antônio Otávio Fernandes[1]

[1] *Computer Science Department*
Universidade Federal de Minas Gerais, MG, Brazil
{jnacif,coelho,otavio}@dcc.ufmg.br

[2] *Jasper Design Automation*
Mountain View, CA, USA
harry@jasper-da.com

[3] *MindSpeed Technologies*
Newport Beach, CA, USA
flavio.depaula@mindspeed.com

[4] *INDT - Instituto Nokia de Tecnologia, AM, Brazil*
edjard.mota@indt.org.br

[5] *Computer Science Department*
Universidade Federal do Amazonas, AM, Brazil
marcia.roberta@gmail.com

Abstract White-box verification is a technique that reduces observability problems by locating a failure during design simulation without the need to propagate the failure to the I/O pins. White-box verification in chip level designs can be implemented using assertion checkers to ensure the correct behavior of a design. With chip gate counts growing exponentially, today's verification techniques, such as white-box, can not always ensure a bug free design. This paper proposes an assertion processor to be used with synthesized assertion checkers in released products to enable intelligent debugging of deployed designs. Extending white-box verification techniques to deployed products helps locate errors that were not found during simulation / emulation phases. We present results of the insertion of assertion checkers and an assertion processor in an 8-Bit processor and a communication core.

Keywords: On-line verification, assertion processor, assertion-based verification

Please use the following format when citing this chapter:
Nacif, José Augusto, M., Coelho, Claudionor Nunes, Jr., Foster, Harry, Flávio, Miana de Paula, Mota, Edjard, Roberta Falcão Mota, Márcia, Fernandes, Antônio, O., 2006, in IFIP International Federation for Information Processing, Volume 200, VLSI-SOC: From Systems to Chips, eds. Glesner, M., Reis, R., Indrusiak, L., Mooney, V., Eveking, H., (Boston: Springer), pp. 101-117.

1. Introduction

There has been a large number of reported design errors detected after the chip has been released, such as in Beatty, 1993; Kantrowitz and Noack, 1996; Taylor et al., 1998. As design complexity increases, it becomes clear that no one can ensure a bug-free design using conventional validation tools using simulation, emulation and formal verification techniques. Probably the most famous bug reported in the literature is the Pentium Floating Point Bug Lowry and Subramaniam, 1998 that was found just after chip deployment.

As design validation clearly becomes one of the most critical issues in chip design today, we propose in this chapter a methodology that pushes design validation beyond chip deployment by using the notion of an assertion processor. An assertion processor is a circuit inserted into the design to monitor synthesized assertions, taking appropriate action in the case of an assertion failure.

This paper is outlined as follows. Section 2 describes the basis for this work, i.e. controllability and observability, and their relation to design validation. Section 3 presents assertion libraries based on PSL and OVL that can be synthesized to facilitate on-chip debug. Section 4, describes the assertion processor framework. Finally, we present conclusion and future work.

2. Controllability and observability

The ability to test a design correlates to the ability of controlling and observing the behavior of a design. The increase of design complexity over the past years has weakened the ability to test a design. Even if a design error can be controlled, it may be very difficult to observe the error using the design I/O pins. This is a widely accepted problem in the integrated circuits industry and academic community, with numerous paper published in this subject Devadas and Keutzer, 1991; Fujiwara, 1990; Fujiwara, 1985; Chen and Breuer, 1985.

Recent improvements of synthesis techniques allowed RTL-based designs to be adopted as the main design capture methodology used by designers. Moreover, the use of RTL-based designs enabled more aggressive validation techniques based on White-box verification as opposed to Black-box verification Foster et al., 2004.

Black-box verification relates to the approach of providing stimulus to the input pins of a design and checking the results in its output pins. This approach offers very poor observability and controllability since a failure inside the design has to propagate to the output pins to be observable. In addition to this problem, the failure may only be noticed several thousands of cycles after it has actually happened, making it difficult to detect and even to recognize the failure.

Consider Figure 1 as an example. This Figure presents an excerpt from a larger sequential design, with registers, represented by the two boxes and gates, represented by combinational logic cloud. In this design, one of the operation modes of the design reduces a portion of the combinational logic to the circuit outlined in the picture, with two AND logic gates, Y and Z and one OR logic gate, X. In this figure, xz and yz represents the internal interconnection wires of these logic gates.

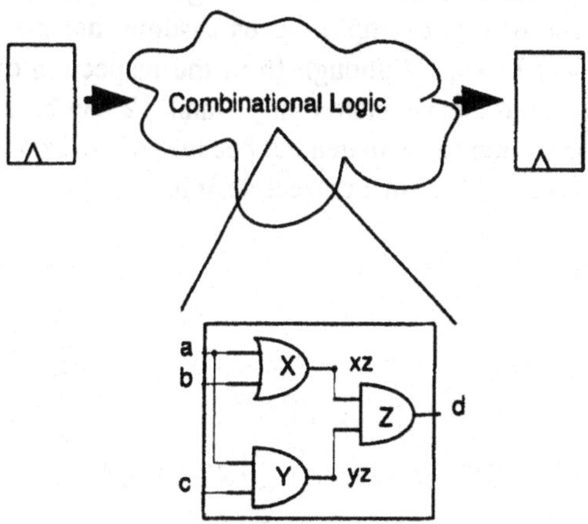

Figure 1. Sequential circuit and its combinational logic.

In this example, let us consider that we want to use the Black-box methodology to test a stuck-at-zero condition at pin b in this piece of logic. First, we apply a set of test vectors to the inputs and compare the output d to the expected value. Even though we can exercise all possible input vectors to this piece of logic, it is not be observable because the logic is redundant on this operation mode.

White-box verification is a technique used to validate a design by inspecting internal wire connections of the design, thus improving the overall observability. When used with monitors, it provides a very powerful tool to aid design validation. An assertion monitor is a piece of HDL code that evaluates specific conditions on the designs' internal wires. Using White-box verification, a designer can locate a failure internal to the design because assertions can trigger immediately after an error occurs.

Assertions are inserted into a design based on the knowledge about legal and illegal behaviors of internal design structures Bergeron, 2000; 0-In Design Automation, Inc., 2002. Usually, the assertions are inferred by a designer according to interface rules or unwanted corner cases of the design.

Assertions can be built from hardware description languages Bergeron, 2000, from some pragmas of a specific tool such as in 0-In Design Automation, Inc., 2002, or from a testbench written using a testbench language, such as Open-

Vera Synopsys, Inc., 2002. White-box verification has become a popular design validation technique, improving the confidence level in a design because assertion monitors, acting like probes inserted into a chip, solve the observability problem of testing chip designs Gupta, 2002; Kazmierczak, 2001.

Consider applying the White-box verification approach in the example in Figure 2. First, we add an assertion correlating the expected behavior of the wires. For the sake of this example, let us assume that the error condition occurs when $f1 + f2 > 1$. Although from the inspection of only this part of the circuit, we could clearly state that $f1$ and $f2$ can be true at the same time, making the assertion false, in general because of enviromental conditions $f1 + f2 > 1$ may never occur in a correct design.

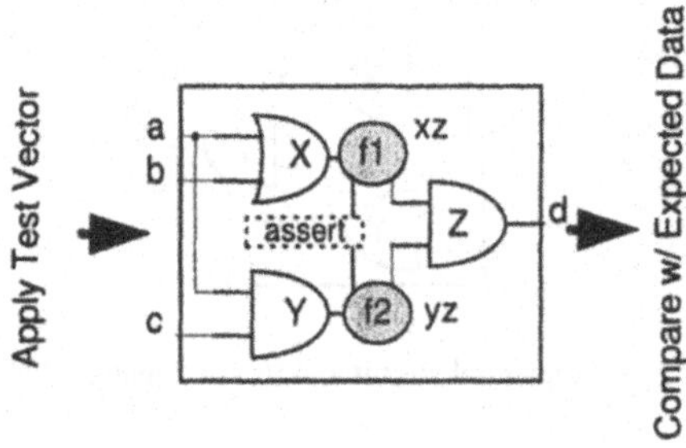

Figure 2. White-box verification approach.

White-box verification has been widely applied during the simulation, formal analysis and emulation Axis Systems, 2002; McMillan, 1993; Shimizu et al., 2000; Switzer et al., 2000 phases of a design. The initial goals of White-box verification are to capture the design intent, to document interface assumptions and to find bugs as the design progresses. However, the use of White-box verification does not guarantee that a design is bug-free because of the overall complexity of the design. In the context of chip-level designs implemented in field-programmable gate arrays (FPGAs), assertion monitors enables reconfigurable designs to be monitored at run-time after deployment of the design. If a bug is ever found in the design, an assertion engine stores the error information that can be later notified to a designer. Because the error information is directly linked to an RTL design, the designer will be able to locate the problem faster, thus being able to provide a new version in a very short time. The previous work on synthesizing assertion checkers Oliveira and Hu, 2002; Drechsler, 2003 didn't address issues like wrong design assumptions or proposed an architecture that could inform which assertion had failed.

3. On-chip verification

An architecture for on-chip verification can be found in Figure 3. This architecture is based on three components: a sea of synthesizable assertions based on Open Verification Library (OVL) Foster and Coelho, 2001 or synthesized from PSL Accellera, 2004; an assertion processor, which is a circuit designed to process the results of the assertions and to take proper action, being as simple as a circuit that raises an error pin or as complex as an embedded processor that dispatches an error correction routine; and a routing mechanism that routes error information from the assertions to the assertion processor.

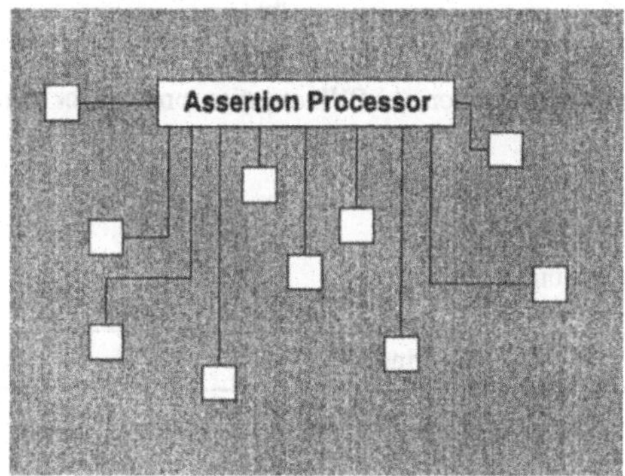

Figure 3. Diagram of assertion processor framework.

3.1 Synthesizable assertions with routing mechanisms from OVL

In Nacif et al., 2003, it was proposed an architecture for an assertion engine to be used in a reconfigurable design by extending the use of the White-box verification beyond the simulation/emulation phases of a design. The main idea was to modify the OVL to support on-chip run-time debug. This modified library was obtained by adding a Boundary-scan IEEE, 2001 chain to the assertions. This library provided support to solve the assertion routing problems, although no assertion processor was used to provide the circuit with an intelligent mechanism to process the error condition.

Figure 4 (a) presents a typical assertion module from OVL. The modified version with scan-chain architecture is presented in 4 (b). Table 2 contains a description of each signal. We refer the reader to a throughout description of the modified library Nacif et al., 2003.

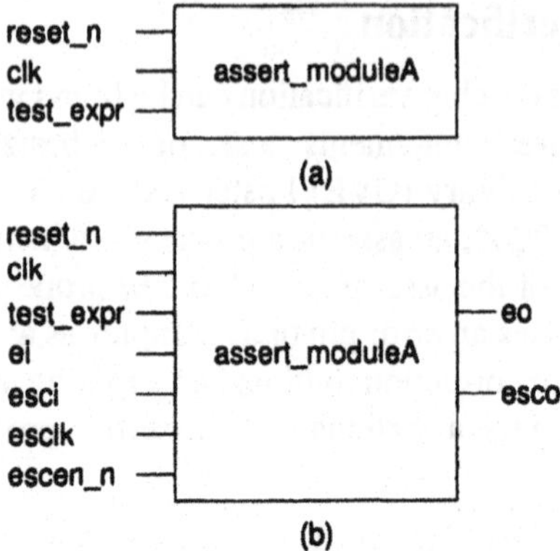

Figure 4. (a) Typical OVL assertion; (b) OVL assertion modified for scan-chain architecture.

Table 1. Signal descriptions for Figure 4(b).

Signal	Description	I/O
reset_n	Reset Active Low	Input
clk	System Clock	Input
test_expr	Any HDL test expression	Input
ei	Error Input	Input
esci	Error Scan Input	Input
eo	Error Output	Output
esco	Error Scan Output	Output
esclk	Error Clock	Input
escen_n	Error Scan Enable Active Low	Input

3.2 Synthesizable assertions from PSL

The Accellera Property Specification Language (PSL) Accellera, 2004 is an ideal language for specifying complex design intent in either linear-time temporal logic or in branching-time temporal logic.

The PSL language definition is segmented into the following layers: Boolean, temporal, modelling, and verification. The temporal layer supports either the linear-time temporal logic or branching-time temporal logic operators. In this section, we consider only a linear-time temporal logic component. For a more complete definition, see Accellera, 2004.

At the Boolean layer, a PSL specification references signals and variables within an HDL description (for example, Verilog or VHDL). Hence, the underlying HDL syntax and semantics for Boolean expressions ensure semantic consistency between the property specification and the HDL model.

Sequences of Boolean conditions that occur at successive clock cycles can be described succinctly using Sequential Extended Regular Expressions (SEREs). Sequences and SEREs can be constructed as follows (where *b* is a Boolean expression):

- `b` : a Boolean expression is a SERE in its simplest form,
- {SERE} : a sequence constructed by a SERE,
- SERE ; SERE : a SERE constructed by concatenating two SEREs,
- `{sequence | sequence}` : a sequence describing alternative sequences,
- `{sequence & sequence}` : a sequence describing parallel non-length matching sequences (that is, two sequences, both hold at the current cycle, regardless of whether they complete in the same cycle or in different cycles),
- `{sequence && sequence}` : a sequence describing parallel length matching sequences (that is, two sequences, both hold at the current cycle, and both complete in the same cycle).

PSL provides various repetition operators ([]) that concisely describe repeated concatenation of the same SERE.

For example, given the SERE `r` and a Boolean `b`:

- `r[*m:n]` : a sequence of `m` to `n` contiguous occurrences of `r`,
- `b[=m:n]` : any sequence containing from `m` to `n` occurrences of `b`,
- `b[->m:n]` : any sequence ending in the mth to nth occurrence of `b`.

The repeat range `m:n` can be replaced by a single constant `n` (for example, `[*2]`). In addition, an unbounded range could be expressed as `[*0:inf]`, where the keyword `inf` represents infinity.

PSL supports all the standard LTL operations. In addition, more readable operators are defined in terms of the base operators. For example, given the PSL temporal formulas `f`, `f1`, `f2`:

- `!f` : `f` does not hold,
- `f1 & f2` : `f1` and `f2` both hold,
- `f1 | f2` : `f1` or `f2` or both hold,
- `f1 -> f2` : `f1` implies `f2`,

- `f1 <-> f2` : `f1 -> f2` and `f2 -> f1`,
- `always f` : `f` holds in every cycle,
- `never f` : `f` does not hold in any cycle,
- `next f` : `f` holds in the next cycle,

PSL also supports operators to build complex properties out of SEREs, such as `{r1} |-> {r2}`, meaning that `{r2}` starts in the last cycle of `{r1}`, and `{r1} |=> {r2}`, meaning that `{r2}` starts in the first cycle after `{r1}`, as wells as a way to define named properties, which facilitates reuse.

A property ensuring that a and b are mutually exclusive can be specified as:

```
property mutex = always !(a <-> b) @(posedge clk);
```

When we synthesize assertions from PSL, we consider only a synthesizable subset of the PSL language, i.e. without any behavior that leads to infinite memory. The synthesis process generates an RTL description of the design that instantiates an OVL assertion. For example, the PSL assertion described below leads to the circuit of Figure 5.

```
assert always {e1; e2; e3} |-> e4 @(posedge clk)
```

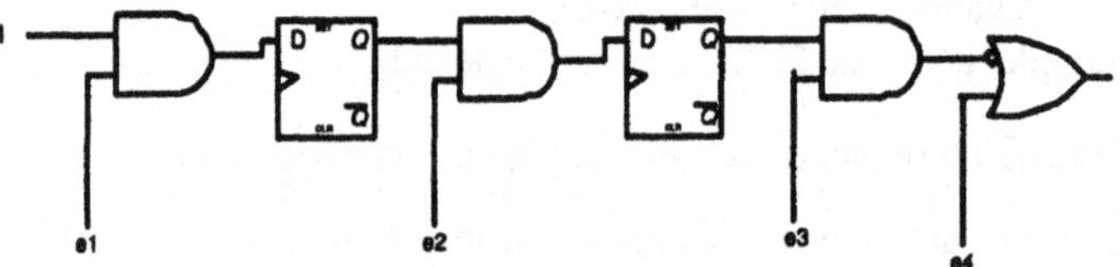

Figure 5. Synthesized circuit from PSL

This circuit corresponds to the following Verilog-HDL description:

```
reg a1, a0;
always @(posedge clk) a0 <= e1 & 1;
always @(posedge clk) a1 <= e2 & a0;
assert_always (reset, clk, e4 | ~(e3 & a1));
```

3.3 Generating chained assertions actual design

One of the major problems in modifying assertion instantiation into chained assertions stems from the fact that the circuit interface must be changed.

Figure 6 depicts an example of an 8051 ALU (Arithmetic Logic Unit) Teran and Simsic, 2002 hierarchical structure with chained assertions. White circles represent the original design hierarchy and dark ones the inserted assertions. Table 2 shows the assertion sequence that must be scanned from the pin *esco* in case an error is signaled by the *eo* pin. A typical timing diagram presenting the behavior of *eo*, *escen_n*, *esco*, and *esclk* pins is depicted in Figure 7.

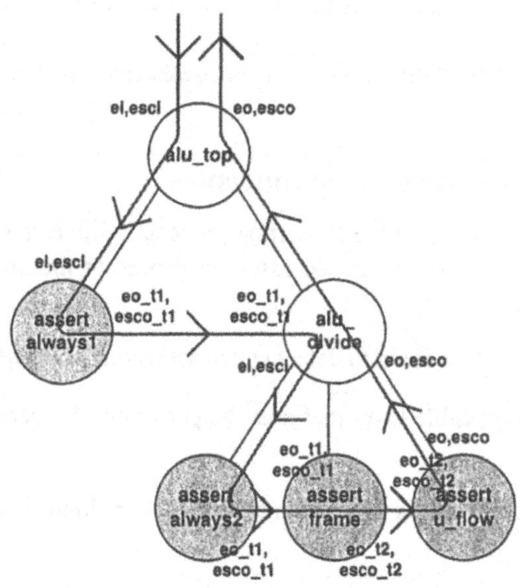

Figure 6. 8051's ALU hierarchy with assertion chaining.

Table 2. Assertion sequence list for Fig. 6.

Sequence Number	Assertion name
1	assert_uflow
2	assert_frame
3	assert_always2
4	assert_always1

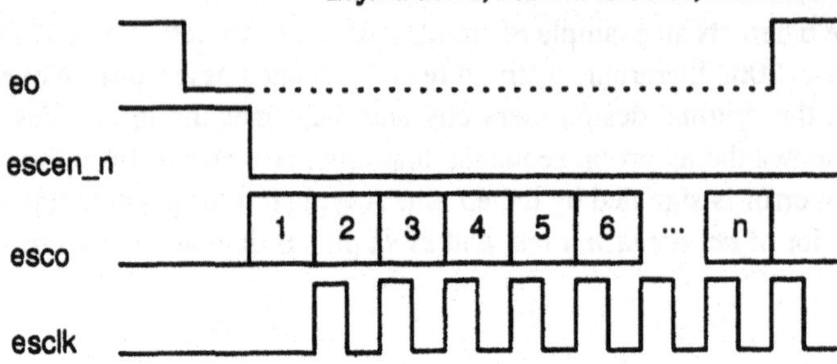

Figure 7. Behavior of the eo, escen_n, esco, and esclk pins.

4. Assertion processor architecture

Figure 8 presents the proposed assertion processor with chained assertions. As it can be seen in the Figure, an assertion processor minimally needs to perform three tasks:

- Scan the assertion chain to detect which assertion has caused the failure;
- Encode the possible tasks that must be performed for each assertion in the circuit;
- Perform specific tasks to overcome the error condition.

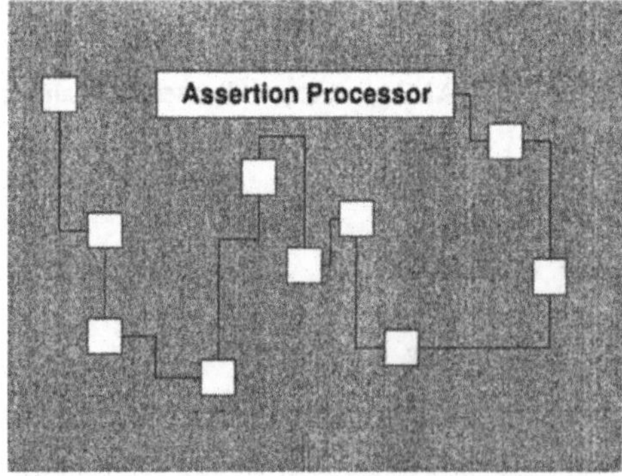

Figure 8. Assertion processor with chained structure.

Figure 9 presents a skeleton for a minimal assertion processor. This assertion processor contains three error processing tasks: halting the processor for more serious errors, resetting the entire chip, or performing a software interrupt to enable a processor core to perform a specific action. The reader should note that the priority for each assertion determining which action must be taken can be obtained directly from the OVL severity level.

```
module AssertionProcessor (...);
reg ['LOGNOFASSERTIONS:0] count;
// SCAN DETECTION
always @(posedge clk) begin
  ...
  if (error detection)
  begin
    count = count + 1;
    if (esci == 1)
      ErrorNo = count;
    ...
  end
end
// PRIORITY ENCODING OF ERROR CONDITION
assign ErrorNo = ErrorEncoding(ErrorNo);
// ERROR CORRECTION
always @(posedge clk) begin
  if (error detected)
  begin
    casex (ErrorPriority)
      3'bxx1:  // HALT INTEGRATED CIRCUIT
      3'bx1x:  // HW RESET
      3'b1xx:  // SW INTERRUPT
    endcase
  end
end
endmodule
```

Figure 9. Assertion processor verilog HDL skeleton.

Although this Figure presents the minimum circuit for an assertion processor, more complex assertion processors can be implemented in the tasks execution part of the assertion processor. For example, if an assertion processor may interact with a network coprocessor if the error must be reported over an ethernet port.

In order to automatize the assertion chaining process XRoach tool has been developed Oliveira et al., 2003. XRoach processes verilog hierarchical designs with OVL and PSL assertions. It compiles the design and links it to the modified synthesis version of OVL. XRoach output files are the verilog design with chained assertions and a list of the assertions in the chained order, that is used by the assertion processor to identify which assertion had failed. Figure 10 presents XRoach user interface. The basic assertion synthesis flow is shown below:

1 Design with PSL and OVL assertions

2 Convertion of PSL properties into RTL code + OVL assertions

3 Selection of assertions to be synthesized

4 Selection of severity level for each synthesized assertion

5 Synthesis of scanning structure for instantiated assertions

6 Synthesis of assertion processor skeleton

7 Generation of new design hierarchy with assertion scan-in/out + assertion processor

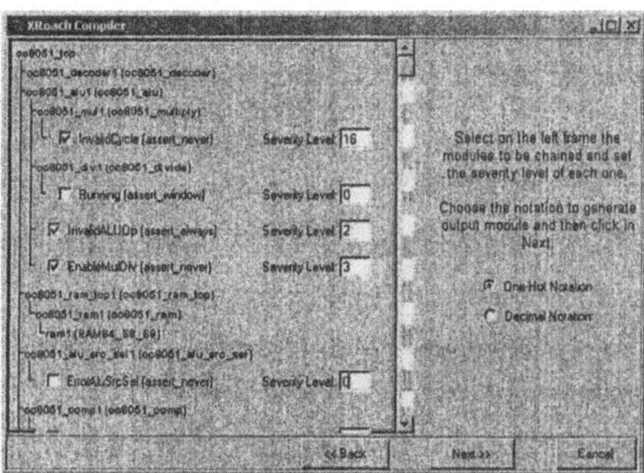

Figure 10. XRoach user interface.

5. Results

This section presents the results of synthesizing assertion processors for an 8051 core Teran and Simsic, 2002 and for an I^{2}C (Inter Integrated Circuit) bus Herveille, 2002. Although the proposed methodology focus in early design stages, the assertions were instantiated based on public domain specifications and the cores' documentation. The synthesis was performed using Xilinx Free Web Pack 5.2i environment. Xilinx Free Web Pack uses Xilinx Synthesis Technology (XST). Better results could be achieved using third party synthesis tools. The designs were synthesized and routed for a Virtex XCV300 FPGA using high area optimization effort.

Figure 11 shows the area in equivalent gate count for an assertion processor monitoring a number of assertions, supposing a priority encoding of five possible actions.

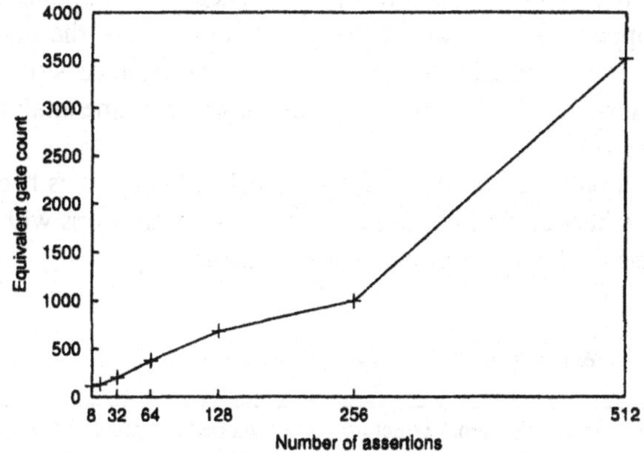

Figure 11. Assertion processor area increase supposing 5 priorities.

5.1 I^{2}C

I^{2}C is a two-wire, bi-directional serial bus that provides a simple and efficient method of data exchange between devices. I^{2}C standard was developed by Philips semiconductors Philips Semiconductors, 2000. Its applications include LCD drivers, remote I/O ports, RAM, EEPROM, data converters, digital tuning and signal processing circuits for radio and video systems, and DTMF generators.

In Table 3 some examples of assertions inserted in the I^{2}C core are shown. The total number of inserted assertions in the original design is 5. These assertions where inserted based on carefully reading and understanding of core's documentation.

Table 3. Assertions inserted in I^2C core.

Assertion type	Functionality
assert_always	Ensures the correct interrupt request operation
assert_never	Ensures that concurrent read and write signals will not occur
assert_one_hot	Ensures the correct operation of control state machines

Table 4 presents the synthesis results for I^2C original design and using the proposed assertion processor architecture. Using total equivalence gate count as an example, we have an overhead of 37.27%. Considering that I^2C core complexity is relatively low, this results are acceptable. Although the circuit speed has dropped 100% from the original design speed, the normal operation of the design could still be maintained, as the chained structure is only active when *escen_n = 0*. As a result, we could run the time analyzer with the constrain *escen_n = 0*.

The reader should note also that the synthesizable OVL has been specified in RTL code. A handcrafted library, or a library with flip-flops with scan chain structures embedded could improve results considerably.

Table 4. Synthesis results for I^2C communication core.

Parameter	Original Structure	Chained Structure + AP	Overhead
Slice Flip Flops	122	129	5.74%
4 Input LUTs	221	357	61.54%
Slices	125	203	62.40%
Equivalent gate count	2,557	3,510	37.27%
Maximum Frequency	89.17 MHz	45.17 MHz	97.41%

5.2 8051 processor core

8051 is an 8-bit processor widely used in many embedded applications. There are several 8051 chip manufactories with different peripherals and memories configurations. The synthesized core has two 16-bit timer/counters, four 8-bit I/O ports, 4K bytes of on-chip program memory, and 128 bytes of on-chip data program (registers). Program memory and registers are inferred by synthesis tool using Virtex Flip-Flops. This memory structure consumes a significative part of design area. Table 5 shows some assertions added to 8051 core. The total number of inserted assertions is 11.

Table 5. Assertions inserted in 8051 processor core.

Assertion type	Functionality
assert_window	Verifies the division operation completion before a new enable signal
assert_time	A four-clock-cycle ACK signal must be produced after an interrupt trigger
assert_overflow	Ensures no stack overflow
assert_always	Ensures that ALU always receives a valid opcode

Table 6 shows synthesis results for original 8051 processor core and using the proposed assertion processor architecture. Because of considerable complexity increase compared with I^2C the assertion overhead is significantly lower. Taking as a parameter the equivalent gate count, we have a 3.19% increase.

Table 6. Synthesis results for 8051 processor core.

Parameter	Original Structure	Chained Structure + AP	Overhead
Slice Flip Flops	838	943	12.53%
4 Input LUTs	4,487	4,698	4.70%
Slices	2,515	2,647	5.25%
Equivalent gate count	68,141	70,131	3.19%
Maximum Frequency	12.99 MHz	12.86 MHz	1.01%

6. Conclusions and future work

We presented a technique that enhances currently validation capability of assertion and property based techniques beyond deployment of chip-level designs. This technique can be applied to inumerous situations, including emulation phase of designs, in which validation testcases execute at full speed and for fault-tolerant chip design, in which an assertion failure can yield to mission failure.

We showed that a decision circuit that monitors assertions in a design can be used to validade the design. This circuit, which was called an assertion processor, upon the detection of an assertion failure can dispatch a set of recovery procedures, ranging from hardware reset, software reset or event put the chip into halt mode.

Along with the assertion processor, we generated a version of the Open Verification Library (OVL) fully synthesizable, and a procedure to generate digital circuits from PSL descriptions by converting them into RTL code along with an OVL assertion.

Since the total number of assertions can be very high, we presented a tool called XRoach which enabled designers to select which assertions he/she wanted to synthesize into the final circuit. These assertions were concatenated to the assertion processor, and a skeleton of the assertion processor was automatically generated.

An example of the tool usage was shown using an 8051 processor, showing that minimum overhead in circuit size was obtained by carefully selecting proper assertions. As future works, we intend to investigate different concatenation procedures between assertions and the assertion processors, and to work in the interaction between software and hardware failures.

Acknowledgments

We thank Leonardo Otaviano, Márcia Oliveira and Fernando Sica for their help in the preparation of an earlier version of this paper. We also thank CNPq for financial support under grants PNM #830107/2002-9 and SensorNet #552111/2002-3.

References

0-In Design Automation, Inc. (2002). Assertion-based verification for complex designs. The Verification Monitor.

Accellera (2004). Proposed Standard Property Specification Language (PSL) 1.1.

Axis Systems (2002). Assertion processor.

Beatty, D.L. (1993). *A Methodology for Formal Hardware Verification with Application to Microprocessors.* PhD thesis, Carnegie Mellon University, School of Computer Science.

Bergeron, J. (2000). *Writing Testbenches Functional Verification of HDL Models.* Kluwer Academic Publishers.

Chen, T. H. and Breuer, M. A. (1985). Automatic design for testability via testability measures. *IEEE Transctions on Computer-Aided Design*, 4(1):3–11.

Devadas, Srinivas and Keutzer, Kurt (1991). A unified approach to the synthesis of fully testable sequential machines. *IEEE Transactions on Computer-Aided Design*, 10(4):39–51.

Drechsler, R. (2003). Synthesizing checkers for on-line verification of system-on-chip designs,. In *Proceedings of IEEE International Symposium on Circuits and Systems*, pages IV748–IV751.

Foster, Harry and Coelho, Claudionor (2001). Assertions targeting a diverse set of tools. In *10 th Annual International HDL Conference Proceedings*.

Foster, Harry, Krolnik, Adam C., and Lacey, David J. (2004). *Assertion-Based Design.* Kluwer Academic Publishers.

Fujiwara, H. (1985). *Logic Testing and Design for Testability.* The MIT Press.

Fujiwara, H. (1990). Computational complexity of controllability/observability problems for combinational circuits. *IEEE Transactions on Computers*, 39(6):762–767.

Gupta, Aarti (2002). Assertion-based verification turns the corner. *IEEE Design & Test of Computers*, 19(4):131–132.

Herveille, Richard (2002). I^{2}C Controller Core. Available at: http://www.opencores.org/projects/i2c.

IEEE (2001). *Standard 1149.1-2001*. IEEE Press.

Kantrowitz, M. and Noack, L. (1996). I'm done simulating; now what? verification coverage analysis and correctness checking of the decchip 21164 alpha microprocessor. In *Proceedings of 33rd Design Automation Conference*, pages 325–330.

Kazmierczak, Marcin (2001). White-box verification techniques in a networking asic design. Technical report, Lund Institute of Technology.

Lowry, M. and Subramaniam, M. (1998). Abstraction for analytic verification of concurrent software systems. In *In Symp. on Abstraction, Reformulation, and Approx.*

McMillan, Kenneth L. (1993). *Symbolic Model Checking*. Kluwer Academic Publishers.

Nacif, José Augusto, de Paula, Flávio Miana, Foster, Harry, Coelho, Claudionor, Sica, Fernando Cortez, da Silva, Diógenes Cec'ilio, and Fernandes, Antônio Otávio (2003). An assertion library for on-chip white-box verification at run-time. In *Proceedings of Latin American Test WorkShop*.

Oliveira, M. and Hu, A. (2002). High-level specification and automatic generation of ip interface monitors. In *Proceedings of 39th Design Automation Conference*, pages 129–134.

Oliveira, Márcia M., Nacif, José Augusto, Coelho, Claudionor N., and Fernandes, Antônio Otávio (2003). XRoach: A Tool for Generation of Embedded Assertions. In *Proceedings of Chip in Sampa Student Forum*.

Philips Semiconductors (2000). The I^{2}C-Bus Specification.

Shimizu, Kanna, Dill, David L., and Hu, Alan J. (2000). Monitor-based formal specification of PCI. In *Formal Methods in Computer-Aided Design*, pages 335–353.

Switzer, S., Landoll, D., and Anderson, T. (2000). Functional verification with embedded checkers. In *9 th Annual International HDL Conference Proceedings*.

Synopsys, Inc. (2002). Assertion-based verification.

Taylor, S., Brown, M. Quinn D., Dohm, N., Hildebrandt, N., Huggins, J., and Ramey, C. (1998). Functional verification of a multiple-issue out-of-order, superscalar alpha processor - the dec alpha 21264 microprocessor. In *Proceedings of 35th Design Automation Conference*, pages 638–643.

Teran, Simon and Simsic, Jaka (2002). 8051 core. Available at: http://www.opencores.org/projects/8051.

RUN-TIME FPGA RECONFIGURATION FOR POWER-/COST-OPTIMIZED REAL-TIME SYSTEMS

Jürgen Becker, Michael Hübner, Michael Ullmann
Institut für Technik der Informationsverarbeitung (ITIV)
Universität Karlsruhe (TH), Germany
http://www.itiv.uni-karlsruhe.de/
{becker, huebner, ullmann}@itiv.uni-karlsruhe.de

Abstract: The paper describes a new approach of a flexible run-time system for handling dynamic function reconfiguration in fine-grain Virtex FPGAs, whereas the fulfillment of given real-time constraints are central. Moreover, the detailed evaluation and measurement of the power consumption situation during this dynamic reconfiguration process is essential for realistically quantifying the power loss of fine-grain FPGAs during dynamic reconfiguration processes. This kind of real-time run-time systems and power analysis give the designer and user the possibility to compare FPGA implementation alternatives and to apply the required functionality reconfigurations during the selected application scenarios. Thus, a qualified decision can be done between fine-grain FPGAs of different sizes and different dynamic reconfiguration frequencies, e.g. using smaller and more cost- as well as power-efficient FPGAs by temporarily outsourcing suitable functionalities.

Key words: Virtex FPGA, power consumption,real-time run-time reconfiguration, function and data management

1. INTRODUCTION

Field programmable gate-arrays (FPGAs) are mainly used today for rapid-prototyping purposes. They can be reconfigured many times for different applications. Modern state-of-the-art FPGA devices like Xilinx Virtex FPGAs [11] additionally support a partial dynamic run-time

Please use the following format when citing this chapter:
Becker, Jürgen, Hübner, Michael, Ullmann, Michael, 2006, in IFIP International Federation for Information Processing, Volume 200, VLSI-SOC: From Systems to Chips, eds. Glesner, M., Reis, R., Indrusiak, L., Mooney, V., Eveking, H., (Boston: Springer), pp. 119-132.

reconfiguration which reveals new aspects for the designer who wants to develop future applications demanding adaptive and flexible hardware. Especially in the domain of mobile computing high-end mobile communication applications will benefit from the capabilities of the new generation of reconfigurable devices.

Actually there exist some recent new approaches deploying Virtex/Virtex II FPGAs in multimedia applications using their capabilities for a dynamic function-multiplex showing new ways for the efficient deployment of partial run-time reconfiguration [2] [4].

A new approach to create systems which are able to manage configuration are run-time systems. These systems use the flexibility of an FPGA by changing the configuration partially. Only the necessary functions are configured in the chip's memory. By demand a function can be substituted by another while used parts stay operative. To solve the problem of substitution and I/O management the configuration needs a main module controlling the tasks.

With such a system it is possible to save resources like output pins and energy because of outsourcing configuration data. The need of chip area becomes smaller and therefore the power consumption can be reduced. Nevertheless the power requirements of such applications will grow with increasing rate of configuration.

Creating such a system has two aspects: Reducing amount of chip area and reducing power consumption by designing a control system which manages the content of FPGAs configuration in an intelligent way to minimize reconfiguration rate. Additionally this management can control the on chip intercommunication bus to prevent an overhead of bus size.

One important aspect is the power consumption during the FPGA's reconfiguration phase. To solve this problem it is necessary to analyze the behavior of FPGA while reconfiguration.

There exist many approaches for the analysis and estimation of the run-time power consumption of designs on FPGAs and they have been able to derive metrics enabling the designer to estimate the design's power consumption at design time [8] [9] [10].

The results show that the most power is dissipated by the on-chip long-wires connecting different functional blocks. Especially the functional multiplex operational mode as mentioned above will demand information on the expected power consumption. In that case not only the operational phase is of interest. Because the frequency of reconfiguration may be increased in such systems it might be of interest how much power is spent during the reconfiguration phase.

Current application scenarios use run-time reconfiguration at a low rate (minute range) so that the power dissipation during reconfiguration can be neglected but that condition can change when the number of applications sharing the same resource FPGA is growing so that the functions' cycle time is reduced to milli-seconds. In that case the energy and time needed for reconfiguration will be of interest for the designer as well.

2. DYNAMICALLY FPGA RECONFIGURATION

Basically a Virtex FPGA consists of two layers and additional configuration and control logic which handles the configuration bitstream loading and the distribution of the configuration data on dedicated positions of the second layer [12]. The first layer contains the reconfigurable hardware like logic-blocks (CLBs), RAM-blocks, I/O-blocks and configurable wiring resources. All blocks of the same type (I/O-blocks excepted) are aligned into columns (see figure 2-1).

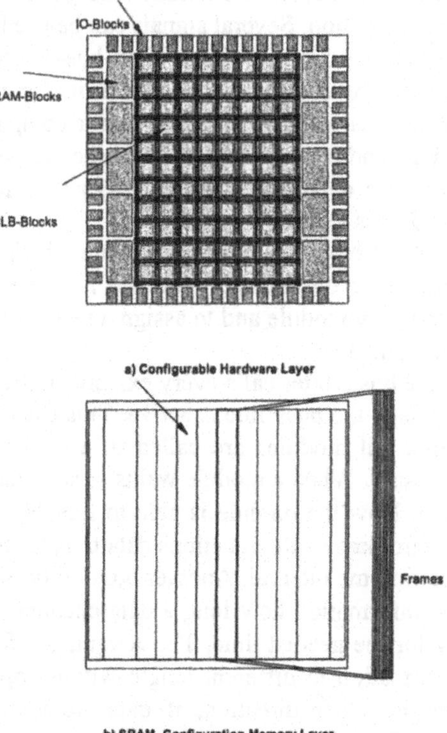

Figure 2-1 Layered FPGA structure (a,b)

The second layer which matches to the first layer is organized into columns as well. Each column whose width depends on the covered block-columns from the first layer consists of a set of one-bit sub-columns called frames.

A frame is the smallest piece of reconfiguration information that can be written on an FPGA and each frame contains fractions of the configuration information needed to configure the blocks assigned to the column. So if one block out of a column is to be reconfigured all other blocks in the same column have to be rewritten as well. This restriction demands that a partial reconfiguration of functions can only be done on groups of consecutive columns. It is possible to write the frames in a random manner out of order, but the configuration files are normally structured in an ordered way.

3. REAL-TIME RUN-TIME SYSTEM APPROACH

Figure 3-1 shows a possible schematic design of the physical FPGA-based part of a simple run-time system. The functions (for example A and B) are positioned on a fixed location. Several signals can be used for controlling and data exchange. The functions control their bus-driver with control signals. This is necessary to configure the drivers in the right way. The modules can be selected via an address bus. Next a comparison with the dedicated address is done and the enable signal becomes active.

By giving the module a default address at design time it is possible to assign a unique logic address to each module.

The design-time default module address is a value which is out of range of the normal address range during operation. So it is easy for the main control to recognize this new module and to assign a new valid logic address within the legal address range.

In operating state the bus arbiter calls every existing address. If a module does not want to send data, the busy line is active. This causes the arbiter to select the next address. All modules are called in a cyclic order without waiting if it's not necessary. When a module wants to send data, the specific request signal is active. Now the module is able to transfer its data via the data bus. The main control knows the functions data transfer time and allows to send data for a specific time interval. Another possibility is to send at first a time stamp with the information how long a data channel is needed. This opens a time window for the needed time. The advantage of this is that the modules can send information in different length without opening the data channel too long. In the other direction, if data from environment are dedicated for a module, the selected module gets the information to receive data via the Data-IN signal. As described in the text before, it is possible to send a time stamp at first to inform the module about the size of data. Should a function which is currently not configured on FPGA start working, a substitution of a not used module is initialized. In this case, the active state and data has to be saved to the local memory of the main control unit. If the function is needed later, the state and data can be re-transferred into this module and the function can start working with the same configuration as before. This context-save is done by using the data and state I/O signal lines.

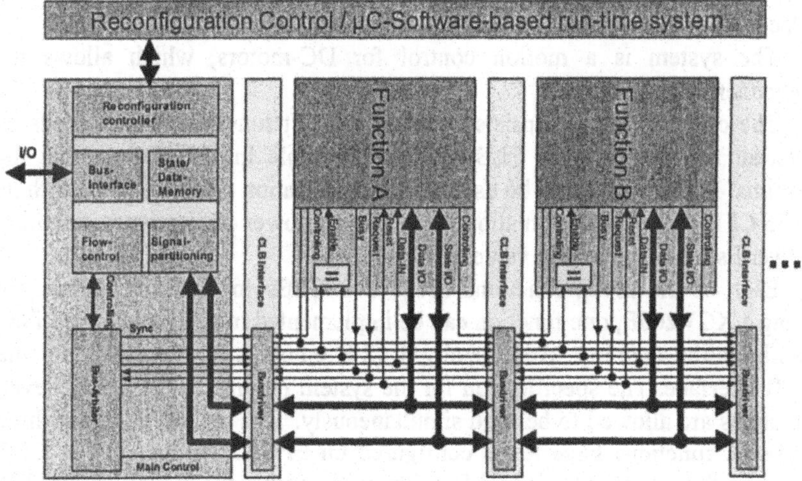

Figure 3-1 Simple run-time system with FPGA partial run-time reconfiguration support

Such a system is conceived to be implemented in future work to get an overview of the run-time and reconfiguration excess power.

We tested a simplified sample scenario using six different sample applications which implement electronic control units from the automotive domain (e.g. seat control, cabin compartment lighting where the results shown in table 1 have been calculated.

By outsourcing functions in this example, 12 I/O-pins were saved. Of course a logic is necessary to connect the shared external multiplexed signal lines in the right way. The saving of 467.6mW shows the capacity of this method in power saving. However, the power consumption of external memory storing the configuration data is not included in this scenario. But the amount of dissipated power of memory is smaller as the FPGAs.

The results were calculated by using real implemented functions on a Xilinx XCV2000E FPGA.

Table 1 Results of sample application

	All functions	Savings
Pins	108	12
Power consumption	1401.2mW	467.6mW

4. SCENARIO AND REAL TIME ANALYSIS

A real industrial scenario was used to analyze the advantage of using run-time systems for outsourcing configuration data.

The system is a motion control for DC-motors, which allows the movement in seven axis.

The overall system consists of seven explicit functions which needs an over-all amount of 1204 CLB blocks. The table in [12] shows, that the minimal FPGA which can be used for this application is a XCV300E with its 1536 CLB blocks. An estimation of dynamical power dissipation with Xilinx Power Estimator gives the value of 176mW.

Each of the seven functions needs 172 CLBs for implementation. By using a XCV200E, one function can be implemented in seven columns of the configuration memory. Additionally 4 CLB columns are necessary for the CLB interface. The specification for the system requires that three of seven functions are allowed to be used simultaneously. This means, that only three of seven functions have to be configured on FPGA. The amount of CLB columns for this case is 3x9 CLB columns for active functional blocks, 3x4 CLB columns for interface. 9 CLB columns for the run-time system (equals to 252 CLB blocks) are available for the control system. Each function causes a dynamic power dissipation of about 41mW on the XCV200E. With three active functions a difference of 53mW to the value for the XCV300E is left for the control system and the interfaces. It is realistic that the amount of power is less to the XCV300E implementation because the average toggle rate of the control system is not as high as the rate of the several functions and the static power dissipation is lower. However the power dissipation caused by reconfiguring has to be considered in this system. A value for this mode of operation is given in chapter 6.

Additionally the costs of XCV200E is less to the XCV300E and also a lower area on PCB is needed because of smaller packaging. An overall advantage by using a run-time system is shown in this small example.

Very important is to abide the real time demand. In the example the worst case situation is if all three functions have to be substituted by another. To reconfigure one functional block it is necessary to transfer seven plus four CLB columns into the configuration memory of the FPGA. The data size for these columns is about 40000 Bytes. Using the SelectMap port for reconfiguring, the amount of time for transferring the data to the FPGA is about 580µs on a transfer rate of 66Mbytes per second. To reconfigure all three functions, a time of 1.74ms is needed. The specification for these functions demands an answer time of <=100ms. 98.26ms are left for the control system managing the data transfer of context-save. With a frequency of 66MHz, 6485160 clock cycles are left for these operations.

The timing values have to be evaluated with the implemented system and can be shown exactly when the complete system is implemented on FPGA.

5. RECONFIGURATION BUS REALIZATION

In figure 3-1 the functions are separated with CLB interfaces. This is necessary to get a precise separation of routing recourses for each function. Bus interfaces are neccessary to seperate modules which can be substituted while dynamic and partial reconfiguration.

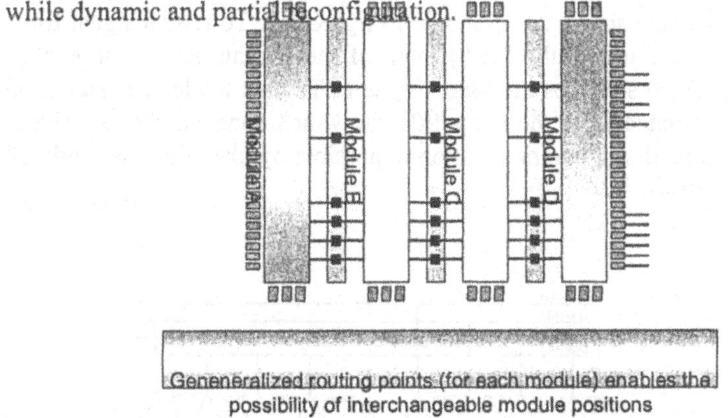
Module A
Module B
Module C
Module D
Genenéralized routing points (for each module) enables the possibility of interchangeable module positions

Figure 5-1 Generalized macros between modules

Figure 5-1 shows the schematic of a system with modules seperated by generalized bus macros. These macros enable the possibility to exchange e.g. module B and module C. Figure 5-2 shows a schematic view on the structure of such an interface. The FPGA's CLB (combined logic blocks) are used to connect the functional blocks. The dashed line is the border between the modules. Four horizontal aligned CLB blocks are used to build a 4 Bit interface.

The cause of using those called double CLB interfaces has its background on the routing in the design phase. To ensure that all connections of the modules are placed on the same position the CLB interfaces are positioned on a fixed place. Because of this, all modules can be implemented on every possible function column.

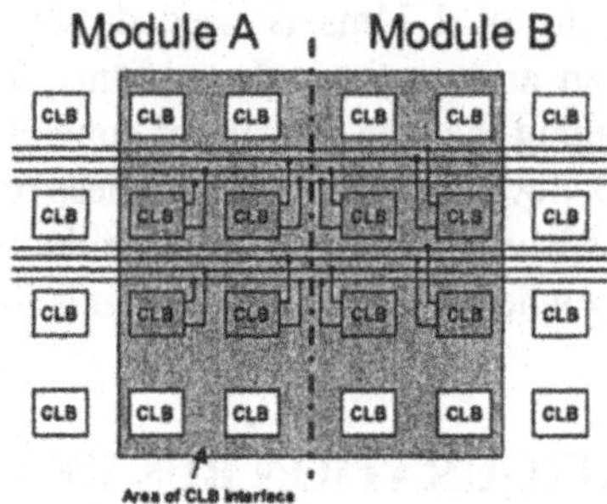

Figure 5-2 Structure of CLB interface

A more detailed schematic is shown in figure 5-3. To send a signal from left to right, for example the LE(0) input of the Tristate gate is on enable level, while RE(0) switches the second gate in Tristate mode. The signal of LI(0) now is connected to the Output(0) line which connects the two sides. In the other direction a communication is possible by changing the mode of the Tristate gates.

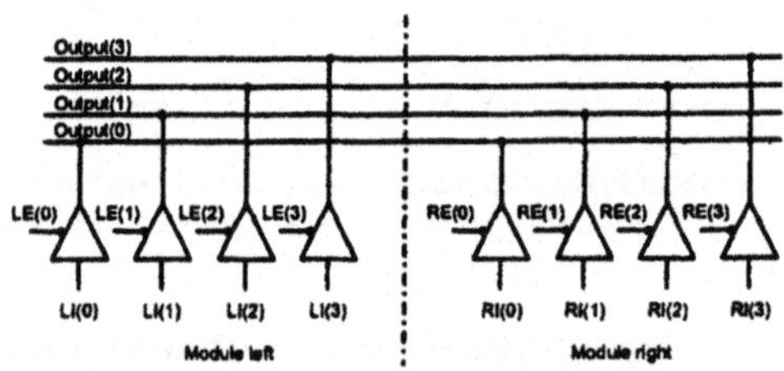

Figure 5-3 Tristate gates of CLB interface

Using Tristate gates for separating modules for dynamic and partial reconfiguration leads to problems while routing. Figure 5-3 shows that the names of the connection lines between the left and right module have the same name (e.g. Output(0)). While routing the autorouter might connect a signal line from the left module to the area of the right module. This causes an open signal line or even a short circuit if this modules ar substituted by another (dynamic and partial reconfiguration).

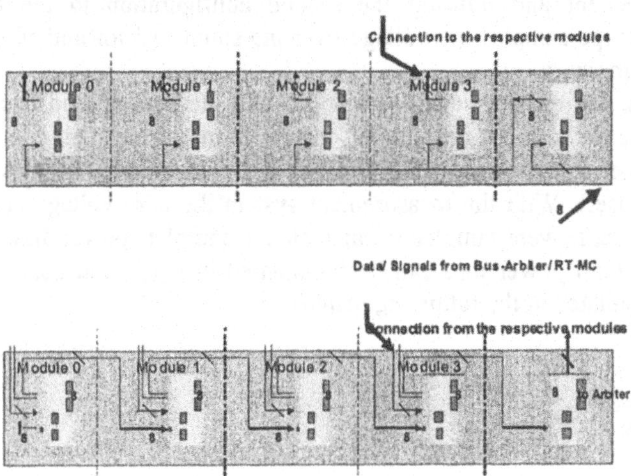

Figure 5-4 Schematic of Input/Output Macro

Figure 5-4 shows the schematic of macros using slices instead of TBUF elements. The slices of the input macro are programmed to route through the signals to the modules. Benefits are, that now the router has defined external pins to connect signals to the macros. This method allows a save and automatic generation of modules without time-consuming checking for signals crossing the modules border. More information about these macros can be found at [15].

6. POWER CONSUMPTION WHILE RECONFIGURATION

With a special measurement system, the amount of current for the FPGA while reconfiguration was measured. The used XCV2000E FPGA needs two power supplies. 1.8V for core and 3.3V supply for I/O elements. Figure 6-1 shows the measurement system. The measurement system consists of a PC with the control software, a Tektronix oscilloscope (Type TDS 220) connected to the PC's RS232 interface. The core and 3.3V supply current is measured with a shunt resistor connected to an opened jumper bridge on the Spyder-Virtex board.

By starting the measurement, the control software initializes the oscilloscope and pre-configures the FPGA. The next step is to start the measurement and transmit the second configuration to the board. The oscilloscope samples the voltage over the shunt resistor and stores the data into its memory.

Then the measure cycle is stopped and the data is read back from the oscilloscope memory into the PC and saved into a file. This file is the basis for further calculations with Matlab. This cycle is iterated n-times (n is given by the user). With the measurement system the core voltage and the 3.3V power supply were sampled to calculate the complete power dissipation.

The total power thus can be fragmented in P_{CORE} and P_{VCC}. The power was calculated in the following manner:

$$\overline{P_{CORE}} = \frac{\int_{T_{start}}^{T_{start}+T_{REC}} P_{CORE}(t)\cdot dt}{T_{REC}} \qquad (6.1)$$

$$\overline{P_{VCC}} = \frac{\int_{T_{start}}^{T_{start}+T_{REC}} P_{VCC}(t)\cdot dt}{T_{REC}} \qquad (6.2)$$

By using equation (6.1) and (6.2) a calculation of the median power of core and 3.3V supply between start and end of configuration is possible. T_{REC} is the total reconfiguration time.

Figure 6-2 and 6-3 show the measured power loss of core and 3.3V supply. To get a general result, these measurements were done with different kinds of implemented functions and repeated 10 times for each configuration.

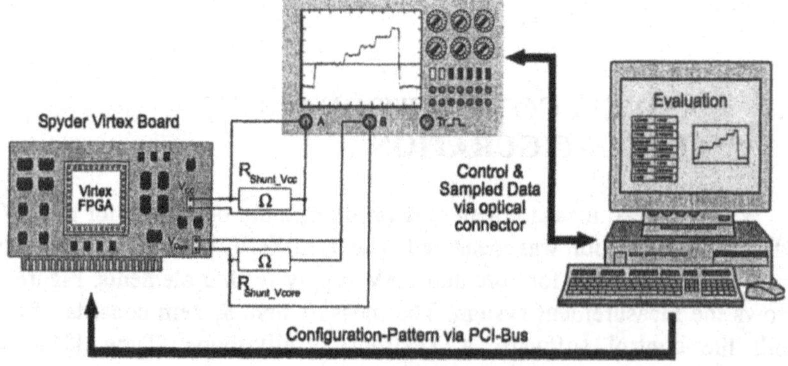

Figure 6-1 Power Measurement System

The rectangular shaped pulse of the 3.3V supply power caused by a partial reconfiguration Bitstream containing 30 CLB columns causes the additional power loss. The amount of core power dissipation stays on the same level. This level depends on the current configuration. We did some extended measurements on this topic. More detailed results were published in [1] showing the behavior of the FPGA while reconfiguration. Table 2 shows the result of the measurement. To get an overall view on the table the results of writing configuration data with a pre-reset deleting all preconfigured data and a value of power dissipation of the reseted FPGA are included. The last line of table 2 shows the values measured in dynamical partial reconfiguration mode. Table 2 and 3 show that the additional power loss may not be ignored. Systems with high reconfiguration rate, especially run-time systems have to be designed with the goal of saving power by managing the reconfiguration in an intelligent way.

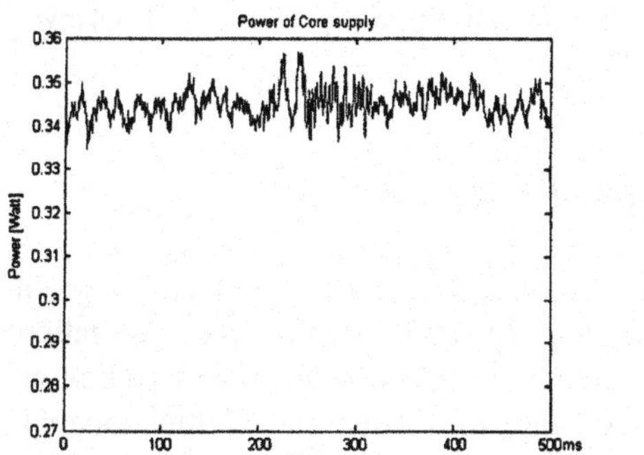

Figure 6-2 Power loss of core while writing partial FPGA configuration

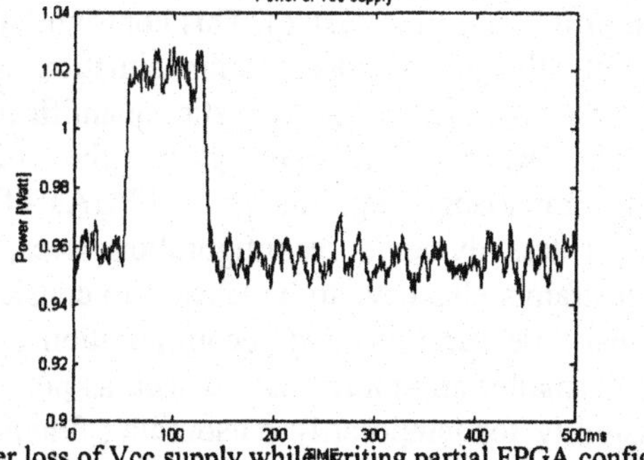

Figure 6-3 Power loss of Vcc supply while writing partial FPGA configuration

Table 2 Power dissipation while reconfiguring

	P_{core}	P_{vcc}	P_{tot}
Reseted empty FPGA (no configuration)	344.8mW	1024.64mW	1369,44mW
Writing configuration data after reset of configuration memory	394.5mW	1075.2mW	1496.7mW
Writing FPGA configuration data without previous reset of configuration memory	347.9mW (model dependent)	1017.2mW	1365.1mW (model dependent)

Table 3 Measured total power dissipation in normal mode

	Measured power dissipation
(1) Seat control unit	1165,1 mW
(2) Window lift unit	1138.4 mW

7. CONCLUSIONS

The paper described in detail a new approach of a flexible run-time system for handling dynamic function reconfiguration in fine-grain Virtex FPGAs. This included also the detailed description and implementation of the needed hardware infrastructure within today's commercial FPGA devices. The feasibility of the suggested approach and the fulfillment of given real-time constraints have been shown. The exemplaric real-time scenario examples provided are based on manual evaluations of the actual implementation of this run-time system. This contribution has evaluated and measured carefully the power consumption situation during the dynamic reconfiguration process. This first realistic quantification of the power consumption of fine-grain FPGAs gives the possibility of fair implementation comparisons, e.g. fine-grain FPGAs of different sizes by applying more or less dynamic reconfiguration steps, whereas cost and performance constraints also have to be taken into consideration. The use of such kind of real-time run-time and reconfiguration systems enables the designer to apply smaller and more cost- as well as power-efficient FPGAs, due to the possibility for temporarily outsourcing functions. Especially the reduction of power dissipation and pinout resources is of great advantage, as shown in the paper.

In addition, it is necessary to design intelligent control systems reducing reconfiguration rate by managing the constellation of modules for the needed scenarios. Thus, next steps will be to provide this kind of run-time-systems with the functionality of dynamically learning during application execution. If a function is more frequently needed as another one, the controller could decide to keep this function available within nearly located configuration memories by substituting a module which is not so frequently needed at the moment. The learned and evaluated scenario information, for example as table based data, has to be given to the main control as specification at design time. This will result in new techniques how real-time control systems manage the dynamic reconfiguration as well as data acquisition between periphery systems and internal modules, including the management of configuration data in internal and external memory topologies.

8. REFERENCES

[1] J. Becker, M. Hübner, M. Ullmann, "Power Estimation and Power Measurement of Xilinx Virtex FPGAs: Trade-offs and Limitations", SBCCI 2003, Brasil.

[2] Y. Ha, B. Mei, P. Schaumont, S.Vernalde, R. Lauwereins, H. De Man; "Development of a Design Framework for Platform- Independent Networked Reconfiguration of Software and Hardware"; Proc. 11th Int'l Conference on Field Programmable Logic and Applications, Belfast, Ireland, 2001.

[3] E.L. Horta, J.W. Lockwood, D.E. Taylor,D. Parlour, "Dynamic hardware plugins in an FPGA with partial run-time reconfiguration", Proceedings of 39th Design Automation Conference, 2002, Page(s): 343 -348

[4] J.-Y. Mignolet, S. Vernalde, D. Verkest, R. Lauwereins: "Enabling hardware-software multitasking on a reconfigurable computing platform for networked portable multimedia appliances"; Int'l. Conf. on Engineering of Reconfigurable Systems and Algorithms; June 25-27 2002, Las Vegas, USA

[5] F. Najm, "Transition density: a new measure of activity in digital circuits", IEEE Transactions on Computer-Aided Design, vol. 12, no. 2, pp. 310-323, February 1993

[6] J.C. Palma, A. Vieira de Melo, F. G. Moraes, N. Calazans, "Core Communication Interface for FPGAs", Proceedings of 15th Symposium on Integrated Circuits and Systems Design (SBCCI), 2002,Porto Alegre BRAZIL, Page(s): 183 –188

[7] http://www.arl.wustl.edu/arl/projects/fpx/parbit/

[8] K. Poon, A. Yan, S.J.E. Wilton, "A Flexible Power Model for FPGAs", 12th International Conference on Field-Programmable Logic and Applications, Sept 2002

[9] K. Poon, "Power Estimation for Field-Programmable Gate Arrays", Master of Applied Science Dissertation, University of British Columbia, 2002

[10] Li Shang, Alireza Kaviani and K. Bathala, "Dynamic Power Consumption in Virtex-II FPGA Family", International Symposium on Field-Programmable Gate Arrays (FPGA'2002), Monterey, CA, Feb. 2002, pp. 157-164.

[11] www.xilinx.com

[12] http://www.xilinx.com/xapp/xapp151.pdf

[13] http://www.xilinx.com/publications/products/software/xc_pdf/xc_xpower.pdf

[14] http://www.xilinx.com/support/techsup/powerest/virtex_power_estimator_v16.xls
[15] M. Huebner, T. Becker, J. Becker: "Real-Time LUT-Based Network Topologies for Dynamic and Partial FPGA Self-Reconfiguration", SBCCI04, Brasil

A SWITCHED OPAMP BASED 10 BITS INTEGRATED ADC FOR ULTRA LOW POWER APPLICATIONS

Giuseppe Bonfini[1], Andrea S. Brogna[1], Roberto Saletti[1], Cristian Garbossa[2], Luca Colombini[2], Maurizio Bacci[2], Stefania Chicca[2] and Franco Bigongiari[2]
[1]*University of Pisa - Dipartimento di Ingegneria dell'Informazione: Elettronica, Informatica, Telecomunicazioni, Via G. Caruso 2, 56122 Pisa (Italy);*
[2]*Aurelia Microelettronica, Via Vetraia 11, 55049 Viareggio (Italy)*

Abstract: *This paper describes an ultra low-power switched opamp-based integrated ADC designed using a cyclic algorithm approach, for cardiac pacemaker applications. The A/D converter shows a typical operating power consumption of 8.18 µW for the analog part and of 9.71 µW for the digital one, whereas the stand by dissipation is about 1 nW and 5 nW, respectively, (measured on 10 chip samples and averaged), considering a typical supply of 2.8 V. The ADC resolution is 10 b, its typical operating clock frequency is 32 kHz (sampling rate is 2.9 kSamples/s) and it is able to reach the same resolution at 2 V, with 0.7 kSamples/s sampling rate, showing a dissipation of 1 µW for the analog part and 1.3 µW for the digital part. Moreover, it is also characterized by low offset and no missing codes.*

Key words: Ultra Low Power, Biomedical Implantable Applications, Switched OpAmp, Analog To Digital Converter

1. INTRODUCTION

The Switched Capacitor (SC) technique needs a particular attention in the design of the switches when used at low supply voltages. In fact, the signal swing applied in these cases is dramatically reduced when a very low supply voltage is used: complementary switches and op amps do not efficiently work because of the insufficient switch overdrive[1]. These problems can be solved by the use of a switched opamp (SOA) technique, instead of SC, to overcome the typical impairments of low-voltage low-power systems[2-7]

Please use the following format when citing this chapter:
Bonfini, Giuseppe, Brogna, Andrea, S., Saletti, Roberto, Garbossa, Cristian, Colombini, Luca, Bacci, Maurizio, Chicca, Stefania, Bigongiari, Franco, 2006, in IFIP International Federation for Information Processing, Volume 200, VLSI-SOC: From Systems to Chips, eds. Glesner, M., Reis, R., Indrusiak, L., Mooney, V., Eveking, H., (Boston: Springer), pp. 133-147.

This paper describes a switched opamp implementation of a cyclic algorithmic ADC that leads to very low power consumption. The architecture complies with the constraints of a biomedical implantable application: ultra low current consumption lower than 4 _A, typical supply voltage of 2.8 V (even if the circuit operates properly in the supply range from 2 V to 3.5 V). Moreover, the ADC can be switched to power-off mode and wakened only when needed. The technology used is a BiCMOS 0.8 μm, with 2 metal and 2 poly layers.

2. THE ADC ARCHITECTURE

The aim of this paper is the design of an integrated ultra-low power consumption A/D converter which operates with a standard battery supply for cardiac pacemaker applications (operating from 3.5 V down to 2 V), with a resolution of 10 bits, conversion rate higher than 2 kSamples/s, input dynamic range of 800 mV and small silicon area.

This performance has been obtained using a cyclic conversion algorithm and the SOA technique. This approach was chosen instead of others because of the following considerations. Both pipeline and sigma delta approaches could be used in applications (biomedical also) where either conversion rate or resolution are requirements more important than consumption and silicon area. In fact, the pipeline architecture is able to achieve a high conversion rate (also in low-voltage applications[8]), but it consists of a series of identical stages that consume additional power, whereas the sigma delta approach can be used in high resolution (more than 16 bits) applications, where hardware simplicity and conversion rate are not the main issues.

Instead, a successive approximation architecture (SAR) using a very low supply voltage allows one to achieve medium speed/medium resolution converter performance with a low power consumption and standard threshold CMOS devices. The results shown in the literature[9] indicate that SAR approach is well suited for operation even below 1 V (around the threshold voltages of the device used), but the very low current dissipation (~30 μA) is achieved with a supply voltage of 1 V. In cardiac pacemaker application, a standard battery is used. It sets the supply voltage value to the typical value of 2.8 V, and thus the same value of dissipation has to be reached at a voltage value larger than 1 V. On the other hand, the SOA technique can be used in our application because of its capability of reducing the supply voltage and overcoming the limits due to the switches overdrive. In fact, the possibility of completely turning "on" opamps only in one of the two phases of the main clock, being the opamp switched off in the other phase, allows us to halve the power consumption of the entire system. This is a great advantage, especially for systems employed in biomedical applications and particularly in pacemakers, where low-voltage and low-power requirements are mandatory.

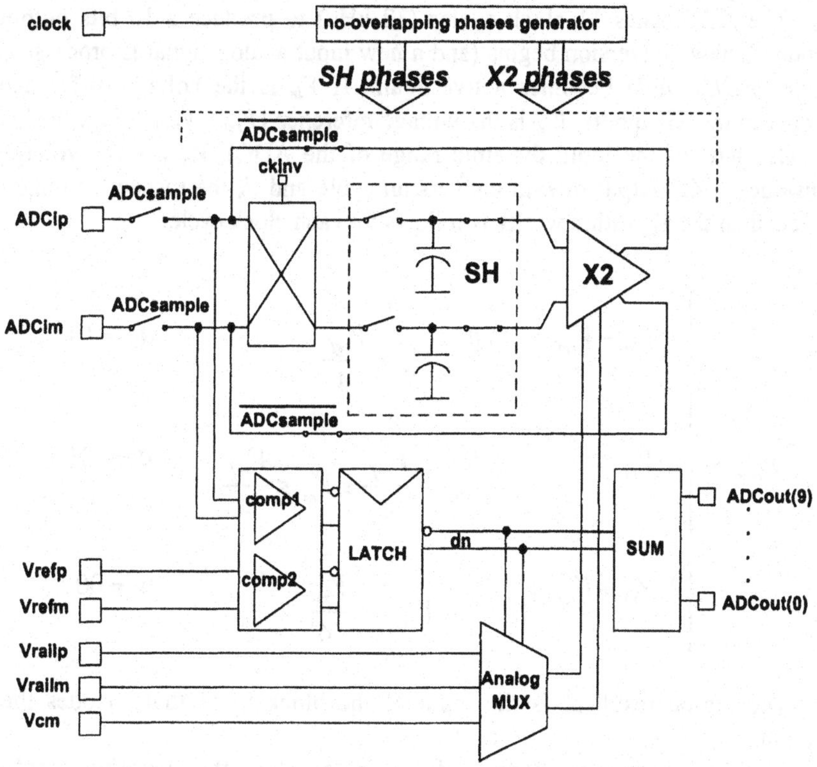

Figure 1. ADC schematic diagram

Figure 1 shows the schematic diagram of the ADC described in this paper. The architecture is a classical Cyclic/Algorithmic topology with 1.5 bit per cycle, consisting of three main blocks: a Sample and Hold (indicated with SH in Figure 1), some comparators and a Multiplying DAC (indicated with X2). The ADC works as follows:

- when *ADCsample* is asserted (see Figure 1), any analog signal present between *ADCip* and *ADCim*, that is sampled through the input switches of SH during the first phase (*phase1* in Figure 2), is converted by the two comparators (that act as a flash sub ADC) in a 2 bit digital number;
- then *ADCsample* is removed, the SH block holds the sampled signal while X2 samples the SH output. At this time the input switches are opened and a loop is created between X2 and SH.

During the following clock cycles (*phase1-phase2*), the operation continues: the SH block samples the X2 output feedback, this signal is compared by the two comparators and the X2 block multiplies by two the SH output. If necessary, a reference voltage is added or subtracted to it, according to the result of the comparison.

The ADC takes 11 clock cycles (32 kHz) to produce a 10 bits output code. A new conversion begins (and a new input analog signal is processed) when *ADCsample* becomes active again. If V_{in} is the voltage difference between the SH inputs, V_{ref} is the voltage reference ($V_{ref} = V_{railp} - V_{railm}$ which is also half of the input dynamic range of the ADC), V_{resn} is the voltage residue at X2 output of n-th conversion cycle and d_n the respective binary code, then the algorithm works as follows for each clock cycle:

$$V_{resn} = \begin{cases} 2 \cdot V_{in} - V_{ref} & \text{if } V_{in} > \dfrac{V_{ref}}{4} \quad d_n = 10 \\ 2 \cdot V_{in} & \text{if } -\dfrac{V_{ref}}{4} \leq V_{in} \leq \dfrac{V_{ref}}{4} \quad d_n = 01 \\ 2 \cdot V_{in} + V_{ref} & \text{if } V_{in} < -\dfrac{V_{ref}}{4} \quad d_n = 00 \end{cases}$$

The digital result d_n is the input of the block SUM that encodes the output.

The SOA technique, traditionally used to reduce the operating supply voltage[8], is mainly used in this case to achieve a reduced power consumption, for a 2 V minimum supply: in fact, if SH is "on", X2 is "off" and vice versa. When a block is off, the SOA output stage is in high impedance and is pulled to *Vcm* (common mode voltage of the ADC) by the related switches.

Figure 2 shows the SH architecture employed in this design. It consists of switches (in a transmission gate configuration), a SOA, two sample capacitors *C*, two hold capacitors *C* (of the same value) and two capacitors having a value of *C*/8, that avoid large spikes on the virtual ground to be traded off with the offset that these capacitors produce.

When *phase1* is low (sample phase), the charge is stored on the sampling capacitors *C* and SOA output stage is in high impedance. In order to minimize the delay caused by the slew rate, that can be high in low-power applications, SOA output stage is pulled to the middle of the dynamic output range (*Vcm*) in this phase. When *phase1* is high, the inputs of this stage are pulled to *Vcm* (while the output of the X2 stage is pulled to *Vcm*), then SOA is turned "on" and all the switches are turned "off". As a consequence, the charge is stored on the sampling capacitors of the X2 stage.

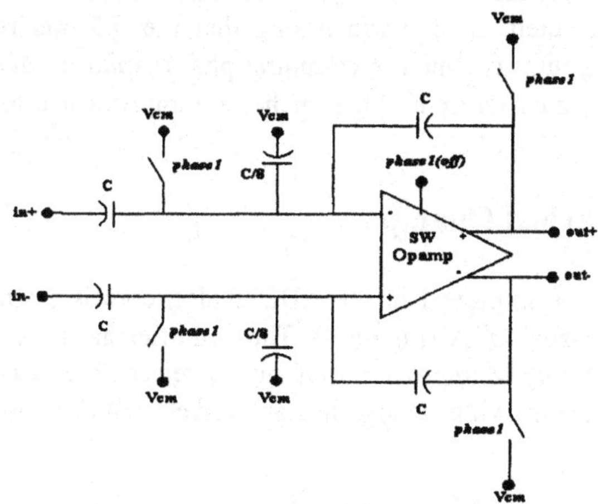

Figure 2. Sample and Hold (SH) detail

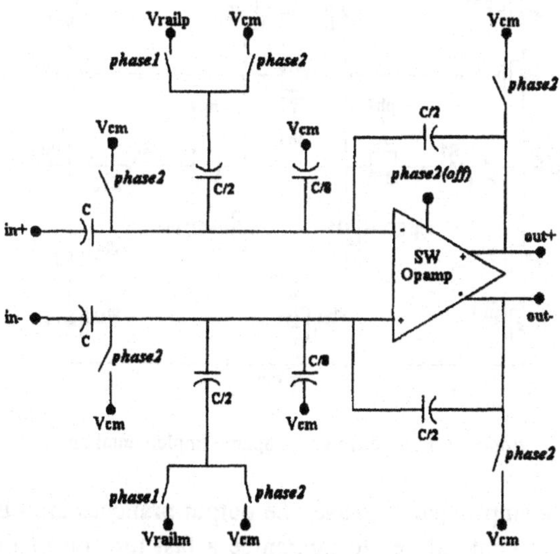

Figure 3. Multiply by two (X2) detail

Figure 3 shows the X2 architecture. There are two capacitors with the same value of the hold capacitors (*C*/2), in order to obtain a 1.5 bits requirement. During *phase2*, the charge is stored "on" the sampling capacitors (through the hold phase of the SH stage), *C*/2 input capacitors are pulled to *Vcm* (as in the output stage of the SOA). At the end of *phase2*, the amplifier is turned "on". During the following phase (*phase1*), the *C*/2 input capacitors are pulled to one of the *Vrail* voltages, depending on the result of the $\pm V_{ref}/4$ comparison. In this way, the subtractions or additions described before are executed. It is worth noting that the 0.5 bits redundancy is obtained using only two non-overlapping phases (*phase1-phase2*) and the complementary ones (since all the switches are implemented as transmission gates).

2.1 Switched Opamp

The amplifier employed in the ADC analog core is a simple 2 stages Miller compensated OTA (Figure 4). The amplifier has a fully differential architecture, mainly to improve power supply rejection ratio (PSRR) and to enhance the signal swing. It also has an active Common Mode FeedBack circuit (CMFB).

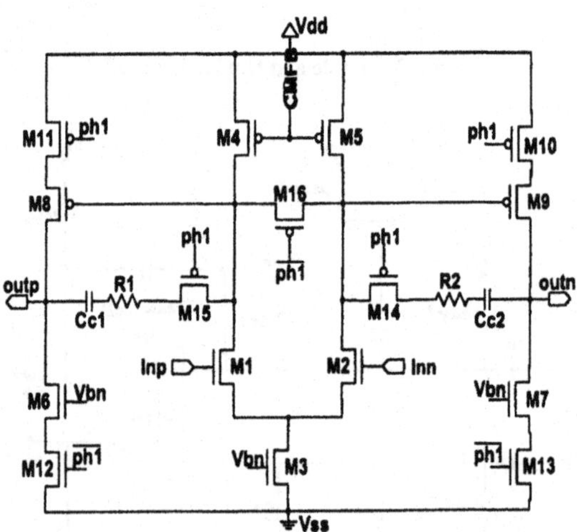

Figure 4. Fully differential opamp implementation

During the opamp inactive phase, the output branches are turned off and the input stage is kept "alive" to guarantee a fast turn on of the amplifier. The dissipation is 150 nA for the input stage and 1 μA for the output stage (500 nA per branch). This means that the current consumption, excluding CMFB circuit, is 1.15 μA during on phase and 150 nA during off phase.

A special attention was paid in designing M4 and M5 in strong inversion (as they work as current mirrors), and the input pair M1, M2 in weak inversion, designing the OTA with suitable dimensions and layout in order to reduce the offset. Switches M14 and M15 prevent the discharge of the compensation capacitors during the off phase of the opamp, thus allowing a fast recovery. Moreover, a switch (M16) between the drains of the input transistors shorts them during the inactive phase, so avoiding the saturation of the input stage caused by the absence of feedback.

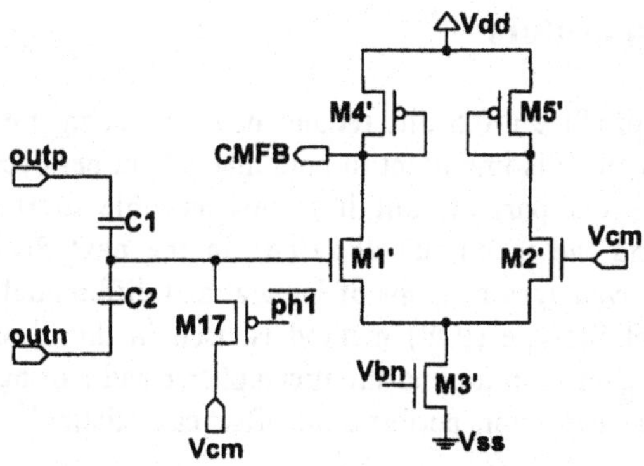

Figure 5. Common Mode FeedBack circuit

The fully differential opamp needs a CMFB (Figure 5) to work properly. A switched capacitor approach was chosen, due to its simplicity and the high linearity it can give.

The outputs are averaged by *C*1 and *C*2 during on phase; this averaged voltage is the input of a simple opamp (replica of the input stage of the main amplifier) and is compared to the wanted Common Mode Voltage (*Vcm*). A feedback signal is generated (CMFB) that controls the current of the input stage of the main opamp (see Figure 4).

Table 1. Opamp simulation results

Parameter	Value	Unit
Supply Voltage	2 ÷ 3.5	V
Supply Current	1.305	μA
Temperature	−15 ÷ 45	°C
Open Loop Gain (Ao)	101	dB
Gain Bandwidth Product (GBW)	387	kHz
Phase Margin (PM)	60	°
Slew Rate (SR)	0.12	V/μs
Silicon Area	0.081	mm^2

During the off phase, capacitors *C*1 and *C*2 are reset (*outp* and *outn* are put to *Vcm*) through switch M17 and the CMFB is kept "on", to be ready for the next phase.

Table 1 summarises the main performance obtained during the simulation of the entire opamp with a load of 4 pF. As shown, the gain is very high (more than what is needed to reach the required linearity) because of the two stages architecture. It is worth noting that slew rate requirements are heavily reduced in this application, thanks to the fact that the output stages are pulled to *Vcm* during the inactive phase.

2.2 Comparators

In order to produce a 0.5 bits redundancy, two comparators have been used in this design. It is important to note that a 1 bit per cycle architecture needs only one comparator, but it is not possible to relax the offset requirement, and the technique described in the next Section does not produce those advantages in terms of integral and differential non-linearity. An Input Offset Storage (IOS) method is used in this design, in which closing a unity gain loop around the preamplifier and storing the offset on the input coupling capacitors performs the offset cancellation[10].

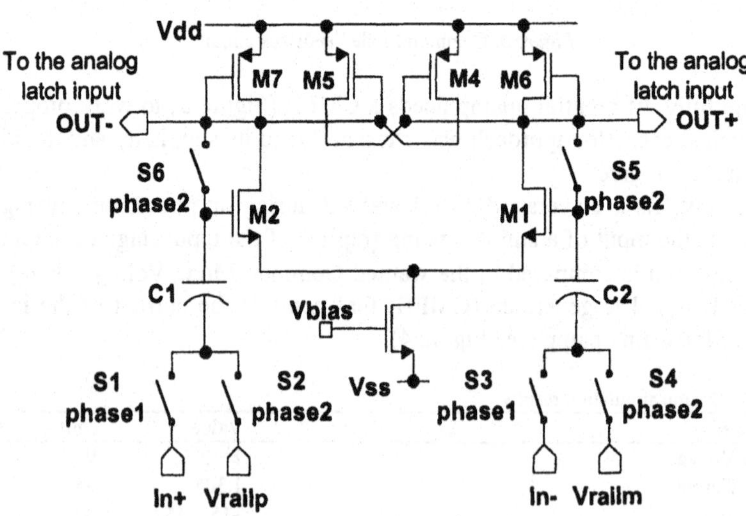

Figure 6. Schematic diagram of the preamplifier

The circuit schematic is shown in Figure 6. During *phase2*, a reference voltage is applied to the inputs of the preamplifier and the offset is stored on the capacitors *C*1-*C*2. Instead, the input signals (coming from SH inputs) are applied and the outputs fed the inputs of the cascaded analog latch stage, during *phase*1. The value of the capacitors *C*1-*C*2 is 600 fF, so that the *kT/C* noise contribution is about 85 _V: a value lower than LSB value (about 780 _V). The power consumption of this stage is 180 nA in the typical case (2.8 V power supply, 2.9 kS/S) and about 50 nA in the "minimum" case (2 V power supply, 0.7 kS/s).

2.3 Offset reduction in the amplifiers

In both SH and X2 blocks, the autozero operation of the amplifier allowed in classical SC architectures cannot occur because in the sample phase (*phase1* for SH for instance) SOA output is "off", so, this effect results in a huge loss of codes and potential linearity problems.

However, in the second clock cycle (the first after *ADCsample* is on), the SH inputs are inverted, through the signal clock *ckinv*. At the third clock cycle *ckinv* (Figure 1) turns "off" and normal conditions are restored. So, the offset stored the first time, that at the end of conversion is multiplied with 2^{N-1} (N = number of bits of the ADC), is subtracted with next offsets (next clock cycles) that are multiplied with 2^{N-k}, where $k = 2..10$ is the number of the clock cycle. This also means that, after *ADCsample* is on, the residual signal of conversion is inverted: the SUM block thus needs additional logic[8].

2.4 2.4. ADC timing diagram

The reset signal asynchronously initialises the digital section of the ADC: the output register is loaded to "0" as conversion result, but *ADCdav* is low so this is not considered a valid result. An initialisation is recommended each time a new data acquisition begins, but the reset pulse width must be as short as possible to minimise power supply consumption.

The ADC timing diagram is shown in Figure 7: the acquisition mode is "free running" and the data stream is collected by a microcontroller which can operate in edge mode or level mode. In fact, *ADCdav* raises each time that a valid data conversion is available (edge triggered) and remains stable until the ending of a new conversion (level triggered). The "one shot" acquisition mode is a trivial sub case.

The ADC starts when it wakes up from stand-by condition. The internal state is bounded to rising edge of the clock; an initial start up time allows the analogue part to reach a steady state and then the conversion starts (the internal signal *ADCsample* is shown for clarity) on the clock falling edge.

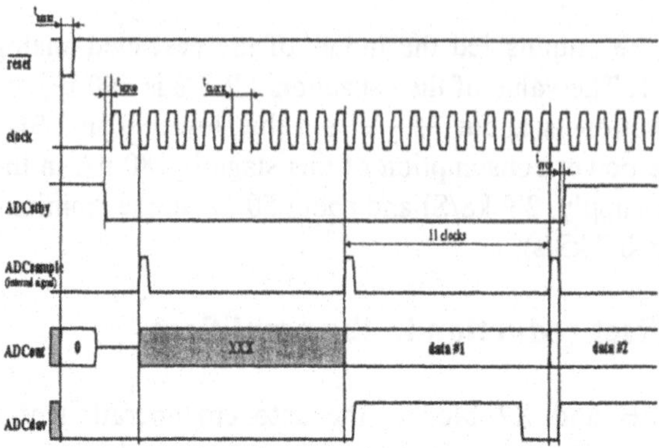

Figure 7. ADC timing diagram

To process the input value, the ADC takes 10 clocks plus an additional clock to reinitialise the internal logic. The data output *ADCout* [9:0] is registered, so the data are stable until the register is updated.

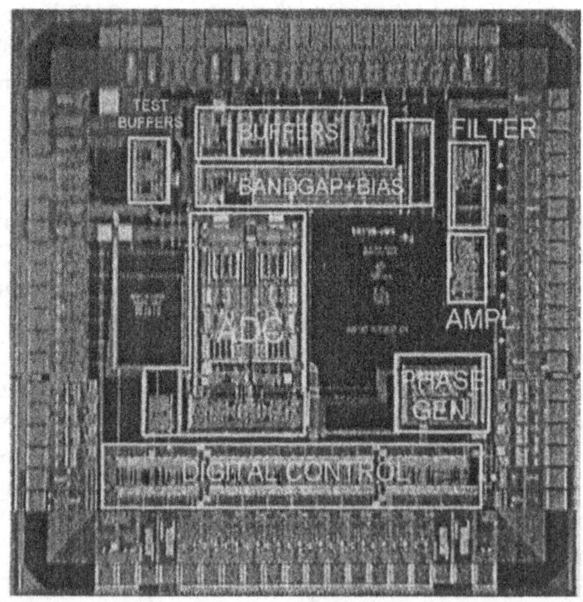

Figure 8. Chip microphotograph (die area about 9 mm^2)

3. EXPERIMENTAL RESULTS

The A/D integrated converter was fabricated in a 0.8 _m BiCMOS technology with 2 metal and 2 poly layers. The ADC described in the previous Section has been designed as a part of a prototype chip developed for implantable pacemakers (the chip microphotograph is shown in Figure 8). The front end section consists of an input amplifier with AC coupling - externally provided - to avoid input offset amplification, a low pass filter, and the 10 bits ADC.

The prototype includes a bandgap voltage reference, which is referred to ground, five voltage buffers to provide the proper references for the ADC and the other devices.

Table 2. Measurement conditions

Power consumption	Supply Voltage [V]	Clock Rate [kHz]	Bias Current
Maximum	3.5	32	full
Typical	2.8	32	full
Minimum	2.0	8	_full

The chip is pad limited and the prototype silicon area could have been reduced (see Figure 8), if no test pads were used. In fact, the ADC cell area is about 0.8 mm^2. Table 2 shows the operating conditions of the A/D converter. Nevertheless, all the prototypes have been characterised with different combinations of power supply, clock rate and bias current, showing a proper functionality in every condition.

Table 3. Summary of the ADC performances

	Minimum Power Consumption	Typical Power Consumption	Maximum Power Consumption	Unit
Resolution	10	10	10	bit
Consumption (analog)	0.56	2.92	3.90	μA
Consumption (digital)	0.67	3.47	4.17	μA
INL	1.13	0.98	0.85	LSB
DNL	0.73	0.67	0.75	LSB
Input Noise	0.43	0.42	0.40	mV (RMS)
Offset	0.72	0.73	0.70	mV
THD	56.6	56.3	57.6	dB
SFDR	60.2	59.6	61.1	dB
ENOB	8.4	8.4	8.4	bit
Offset drift	-	21.4	-	μV/°C
Gain drift	-	100	-	ppm/°C
Active Area		0.8		mm^2
Technology		AMS 0.8μm BiCMOS		

Very low power consumption has been measured for the analog core when reduced power supply, clock rate and bias current were used: 0.56 _A in the minimum case. Digital core dissipation (see Table 3) is rather high for the target application (total consumption exceeds 4 _A in typical case), but no particular attention has been paid to this point, because of the following considerations:

- in implantable device applications, an embedded processor that can implement the main part of the digital section of the ADC is commonly available.
- no low power digital library was available for the considered technology, so a standard digital library was used.

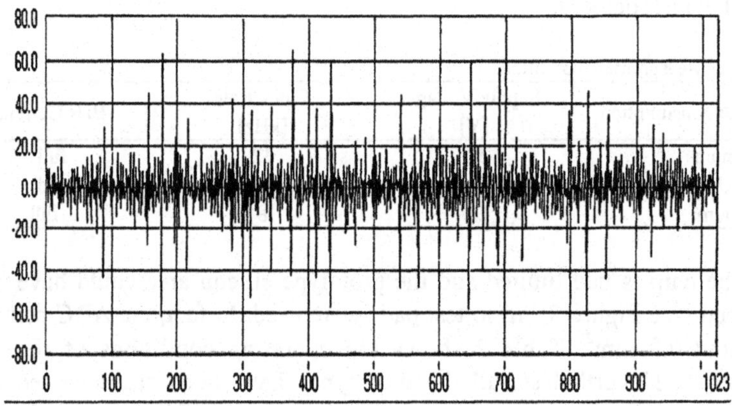

Figure 9. Measured DNL (LSB*100) in typical conditions

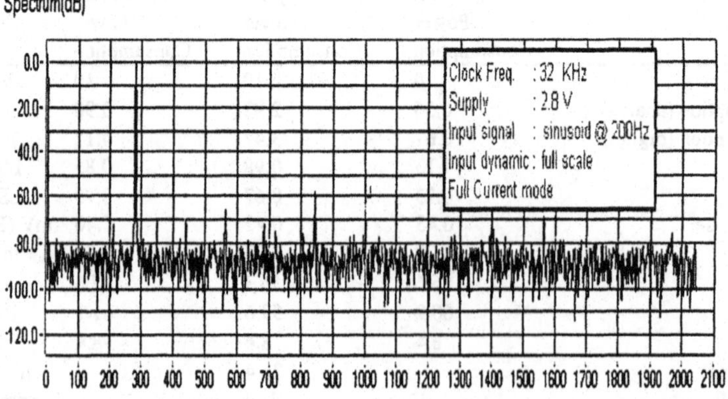

Figure 10. Output spectrum for an input tone at 200 Hz

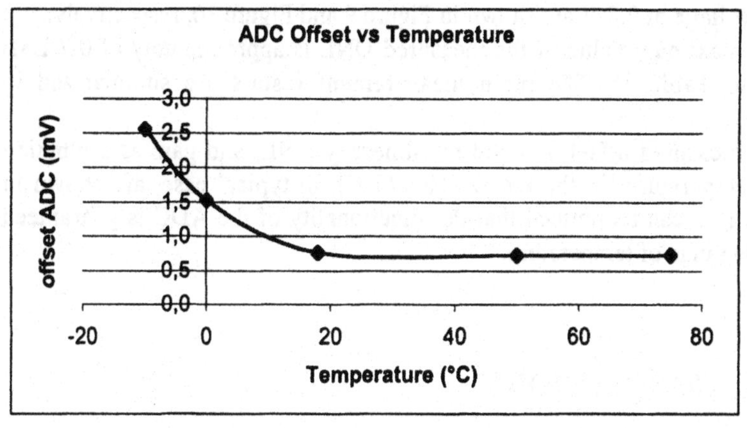

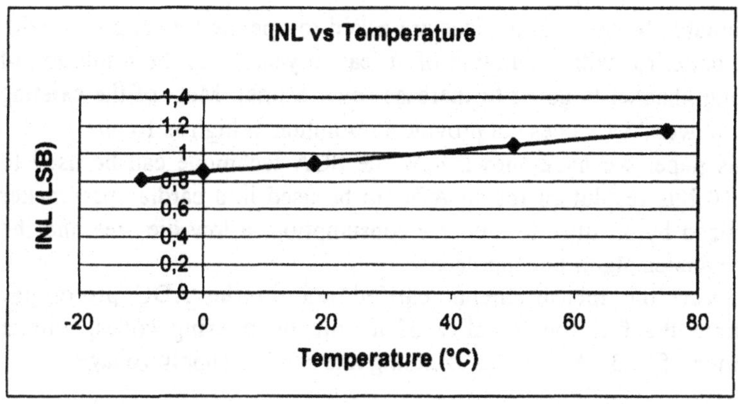

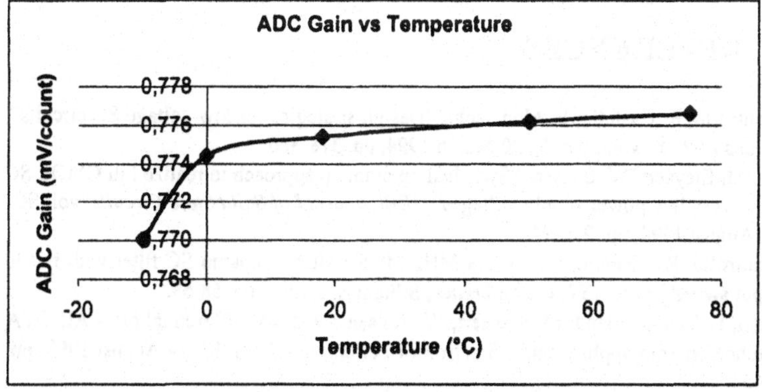

Figure 11. Offset, INL and Gain versus Temperature

The measured differential non-linearity (DNL) curve (typical case) and the measured output spectrum (2048 point FFT spaced 1.42 Hz each other) of a reconstructed 200 Hz full scale sine wave sampled at 2.9 kS/s with a supply voltage of 2.8 V are shown in Figure 9 and Figure 10, respectively.

The maximum value of the measured DNL is approximately of 0.7 LSB (see also Table 3). The main measurement results are summarized in Table 3.

The measured offset, integral non-linearity (INL) and gain as a function of the temperature (in the range −10/+75°C), in typical case, are shown in Figure 11. It can be noticed that the functionality of the ADC is guaranteed in a wide range of temperature.

4. CONCLUSIONS

Implantable biomedical devices are asked to operate for long time with long life batteries (with a duration of at least 6 years) and the resolution of the sensing channel is going to increase (over 8 bits). Most of the existing ultra low-power ADCs does not provide a resolution as high as 10 bits.

In this paper we have shown how the SOA technique can be used to achieve 10 bits resolution for an ADC to be used in a cardiac pacemaker characterized by an ultra low power consumption, a low die area and the cyclic conversion algorithm approach.

Moreover, the measurements carried out on the ADC prototypes demonstrate the full functionality of it with no missing codes, with a consumption of 1.23 _A, 0.7 kS/s sampling rate and 2 V supply voltage.

5. REFERENCES

1. A. Baschirotto, R. Castello, F. Montecchi, "Design strategies for low voltage SC circuits", *Electronics Letters.*, vol. 30 n. 5,, 30 March 1994, pp. 378-380.
2. J. Crols, M. Steyaert, W. Sansen, "Switched op amp: an approach to realize full CMOS SC circuits at very low power supply voltages", *IEEE Journal of Solid State Circuits*, vol. SC-29, n.8, August 1994, pp. 936-942.
3. A. Baschirotto, R. Castello, "A 1V 1.8 MHz MOS switched opamp SC filter with rail to rail output swing", *Proc. ISSCC Conference*, S.Francisco,1997, pp. 58-59.
4. V. Peluso, P. Vancorenland, M. Stayaert, W. Sansen, "900 mV differential class AB OTA for switched op amp applications", *Electronics Letters*, vol. 33 n. 17, 14 August 1997, pp. 1455-1456.
5. G. Ferri, A. Costa, A. Baschirotto, "A 1.2V rail to rail switched buffer", *Proceedings ICECS*, Lisboa, Portugal, September 1998.

6. G. Ferri, A. Baschirotto, "Low voltage rail to rail switched buffer topologies", *International Journal of Circuit Theory and Applications*, vol.29, n.4, July 2001, pp.413-422.
7. A. Gerosa, A. Novo, A. Neviani, "Low power sensing and digitization of cardiac signals based on sigma delta conversion", *Proc. ISPLED*, Rapallo, Sept.2000, pp.386-389.
8. M. Waltari, K. Halonen, "1 Volt, 9 bit Pipelined CMOS ADC", *Proc. ESSCIRC*, Sept. 2000, pp. 360-363.
9. J. Sauerbrey, D. Schmitt-Landsiedel, R. Thewes, "A 0.5 V, 1_W Successive Approximation ADC", *Proc. ESSCIRC*, Florence, Italy, 24th-26th September 2002.
10. B. Razavi, B. A. Wooley, "Design Technique for High Speed, High Resolution Comparators", *IEEE Journal of Solid State Circuits*, vol. 27, n. 12, December 1992.
11. B. Ginetti, P. G. A. Jespers, A. Vandemeulebroecke, "A CMOS 13 bits Cyclic RSD A/D Converter", *IEEE Journal of Solid State Circuits*, Vol. 27, n.7, July 1992, pp. 957-964.

EXPLORING THE CAPABILITIES OF RECONFIGURABLE HARDWARE FOR OFDM-BASED WLANS

Thilo Pionteck, Lukusa D. Kabulepa, and Manfred Glesner
Darmstadt University of Technology, Institute of Microelectronic Systems, Karlstr. 15, 64285 Darmstadt, Germany
pionteck@mes.tu-darmstadt.de

Abstract In this paper, the potential of reconfigurable hardware in wireless communication systems is evaluated. As most of the published works aim at the usage of reconfigurable architectures as a universal platform for different standards, the optimization capabilities of reconfiguration techniques for hardware modules within one standard are often not considered. This work focuses on the hardware optimization within one transmission scheme, namely OFDM (Orthogonal Frequency Division Multiplexing). By making use of special characteristics of packet-based WLANs and standard specifications, functional blocks of an OFDM receiver can be mapped on the same hardware. Even performance enhancements can be achieved without any additional hardware.

Keywords: Reconfigurable hardware, OFDM, digital baseband

1. Introduction

The usage of reconfigurable architectures as a hardware platform for wireless communication systems is often considered to be very suitable for solving the increasing requirements of modern communication systems. Critical design issues like performance, low-power consumption and flexibility seem to be facilitated by reconfigurable hardware. Especially the flexibility of reconfigurable architectures raises the idea of a universal architecture supporting different communication standards. Some works even expand the idea to a support of standards based on different transmission schemes [Helmschmidt et al., 2003]. Thereby designers inevitably encounter the fundamental trade-off between flexibility and efficiency. Flexibility can only be achieved at the cost of performance, chip area and power consumption. As different transmission schemes significantly differ in their transceiver structure, a universal hardware platform results in a significant hardware overhead and performance degrada-

Please use the following format when citing this chapter:
Pionteck, Thilo, Kabulepa, Lukusa, D., Glesner, Manfred, 2006, in IFIP International Federation for Information Processing, Volume 200, VLSI-SOC: From Systems to Chips, eds. Glesner, M., Reis, R., Indrusiak, L., Mooney, V., Eveking, H., (Boston: Springer), pp. 149-164.

tion. Hence, many research efforts are devoted to reconfigurable architectures supporting different standards based on the same transmission scheme. Such architectures require less flexibility as the basic operations of a transceiver for one specific transmission scheme are mostly the same. For example, OFDM-based (Orthogonal Frequency Division Multiplexing) WLAN standards like IEEE 802.11a and HiperLAN/2 show a huge similarity in the physical layer. The usage of reconfigurable architectures for supporting both standards seems to be a promising application field.

The flexibility required for supporting different standards of the same transmission scheme can also be used to improve the performance within one standard. WLAN standards show certain characteristics which can be used to reduce the hardware effort of receivers. Dynamic reconfiguration can be applied as certain functions within a receiver are not required permanently. By mapping these functions onto the same hardware components, the released resources can be used for the sake of optimizing the receiver performance.

The rest of the paper is organized as follows: in section 2, the basic principles of OFDM-based WLANs are presented. Based on these analyses, reconfiguration capabilities of OFDM receivers are shown in section 3. Section 4 gives a design example for a function-specific dynamically reconfigurable hardware design for the blocks depicted in the last section. Finally, a conclusion is given in section 5.

2. OFDM-based WLANs

The usage of OFDM as a transmission scheme for emerging WLAN standards like HiperLAN/2[Hip, 2001] and IEEE 802.11a[IEEE802.11a, 1999] has proved its suitability for high data-rate wireless communication systems. The basic principle behind OFDM consists in splitting the available bandwidth into many narrowband subchannels. A high transmission rate can be achieved by sending a large number of subcarriers in parallel as lower rate data streams. The discrete-time equivalent of a complex OFDM signal can be obtained by means of the IDFT (Inverse Discrete Fourier Transform) [Nee and Prasad, 2000]. Therefore the receiver can demodulate the transmitted signal by applying the DFT (Discrete Fourier Transform). Thus the FFT is essential for an OFDM receiver and is also one of the most computation-intensive blocks besides the Viterbi decoder. The complete structure of an OFDM receiver is shown in figure 1.

The first functional block in the baseband processing for an OFDM receiver is the synchronization unit. Its aim is to detect an incoming data packet and to compensate for a potential frequency offset. Therefore special preambles at the beginning of a transmission are used. After removing a cyclic prefix which was introduced at the transmitter side in order to avoid intersymbol interference,

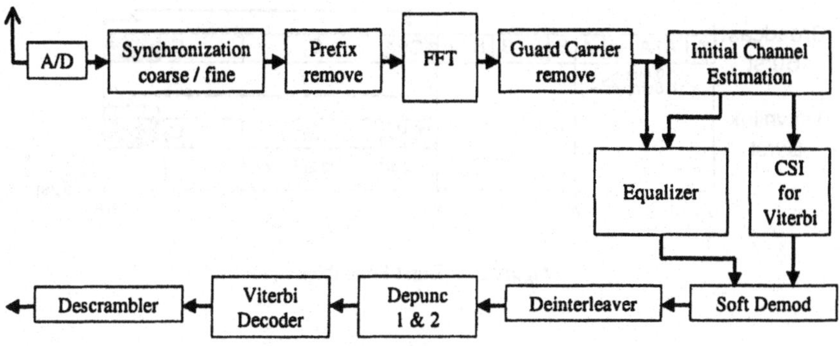

Figure 1. Block Diagram of an OFDM Receiver

the signal is transformed from the time-domain back to the frequency-domain by applying a FFT. For the compensation of the channel distortion, a channel estimation is done at the beginning of each transmission burst. The estimates are used by the equalizer for the compensation of the channel distortion by means of a zero-forcing equalization. In addition, channel state information are generated by using the channel estimates. The channel state information are used to weight the equalized soft values from the equalizer. Prior to the Viterbi processing of the soft values, the interleaving and puncturing have to be reversed. Finally the data bits are descrambled.

Transmission Format

The HiperLAN/2 standard defines a basic frame with a length of 2 ms, which comprises five different burst types. For a downlink connection, two types are defined, a BCB (BroadCast Burst) and a DLB (DownLink Burst) [Hip, 2001]. Figure 2 shows the structures of these two burst types. The preamble of the BCB is divided into three symbol groups. The A and B symbols enable frame synchronization, automatic gain control and frequency synchronization. The C symbols are used for channel estimation as well as for fine frequency synchronization. While the C symbols are included in both burst types, the A and B symbols are only used for the BCB. As a DLB is preceded by a BCB, a new frame and frequency synchronization is not required for the DLB.

For the WLAN standards IEEE 802.11a and HiperLAN/2, a channel raster of 20 MHz is used. With a subcarrier spacing of 312.5 kHz, this results in 64 subchannels. For one OFDM symbol channel, 11 guard-carriers are defined in order to achieve a sufficient adjacent channel suppression. Moreover, the DC-carrier is not used either. The remaining 52 subchannels are used for

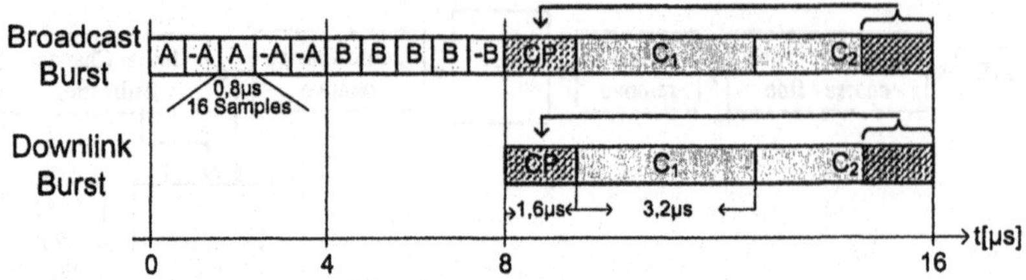

Figure 2. Burst Preambles

transmission. Out of these 52 subchannels, 4 pilot-carriers are scattered among the frequency spectrum and thus 48 subcarriers are used for data transmission. Figure 3 shows the frequency spectrum allocation.

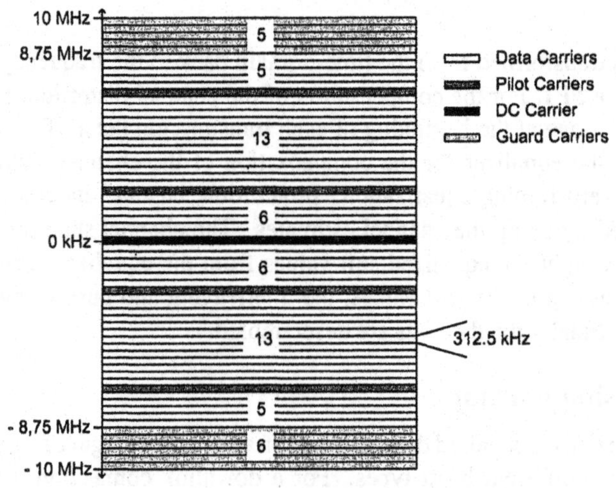

Figure 3. Subcarrier Frequency Spectrum

3. Reconfiguration Capabilities

Reconfiguration always comes at cost of hardware efficiency. Due to the required high computational power of modern wireless communication systems, the use of reconfigurable architectures for wireless systems based on different transmission schemes seems not to be advisable. By restricting the flexibility to function-specific reconfigurable designs, the hardware overhead can be reduced significantly [Zhang and Bodersen, 2000]. Such hardware architectures will be specific to a transmission scheme but not to a special standard. In the following, reconfiguration capabilities based on WLAN and standard characteristics are analyzed. Notice that computation-intensive tasks like FFT and Viterbi decoding are not considered in the following. Due to the high data rates up to 54 MBits/s, these tasks are often realized with specialized hardware designs. This specialization hampers the utilization of these hardware blocks for reconfiguration.

Reconfiguration Options based on WLAN Characteristics

The main option enabling reconfiguration is the fact that not all functional blocks within an OFDM receiver are active while processing a burst. During timing and frequency synchronization, no payload processing is done (neglecting a pipelined processing of the last burst). On the other side, while decoding the payload, synchronization has already been achieved. This raises the idea of mapping both tasks onto the same hardware.

Figure 4. Synchronization Block Diagram

Figure 4 shows the block diagram of a synchronization unit. The synchronization process can be divided into two steps. In the first step, the time and frequency values are determined. Therefore the repetition peambles within the received data sequence r have to be detected. This process is based on the computation of the complex correlation

$$S_n = \sum_{m=0}^{N_s-1} r_{n+m} r^*_{n+m+N_s} \tag{1}$$

and the power sum of a synchronization windows of length N_s

$$P_n = \sum_{m=0}^{N_s-1} |r_{n+m}|^2. \tag{2}$$

The hardware structure for the symbol correlator is given in figure 5.

There exist several metrics which use the complex correlation and the power sum to determine the synchronization point and the frequency offset. In this work, a combination of the MNC (Maximum-Normalized-Correlation) scheme [Schmidl and Cox, 1997] and the MMSE (Minimum-Mean-Square-Error) scheme [Chevillat et al., 1987] is used, which was proposed in [Kabulepa et al., 2002]. While the MMSE scheme has shown to be efficient in continuous OFDM applications, it tends to increase false alarm probabilities in burst-oriented OFDM systems when no data frame is transmitted. By applying the MNC scheme to detect the existence of a data frame and using the

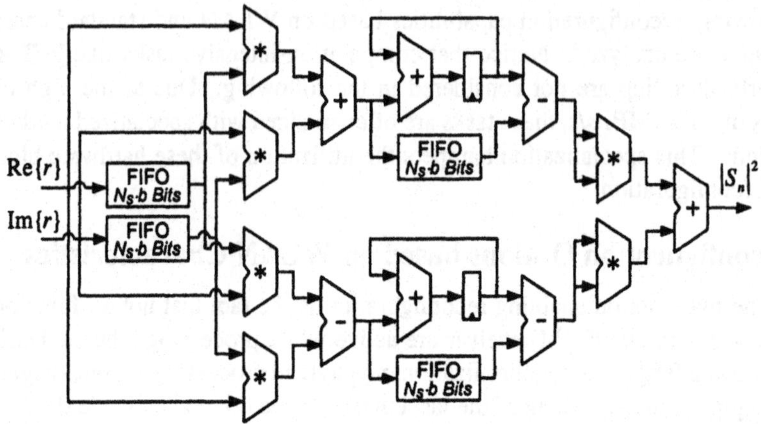

Figure 5. OFDM Symbol Correlator

MMSE scheme to determine the exact synchronization point, it was possible to combine the advantages of both schemes. The hardware structure for this synchronization scheme is given in figure 6.

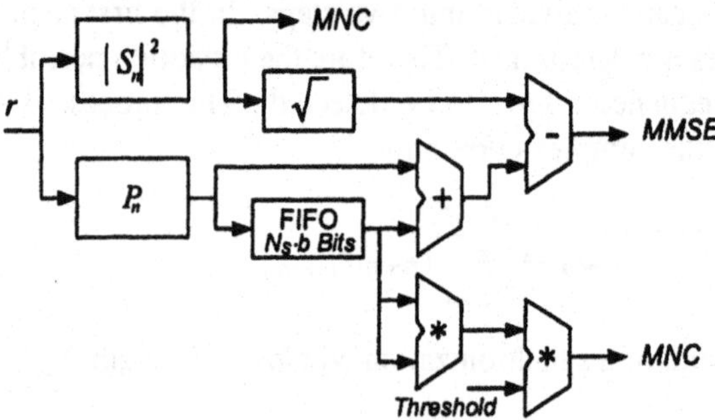

Figure 6. Hardware structure of synchronization scheme

When the synchronization point has been detected, the received data are aligned in a second step. Therefore, FIFOs are used for the time alignment and a CORDIC (COordinate Rotation DIgital Computer) is used for the frequency correction. The CORDIC is also required for angle calculation during the frequency offset estimation during the first phase. Thus, only the hardware components realizing the synchronization scheme as depicted in figure 4 can be reused as these modules are not required all the time.

Once synchronization has been achieved, payload processing can be started. Focusing on the functional blocks succeeding the FFT, the payload processing starts with equalization. Equalization (zero-forcing) can be performed by multiplying the received signal $\hat{d}_n(i)$ with the estimated inverse channel transfer function $\hat{C}_n^{-1}$.

$$\tilde{d}_n(i) = \hat{d}_n(i) \cdot \hat{C}_n^{-1} = d_n(i) + \frac{N_n}{\hat{C}_n} \tag{3}$$

resulting in the transmitted signal $d_n(i)$ and a noise term N_n. The channel estimates $\hat{C}_n$ can be achieved by using the two C-symbols at the beginning of each burst. Equation 4 shows the computation of the channel estimates according to the preamble structure of IEEE 802.11a and HiperLAN/2.

$$\hat{C}_n = \frac{\hat{d}_n^P(0) + \hat{d}_n^P(1)}{2 \cdot d_n^P} \tag{4}$$

In case of a non perfect synchronization, the equalized data may still show a remaining phase rotation. In order to compensate for this impairment, the phases of the pilot-carriers can be monitored. By averaging the phase rotations of the pilot-carriers in an OFDM symbol, a correction value $\widetilde{\varphi}(i)$ can be determined which is used to rotate the equalized data. This value can also be used to update the channel estimates, resulting in an improved equalization for succeeding OFDM symbols. Equation 5 presents a method for averaging the phase rotations of the pilot-carriers under consideration of the channel state. N_p gives the number of pilot-carriers, $\tilde{d}_j^P(i)$ are the equalized pilots and $\hat{C}_j^P(i)$ are the channel estimates for the pilot-carriers. A more detailed description of this method can be found in [Pionteck et al., 2003a].

$$\widetilde{\varphi}(i) = \frac{\sum_{j=0}^{N_p-1} arg\left(\tilde{d}_j^P(i) \cdot \left|\hat{C}_j^P(i)\right|^2\right)}{\sum_{j=0}^{N_p-1} \left|\hat{C}_j^P(i)\right|^2}. \tag{5}$$

Figure 7 shows the hardware structure of a zero-forcing equalizer with phase rotation compensation. The grey area highlights the components which are used for zero-forcing equalization only. It was possible to avoid the division of the equalized data by the squared absolute values according to equation 3, as the succeeding functional block, the soft value generation, requires a multiplication of the data with this factor.

For the soft value generation, the LLR (Log-Likelihood Ratio) is used as a measure for the reliability of a decision. The sign of the LLR denotes the symbol (±1) while the absolute value represents the realibility of the decision. Especially for higher modulation schemes, the computation of the LLR function

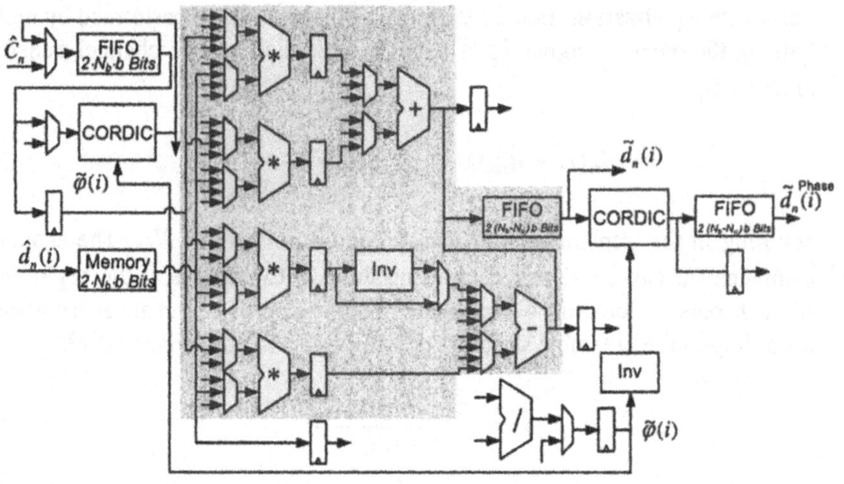

Figure 7. **Block structure of the equalizer**

is very complex. Therefore, a simplified version is used, which was presented by [Tosato and Bisaglia, 2002]. Let k be the dimension of the modulation scheme and $|G_n(i)|^2$ be the channel state information of subcarrier n. The half distance between the partition boundaries of the constellation for $k > 1$ is given by $a_{k,n}$. According to [Tosato and Bisaglia, 2002], an approximation of the LLR function for a square QAM constellation is given by:

$$D_{k,n}(i) \approx \begin{cases} \tilde{d}_n(i) & k = 1 \\ -|D_{k-1,n}(i)| + a_{k,n}(i) & k > 1 \end{cases} \tag{6}$$

$$LLR\left(b_{k,n}(i)\right) = |G_n(i)|^2 \cdot D_{k,n}(i). \tag{7}$$

The corresponding hardware structure for $k \leq 3$ is given in figure 8. The multiplication of $\tilde{d}_n(i)$ with $|G_n(i)|^2$ can be avoided by adapting the equalizer as mentioned before.

The hardware structures of the presented functional blocks show similarities to the symbol correlator and the power sum computation module. Due to the non-overlapping execution times, the use of a common hardware platform for these functional blocks seems to be advisable. The usually introduced hardware overhead through dynamic reconfiguration can be avoided as the reconfiguration capabilities are very function-specific. A hardware platform for these functional blocks is presented in section 4.

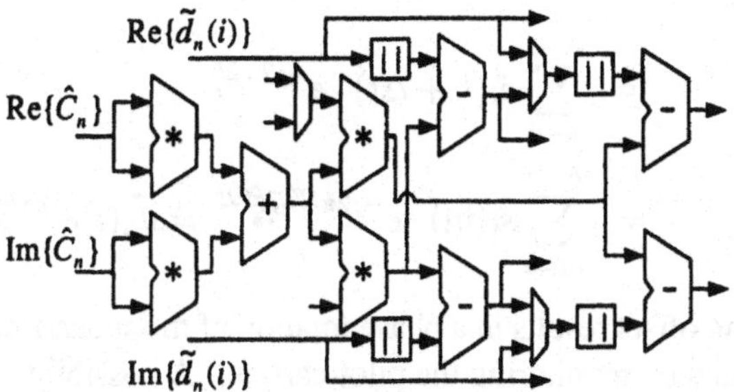

Figure 8. Datapath for generating soft decision values

Reconfiguration Options based on Standard Characteristics

Another way of finding reconfiguration options is the analysis of the standards to be implemented. Focusing on the HiperLAN/2 standard, the different characteristics of the two burst types (BCB and DLB) for a downlink connection can be exploited. The BCB is used to transmit control information to all users while the DLB contains the payload for one specific user. Due to the importance of the BCB, simple modulation schemes are used to make sure that all users are able to decode the control information. In HiperLAN/2, the BPSK modulation scheme is exclusively defined for the BCB, while for the DLB more complex modulations schemes like 64QAM can be used. It is evident that higher modulation schemes require more complex decoding algorithms. As for the BCB only BPSK is used, hardware required for the demodulation of 64QAM is idle. This hardware can be used to improve the performance of the receiver without any costs.

Based on the functional blocks introduced in section 3, the soft values generation can be simplified. For BPSK, the soft information is proportional to the distance from the decision boundary [Tosato and Bisaglia, 2002]. Thus, the hardware blocks for generating the channel state information and soft values are not required for the BCB. All further hardware blocks presented in section 3 are independent of the modulation scheme and thus are required for both burst types.

The freed hardware resources can be used to improve the timing synchronization. OFDM systems are very sensitive to timing synchronization errors [Bølcskei, 2001]. Even a small misalignment leads to a degradation of the system performance. The sensitivity increases with the complexity of the modulation scheme. In the presence of a timing offset Δt, where Δt is an integer multiple of the sampling period, the received signal can be written as

$$\hat{d_n}^{\Delta t}(i) = \sum_{k=0}^{N-1} \hat{s}_i(k+\Delta t) \cdot e^{-j2\pi \frac{nk}{T_s}} \tag{8}$$

$$= \sum_{m=0}^{N-1} \hat{s}_i(m) \cdot e^{-j2\pi \frac{(m-\Delta t)k}{T_s}} = \hat{d}_n(i) e^{j2\pi \frac{\Delta t n}{T_s}} \tag{9}$$

Thus a timing offset results in a phase rotation of the subcarriers, which can easily be detected by monitoring the pilot-carriers. A possibility to estimate a timing offset is to determine the best fit line of the phase rotations of the pilot-carriers. With linear regression, the influence of the channel can be reduced. Figure 9 shows the best fit line for a timimng offset of -1. By monitoring the slope of the best fit line during the BCB, a residual timing offset can be detected reliably [Pionteck et al., 2003b]. A receiver can compensate for the estimated timing offset before starting the processing of the DLB. The required hardware effort is marginal. The calculation of the phase of the pilot-carriers, which is the most complex task, is already done in the equalizer when performing phase rotation compensation. For correcting the offset, no additional hardware is required as only memory positions inside the synchronization module have to be updated.

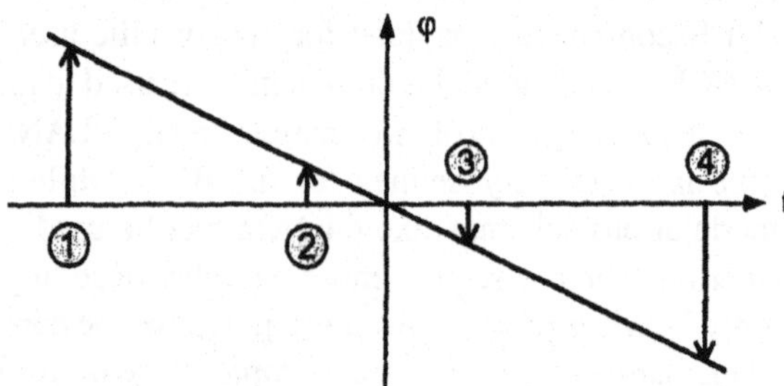

Figure 9. Timing offset estimation with best fit line

The timing offset estimation based on pilot-carriers phase shift can be done according to the following equation:

$$m = \frac{\sum_{i=0}^{N_{BCB}-1} \sum_{j=0}^{N_p-1} (\overline{x} - x_j) \left[\left(\frac{1}{N_p} \sum_{k=0}^{N_p-1} \varphi(i) \right) - \varphi_j(i) \right]}{N_{BCB} \cdot \sum_{k=0}^{N_p-1} (\overline{x} - x_k)^2} \tag{10}$$

where N_{BCB} denotes the number of considered OFDM symbols, N_P gives the number of pilot-carriers, x_j and x_k are the subcarriers at position i,j and

$\varphi_j(i)$ is the phase of jth pilot-carrier of the ith OFDM symbol. As some of the pilot-carriers are rotated, an addition of the phase of these pilot-carriers with π is required before determining the slope of the best fit line. The timing offset estimator can be realized with only a few hardware components. This allows the realization of this optimization method onto the hardware blocks of the channel state information and soft value generation. By means of a dynamic reconfiguration between the BCB and DLB, no additional hardware for the timing offset estimator is required. The division in equation 10 can be avoided by adapting the reference values which are used to determine the size of the timing offset.

4. Hardware Design

The previous section indicates that several possibilities exist where dynamic reconfiguration can reduce the hardware requirements of a receiver within one transmission scheme. These function-specific reconfigurations can be used either as optimizations within a dynamically reconfigurable architecture or for a stand-alone implementation. A possible stand-alone implementation for the usage inside an ASIC is presented in the following.

Figure 10 shows the datapath of the proposed hardware architecture. For the sake of simplicity, not all feedback connections and pipeline registers are shown. Memory blocks which can be used at different positions in the datapath are drawn separately. Each of these memory blocks has a capacity of 52 Bytes, capable of storing the real or imaginary part of a complete OFDM symbol. The remaining memory blocks provide a capacity of 104 Bytes. For the CORDIC elements a word-parallel architecture was chosen. This increases the hardware requirements but leads to a reduced latency for the complete system. The same holds true for the architecture of the divider.

The design comprises all arithmetic blocks and datapaths for realizing the functional blocks discussed in the last section, including the timing offset estimator for the BCB. Dynamic reconfiguration enables the switching between different functional blocks at runtime. By making use of timing and scheduling properties of the different tasks, the functional blocks can be grouped into three different configuration mappings. Figure 11 shows the mapping of the functional blocks onto the proposed hardware design.

mapping A The first mapping realizes the combination of the MMSE and the MNC scheme as described in section 5. According to the numbering in figure 10, the result of the MMSE scheme is provided at position ⑤ and the result of MNC scheme at position ④ and ⑥ of the circuit. Besides memory block 4 and both CORDICs, all hardware and memory elements of the architecture are used.

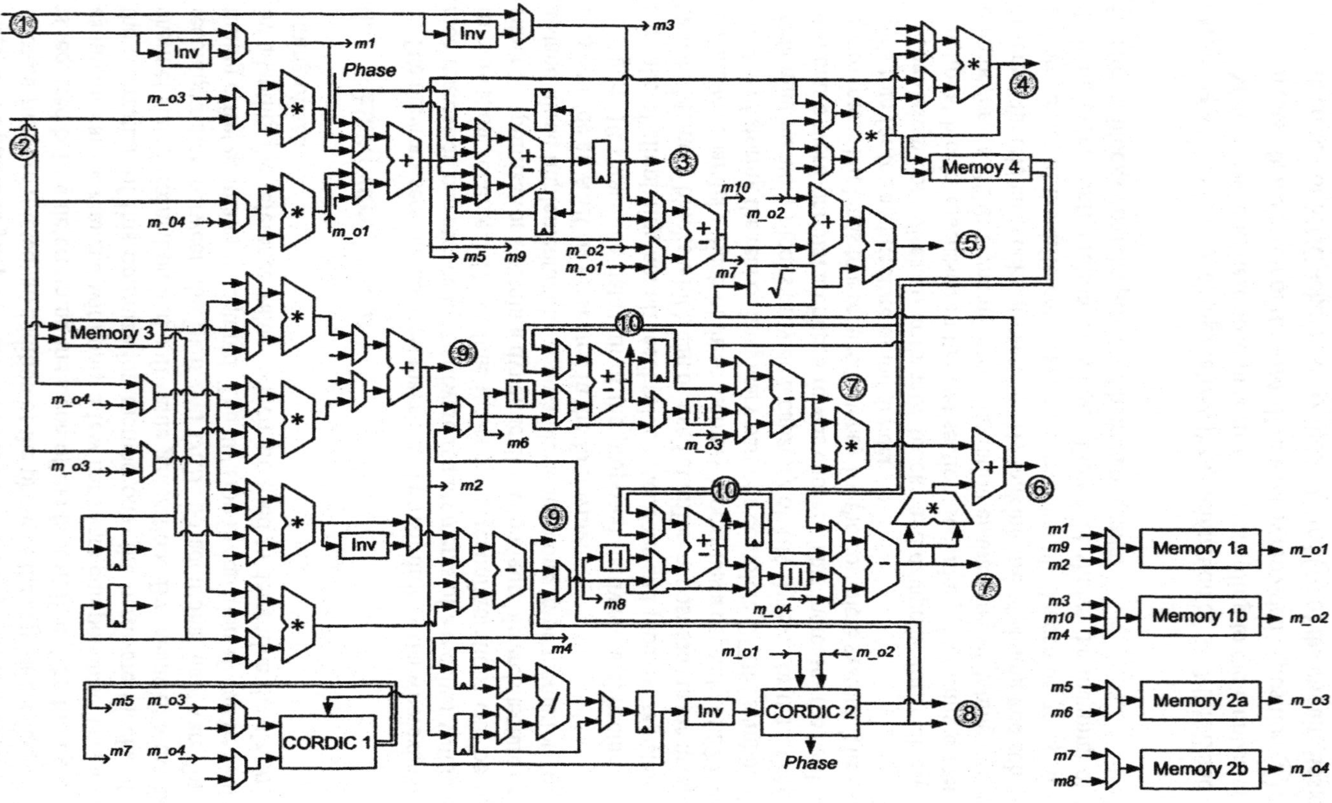

Figure 10. Dynamically Reconfigurable Architecture

mapping B The second mapping realizes the channel estimater, equalizer, channel state information and soft values generator. This mapping can be used for payload processing in HiperLAN/2 (DLB and BCB) and IEEE 802.11a. Modulation schemes up to 64QAM are supported. Input (1) provides the C symbols which are used to calculate the channel estimates. These values are stored in memory block 2 for the equalization of the data symbols. The incoming data symbols $\hat{d}_n(i)$ at input (2) are stored in memory block 3. In parallel the first data symbols are equalized. As soon as all pilot-carriers have been received, the computation of the correction value for the phase rotation is started. In the meantime, memory block 1 is used as a buffer for the data symbols. The phase compensation of the equalized data is done in CORDIC block 2, while CORDIC block 1 is used to update the channel estimates. The equalized and rotated data symbols are available at position (8). If only zero-forcing equalization is to be performed, the result is available at position (9). For the soft value generation, the equalized data is fed back to the architecture. According to the used modulation scheme, the soft decision values are provided at position (7) or (10).

mapping C The third mapping realizes the channel estimator, equalizer and timing offset estimator as described in subsection 3.2. Generation of soft values and channel state information are not required as BPSK is used as modulation scheme. Thus this mapping is specific to HiperLAN/2 as it exploits standard characteristics for the BCB of the downlink connection. The result of the timing offset estimation is available at position (3).

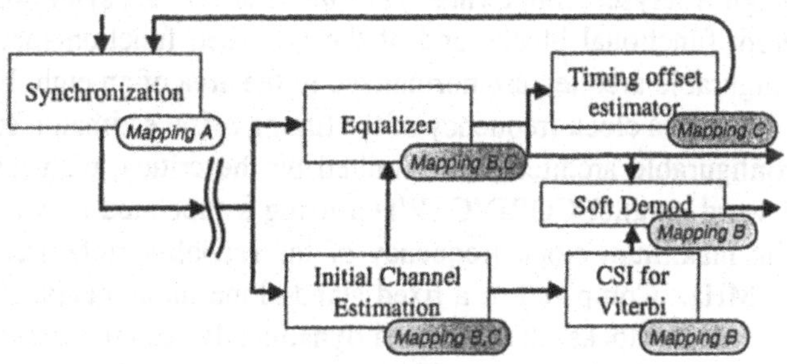

Figure 11. Functional mapping

The configuration memory comprises 48 Bytes and is divided into two sections. In the first section, three different configuration sets can be stored. The

Table 1. Synthesis Results

Modules	*Frequency [MHz]*	*Area (normalized)*
synchronization metric	171	15.1
timing offset estimator	31	19.9
channel estimator	534	16.5
equalizer	20	121
channel state inform.	110	3.4
soft value generation	223	2.6
reconfigurable architecture	30	143

second section is used to provide some general settings which are applicable to all of the three configuration sets. These settings include information about the preamble structure (HiperLAN/2 or IEEE 802.11a) and the position of rotated pilot-carriers. The architecture allows a single cycle context-switch between different mappings, as it is required for a continuous data processing during a burst. The hardware architecture is controlled by an external device (e.g., microcontroller), which also manages the reconfiguration process. This external device has to implement functions of the data link layer according to the OSI layer model, as only this layer has knowledge of the type of a burst. The switching between the preamble processing and the payload processing of a burst is controlled by the architecture itself.

The proposed dynamically reconfigurable design was implemented using a $0.35\mu m$ standard CMOS process with 3 metal layers. Synthesis was done with the Design Analyzer from Synopsys. Table 1 shows the synthesis results of the different functional blocks and of the proposed function-specific dynamic reconfigurable architecture normalized to the area of an eight bits multiplier. The maximum clock frequency of the timing offset estimator, equalizer and the reconfigurable architecture is limited by the critical path inside the unpipelined word-parallel CORDIC. When using a pipelined or word-seriel CORDIC, the maximum clock frequency of the reconfigurable architecture is about 100 MHz. Compared to a fixed stand-alone implementation of the presented functional blocks, the proposed dynamically reconfigurable design significantly reduces the hardware effort while providing a higher flexibility than a fixed design. The proposed architecture has a normalized area of about 153. Compared to the normalized area of 178.5 for implementing all functional blocks separately, hardware savings of about 20% could be achieved by using the reconfigurable platform. The hardware overhead introduced by dynamic reconfiguration is very low (about 6%), as only three configurations have to be stored. Even performance optimizations by improving the timing synchronization were to be realized without increasing the hardware effort.

5. Conclusion

In this work, reconfiguration capabilities for OFDM-based WLANs are explored. By focusing on WLAN and standard characteristics, a set of functional blocks suitable for a dynamic reconfiguration were extracted. The blocks were chosen according to the division of the data processing into preamble and payload processing. The reconfiguration possibilities could be used either for optimization purposes within a dynamically reconfigurable architecture or for a stand-alone implementation. As an example for a stand-alone implementation, the extracted functional blocks were realized onto a function-specific dynamically reconfigurable hardware design, offering significant savings in terms of area. Making use of characteristics in the HiperLAN/2 standard, it was also possible to use the presented hardware design to improve the receiver performance by optimizing the timing synchronization. The work shows that the use of reconfiguration techniques is not only worthwhile for hardware designs for different standards but also for optimizations within a standard. By offering a restricted set of reconfiguration capabilities reduces the overhead introduced by dynamic reconfiguration and eases the use of reconfigurable hardware designs.

6. References

[Belcskei, 2001] Belcskei, H. (2001). Blind Estimation of Symbol Timing and Carrier Frequency Offset in Wireless OFDM Systems. *IEEE Trans. on Communiction*, pages 988–999.

[Chevillat et al., 1987] Chevillat, Pierre R., Maiwald, Dietrich, and Ungerboeck, Gottfried (1987). Rapid Training of a Voiceband Data-Modem Receiver Employing an Equalizer with Fractional-T Spaced Coeddicients.

[Helmschmidt et al., 2003] Helmschmidt, J., Schüler, E., Rao, P., Rossi, S., Matteo, S., and Bonitz, R. (2003). Reconfigurable Signal Processing in Wireless Terminals. In *Date'03 Designers Forum*.

[Hip, 2001] Hip (2001). Broadband Radio Access Networks (BRAN); HIPERLAN Type 2; Physical (PHY) Layer, ETSI TS 101 475.

[IEEE802.11a, 1999] IEEE802.11a (1999). *Part 11: Wireless LAN Medium Access Control (MAC) and Physical Layer (PHY) Specifications: High-Speed Physical Layer in the 5 GHz Band*. Institute of Electrical and Electronic Engineers (IEEE).

[Kabulepa et al., 2002] Kabulepa, Lukusa D., Ortiz, Alberto Garcia, and Glesner, Manfred (2002). Design of an Efficient OFDM Burst Synchronization Scheme.

[Nee and Prasad, 2000] Nee, R. Van and Prasad, R. (2000). *OFDM for Wireless Multimedia Communications*. Artech House Publishers.

[Pionteck et al., 2003a] Pionteck, Thilo, Kabulepa, Lukusa D., and Glesner, Manfred (2003a). On the Rapid Prototyping of Equalizers for OFDM Systems. *Design Automation for Embedded Systems*, 8(4):283–295.

[Pionteck et al., 2003b] Pionteck, Thilo, Kabulepa, Lukusa D., Koppel, Stefan, Garcia, Alberto, and Glesner, Manfred (2003b). Hardware-Efficient Detection and Correction of Timing Offsets in OFDM-based WLANs. *8th International OFDM-Workshop.*

[Schmidl and Cox, 1997] Schmidl, T. and Cox, D. (1997). Robust Frequency and Timing Synchronization for OFDM. In *IEEE Trans. on Communication.*

[Tosato and Bisaglia, 2002] Tosato, F. and Bisaglia, P. (2002). Simplified Soft-Output Demapper for Binary Interleaved COFDM with Application to HiperLAN/2. In *IEEE International Conference on Communications.*

[Zhang and Bodersen, 2000] Zhang, N. and Bodersen, R. W. (2000). Architectural Evaluation of Flexible Digital Signal Processing for Wireless Receivers. In *Asilomar Conference on Signals, Systems and Computers.*

SOFTWARE-BASED TEST FOR NON-PROGRAMMABLE CORES IN BUS-BASED SYSTEM-ON-CHIP ARCHITECTURES

Alexandre M. Amory[1], Leandro A. Oliveira[1] and Fernando G. Moraes[2]

[1]Instituto de Informática - Universidade Federal do Rio Grande do Sul (UFRGS) - av. bento gonçalves, 9500 - prédio 43412/bloco IV - Porto Alegre - Brazil - CEP 91501-970; [2]Faculdade de Informática - Pontificia Universidade Católica do Rio Grande do Sul (PUCRS) - av. ipiranga, 6681 - prédio 30/bloco 4 - Porto Alegre - Brazil - CEP 90619-900 - moraes@inf.pucrs.br

Abstract: With the advance in hardware integration, system-on-a-chip (SoC) test activities using only automatic test equipments (ATEs) result in an expensive option. Hardware-based test may reduce the ATE dependency. However, hardware-based test imposes some constraints like area overhead and processing speed degradation. The main objective of this work is to investigate and evaluate a less intrusive test approach called software-based test. Software-based test uses an embedded processor as source and sink of the test, sending the test patterns and reading the responses. A new integrated design and test environment has been developed to automatically synthesize test programs to test non-programmable cores of SoCs. Some benchmarks ISCAS85 and ISCAS89 are used to evaluate the proposed methodology.

Key words: SoC test; Software-based test; computer aided test (CAT); LFSR reseeding.

1. INTRODUCTION

In recent SoC based systems the amount of test data transferred between automatic test equipments (ATEs) and devices under test is becoming too large. Even expensive state-of-the-art ATEs restrict the SoC test as a result of limited memory resources, narrow channel bandwidth and low speed.

Please use the following format when citing this chapter:

Amory, Alexandre, M., Oliveira, Leandro, A., Moraes, Fernando, G., 2006, in IFIP International Federation for Information Processing, Volume 200, VLSI-SOC: From Systems to Chips, eds. Glesner, M., Reis, R., Indrusiak, L., Mooney, V., Eveking, H., (Boston: Springer), pp. 165-179.

One known approach to overcome ATE limitations is to use hardware-based test (i.e. built-in self-test BIST) to generate patterns and to analyze the results at -speed. This approach reduces the ATE constraints and the test cost. However, BIST has some drawbacks[1]: (i) some circuits are resistant to random patterns, resulting into a low fault-coverage; (ii) since new modules are inserted into the system, the total area, operation frequency and power consumption are negatively affected.

Software-based test is an alternative approach to BIST and ATE. SoC devices usually contain, at least, one embedded processor and use bus-based interconnection to integrate several IP cores. We propose a test methodology to test non-programmable IP cores using an embedded processor. Since there are no new test modules added to the system, figures as area usage, speed and power are not changed. A possible drawback of the software-based test, when compared to the hardware-based test is the longer time needed to apply the patterns and/or analyze the results[1].

This work has two main objectives. The first one is the evaluation of software-based against hardware-based test. The second objective is to present the developed Computer Aided Test (CAT) tool to synthesize the test program and to integrate IP cores in the SoC.

Unlike previous approaches[2,3,4], this paper presents a new tightly integrated design and test methodology, including commercial and in-house tools. Moreover, the evaluation compares software-based test with hardware-based test, unlike[3] which compares software-based test with boundary scan.

This paper is organized as follows. Section 2 presents the state-of-the-art in software-based test. Section 3 presents the bus-based SoC architecture. Section 4 details the developed CAT environment. Section 5 focuses on the software-based test evaluation based on some ISCAS85/89 benchmarks. Section 0 concludes this paper and presents some directions for future work.

2. SOFTWARE-BASED TEST

Software-based test can minimize some of the BIST drawbacks discussed before. The following is a list of advantages of this approach:

- Ease to reuse and to modify the test strategy as it is implemented in software;
- No specific test controller is required, since a generic embedded processor is responsible for the test control and execution;
- Reduced (or none) area overhead due to the test patterns generation and response compaction implemented in software;
- Reduced (or none) speed/power degradation, as there are no additional test modules;

- Can be applied to test processors, memories, general cores and interconnection (bus);
- Reduced design time compared to BIST even considering automated tools since there is an additional manual process required to make the target core become BIST-ready[1];
- Test occurs in normal operational mode, eliminating the extra power consumption of BIST[5];
- Can apply and analyze at-speed test signals[5] to detect delay faults, alleviating the need of high-speed tester. Moreover, since the test is applied in normal operational mode, the system is not over-tested.

Possible drawbacks of the software-based test are:

- The SoC must have an embedded processor;
- Additional time to create the test program;
- Extra memory needed to store the test program and the deterministic test patterns;
- Increased test time when comparing to hardware-based test. BIST usually generates patterns/ compact responses in just one clock cycle. However, in software-based test patterns are provided by the processor to the cores, taking longer to execute the same task than BIST modules. On the other hand, software-based test may be faster than ATE based test due to the limited bandwidth[3].

Thus, the goal of this work is to automate the test program generation, reducing the project time and to evaluate quantitatively the last two drawbacks (i.e. the increased test time and memory requirements).

2.1 Processor Test

The first works on software-based test for processors were conducted in the 70s by Thatte and Abraham[6]. Software-based self-test of processors is a challenging issue due to: (i) the number of different hardware modules to be tested; (ii) the limited number of pins to access the processor; (iii) the existence of modules that can not be accessed directly, as interrupt controller, flag, exception handler; (iv) the huge variation of implementations and architectures. Recent works on software-based processor test can be found in [1,7].

2.2 Memory Test

Memory test is very simple to be implemented by a processor due to its regular structure. Well-know algorithms[8] using deterministic patterns, e.g. 55h and AAh, are used. Zorian and Ivanov[9] consider a ROM as a combinational circuit, where the address is the input and the memory content is the output. An exhaustive test is performed reading all memory contents and compacting the output.

2.3 Non-Programmable Core Test

The embedded processor can execute a specific program to test each core of the SoC, which is the goal of this paper. This program has the following functions: (i) generate the test patterns; (ii) send these patterns to the cores through the communication path (e.g. bus or TAM); (iii) read the test responses; (iv) compact these responses. Since it is a software-based test, it is possible to use different algorithms to test each core.

The test program can emulate the behavior of a LFSR and MISR in software, generating pseudo-random patterns and compacting responses[10]. A disadvantage of emulating a LFSR or a MISR in software is the increasing test time, since it may need several clock cycles to produce one pattern. Other approaches using adders, subtractors and multipliers to generate patterns and/or to compact responses can be used. In [11] it is shown that these modules can be used to substitute LFSR, generating comparable fault coverage, with the advantage of generating a pattern for each instruction (e.g. add).

Huang *et al.* [2] developed a bus-based architecture with MIPS processor, PCI bus and VCI interface. Using this architecture, they evaluated the test time and fault coverage of some ISCAS89 benchmarks. Lai and Cheng[4] used the same architecture presented in [2] to evaluate test programs generated for four ISCAS89 benchmarks, using the DLX processor. The test program length is from 40 to 27000 bytes. The test time is from 94 to 30430 clock cycles. The results point to important test memory requirements and test time compared to hardware-based test.

Hwang and Abraham[3] developed a bus-based architecture with an ARM processor and Wishbone interface. The authors compared the test time and area overhead between software-based test and boundary scan. In both cases, the software-based test presented better results.

3. SOC ARCHITECTURE AND TOOLS

This work targets SoCs employing a bus to connect a processor to memory and IP cores. The proposed approach is prototyped in FPGAs using the Excalibur™ environment[12]. Figure 1 presents the target SoC architecture. A PC is used as the external tester, responsible for loading the test program and deterministic patterns into the memory through the serial interface.

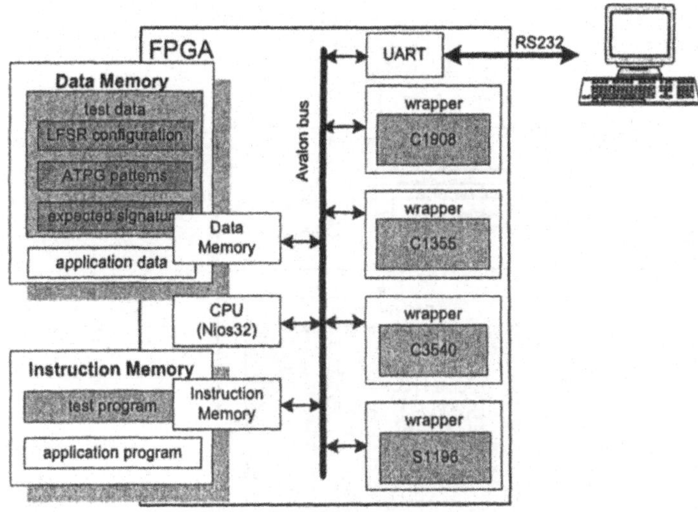

Figure 1. Target SoC architecture.

The embedded processor used is the 5-stage pipeline Nios32[12]. The 32-bit Avalon™ bus is used to connect the SoC components. The c1908, c1355, c3540 and s1196 benchmarks, randomly chosen, are used to evaluate the proposed test method and the CAT tool. Each core connected to the bus is wrapped. Using a uniform wrapper, the same test procedures can be applied to all cores under test.

In the data memory, each core has its LFSR configuration (i.e. the multiple seeds and polynomials) chosen in a manual process, the ATPG patterns generated by a commercial tool, the expected signatures and the application data. The instruction memory contains the synthesized test program used to test all non-programmable cores and the application program.

4. CAT TOOL ENVIRONMENT

The CAT environment developed has four main functions: (i) automatically synthesize test programs in C language; (ii) automatically synthesize the VHDL wrappers to each core; (iii) insert scan chains when necessary; (iv) automatically synthesize the SoC interface to integrate the cores. Figure 2 and Figure 4 present a design flow divided in two parts: the SoC generation and the test program generation.

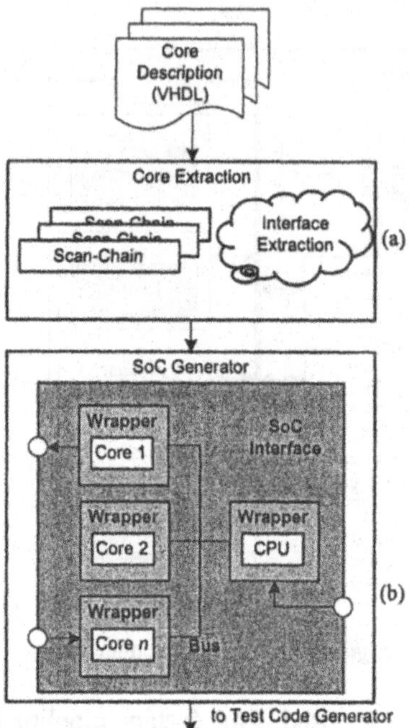

Figure 2. SoC generation CAD flow.

QT (http://www.trolltech.com) is used to build the graphical interface of this environment. The CAT tool synthesizes 'C' and 'VHDL' descriptions from pre-validated templates, integrates commercial CAD tools (Leonardo, ModelSim, FlexTest, FastScan, DFTAdvisor) and guides the user through the design and test flow.

4.1 SoC Generation

The SoC generation flow starts with a tool called *core extraction* –Figure 2(a). This tool extracts the core interface (input/outputs) and may insert scan chains when required. Scan chains are inserted using the DFT Advisor tool. In the interface extraction, some of the extracted data are: (i) the name of the core; (ii) the type of the core - *core, processor* or *memory*; (iii) the files composing the core; (iv) the top entity; (v) the ports and generics of the top entity. The user must specify the port usage for each port of the core. Port usage means how the port is used in the system (e.g. *data, clock, reset, enable, external*). This is a key piece of information, since it enables the *SoC generator* tool to automatically synthesize the wrapper for this core and the SoC interface.

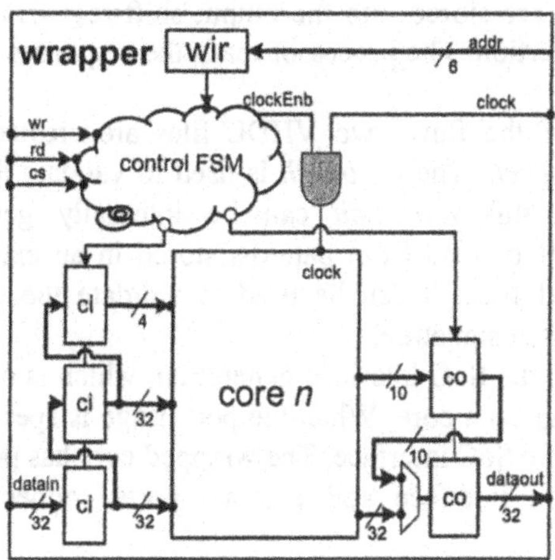

Figure 3. Wrapper scheme.

The next step is the SoC generation, Figure 2(b). The following actions are executed during this step: (i) selection of the cores and the processor; (ii) selection of the communication media; (iii) wrapper synthesis for each core; (iv) SoC interface synthesis.

The cores are selected from a library created in the first step (*core extraction*). After that, the user defines the communication architecture. Different communication architectures can be easily integrated to the environment, but presently only bus-based is allowed.

Once the communication architecture is defined, the wrapper is automatically synthesized for each core. Wrappers are synthesized after the communication architecture selection since the communication protocol and the wrapper interface are associated to the communication media, in this case, a bus.

Figure 3 presents an example of wrapper for a bus-based architecture. The wrapper has shift-registers adapting the bus width (32-bit) to the core input/output widths, represented by *ci* and *co* registers. The test is executed in two steps: patterns loading and response reading. The six least significant bits of the *addr* port are used to set the test mode, i.e. pattern loading or response reading.

When the wrapper is in test mode and a write operation is required (*wr* port), the internal *clockEnb* is asserted during one clock cycle to process the incoming test patterns. Note that the *clockEnb* assertion occurs only when the last part of the incoming pattern is written into the input shift-register.

The responses are stored into the output shift-registers one clock cycle after *clockEnb* assertion. The processor reads the responses when the *rd* port is asserted.

At this step of the flow, two VHDL files are created for each core: *wrapper* and *testbench*. The *testbench* is used to validate the *wrapper*. The test patterns for this *testbench* can be internally generated from a configurable LFSR or read from patterns stored in an external file. When read from external files, it can be used to validate the test patterns and generate the expected signature.

The last step is the SoC interface generation, which is obtained from the data extracted from each core. When the port usage is specified as external, it is connected to the SoC interface. The wrapped core has ports connected to the communication interface and external ports connected to the SoC interface.

These external ports impose modifications (addition of ports) in the wrapper and in the SoC interface.

4.2 Test Program Generation

Once the SoC is built, the next step is to create the *test program*. The test program generates the test vectors, compacts the responses and evaluates the signature of each core. The selected test pattern generation approach is based on Hellebrand's approach[10], which uses different LFSR configurations to increase the fault coverage obtained from pseudo-random patterns. Therefore, fewer deterministic patterns are stored in the memory. The main advantage of this approach is the minimization of technological requirements of the external tester, such as bandwidth, memory and frequency. However, the challenge is to find a tradeoff between run-time for pseudo-random test and the memory requirements for deterministic test. The test program generation flow is presented in Figure 4.

The first action of the *test program generator* (Figure 4(a)) is to select the polynomials and seeds to generate patterns to each core. Multiple polynomials can be selected and each polynomial may have several seeds. The environment supports generic modular LFSR and MISR descriptions. A reseeding tool like[13] could present better encoding efficiency, i.e. more compact test pattern set to achieve higher fault coverage. However, the LFSR configuration was randomly chosen since this kind of tool is not available, at the moment.

Each LFSR configuration is simulated, creating the test patterns for each core - Figure 4(b). The generated patterns are translated to the fault simulator format (Flextest™/Fastscan™) and the fault coverage is evaluated, using the synthesized description of the cores. If the resulting fault coverage is not sufficient, the user may run the ATPG tool to generate the remaining test patterns or choose other LFSR configuration, restarting the process. At the end of this process, the user has all the necessary patterns to test each core. Thus, the expected signature can be generated using a logic simulator - Figure 4(c).

Finally, these patterns (pseudo-random and/or deterministic) are translated to the test program in C language - Figure 4(d). After the program generation, the user can evaluate the required test time using a hardware/software co-simulation tool - Figure 4 (e). The co-simulation tool is provided by the Excalibur environment. Figure 5 presents an example of pseudo test program.

The core under test in Figure 5 is memory mapped, which means that the processor accesses cores with load and store instructions. This can be observed by the **cutPtr* pointer, in the third line.

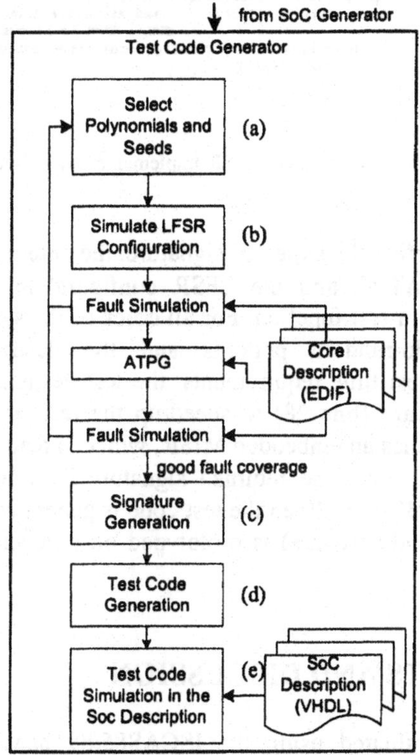

Figure 4. Test program generation CAT flow.

```
1)  #define nCores 5
2)  const coreTyp coreCfg[nCores]={{expectedSig,corePtr},{},{}};
3)  const unsigned detVet[nCores]={{},{},{},{},{}};
4)  const unsigned lfsrCfg[nCores]={{},{},{},{},{}};
5)  void lfsr( unsigned polynomial, int *state);
6)
7)  void ApplyPseudoRandomPattern(volatile void *hwif){
8)      for each polynomial {
9)              poly = lfsrCfg[ind];
10)             n_seed = lfsrCfg[ind+1];
11)             ind+=2;
12)             for each seed {
13)                     value = lfsrCfg[ind];
14)                     n_patterns = lfsrCfg[ind+1];
15)                     ind+=2;
16)                     for each pattern {
17)                     *hwif = value;        // send pattern to CUT
18)                     lfsr(poly;&value);    // generate new pattern
19) }}}}
20) void ApplyDeterministicPattern(volatile void *hwif){
21)     for each pattern {
22)             *hwif = detVet[i]; // send det. pattern to CUT
23) }}
24) int ExecuteTest(void){
25)     volatile void *cut;
26)     for each core {
27)             cut = coreCfg[core].cutPtr;
28)             ApplyPseudoRandomPattern(cut);
29)             ApplyDeterministicPattern(cut);
30)             genSignature = *cut;  // read signature from the
31)                                   // CUT - MISR in hardware
32)             if (genSignature != coreCfg[core].expectedSig)
33)                     abortTest();
34)     }
35) }}
```

Figure 5. Pseudo test program example. LFSR implemented in software (lfsr function) and MISR in hardware.

The first lines define the expected signature, the deterministic test vectors generated by the ATPG and the LFSR configuration. The *ExecuteTest* procedure is the main test function. For all cores of the system this function applies the pseudo-random patterns and the deterministic patterns. Depending on the test time requirements, the test designer may implement the MISR in hardware (line 28) to speed-up the test, as presented in this example. Each core has an embedded MISR, which is read by the software at the end of the core test. The returned signature is compared against the expected one (lines 30-31). When the test code is generated to each core, the system (hardware and software) is prototyped on a development board for the final validation.

5. RESULTS AND DISCUSSION

The features evaluated using the ISCAS85/89 benchmarks are fault coverage, test time and memory requirements to store the test program (instruction memory) and deterministic patterns (data memory). Table 1 presents some additional information of the benchmarks.

Table 1. Benchmarks general information.

Core	Inputs	Outputs	Core area	Core + Wrapper Area
c1908	33	25	286	843
c1355	41	32	271	948
c3540	50	22	756	1499
s1196	14	14	476	854

The benchmarks are mapped to a 0.35 library and the area is presented in nand2 equivalent gates. The cores evaluated are small and do not represent real case cores. Using complex benchmarks, the wrapper overhead will be reduced since it is a function mainly of the number of inputs and outputs.

Table 2 presents the fault coverage (FC) obtained and the number of patterns for the mixed patterns approach. For example, core c1908 obtained 67.94% fault coverage using only pseudo-random patterns. Deterministic patterns, generated by a commercial ATPG, increased the fault coverage to 99.52%. The last two columns present the number of patterns (pseudo-random and deterministic) used.

Table 3 presents the data memory requirements to store the LFSR configuration (*random size*) and the ATPG patterns (*deterministic size*) in bytes. As can be observed, the number of bytes to store a LFSR configuration is much smaller than deterministic patterns, as stated in [10]. However, a significant number of deterministic patterns had to be used to reach acceptable fault coverage. Those remaining deterministic patterns can be reduced even more using embedded compression algorithms or using a tool to select the best seeds and polynomials for each core. Using compression algorithms will obviously reduce the test data and the time to download them. However, this can have a negative impact in the test time, since the patterns had to be decompressed before being sent to the core. On the other hand, using a tool to select seeds can increase the quality of the pseudo random patterns (i.e. increase the number of useful patterns) reducing the need of deterministic patterns[13].

The mixed test pattern approach (i.e. deterministic and pseudo-random) was partitioned into four different hardware/software test configurations. In the first partition, the LFSR, the MISR and the control structure used to apply deterministic patterns are implemented in hardware (*hardware partition*). In the second partition, all test modules are implemented in software (*software partition*). The third partition implements the LFSR in software and the MISR in hardware (*misr hardware partition*). In the fourth partition, the test patterns are generated using arithmetic functions (*adders software partition)* provided by the processor and the MISR is implemented in hardware[11]. The fault coverage using adders is not evaluated.

Table 2. Fault coverage (FC) and number of test patterns for mixed test patterns approach.

Core	Random FC	Deterministic + Random FC	# Random Patterns	# Deterministic Patterns
c1908	67.94	99.52	40	68
c1355	88.89	100.00	66	29
c3540	75.22	97.48	105	184
s1196	53.41	94.57	159	459

Table 3. Data memory requirements, in bytes, for each core.

Core	Random Size	Deterministic Size
c1908	9	137
c1355	17	59
c3540	21	369
s1196	15	460

Table 4 compares the instruction memory requirements to store the test program (each instruction uses two bytes) for three partitions. The test program is the same for all cores. As can be noted, since the test program is simplified (i.e. moved from LFSR to adder to generate patterns) the test program obviously is reduced in size. As can be seen in Table 4 the instruction memory requirements, i.e. bytes necessary to store the test program, are relatively small. The instruction memory requires less than 1Kbyte to store a test program to test all the non-programmable cores. This shows that the memory requirements to store the test program are not a constraint for software-based test applied to non-programmable core.

Table 5 presents the number of clock cycles between the generation of two patterns. In the hardware partition, the pseudo-random patterns are generated cycle by cycle and the deterministic patterns are estimated to 10 clock cycles, due to the time to access the test memory. The software partition takes 185 cycles to apply a pseudo-random pattern and 175 cycles to apply a deterministic pattern, due to the test serialization. Since part of the test has been removed from the software (i.e. response compaction moved to hardware), the time is reduced. In the last partition, it is possible to see that the test time is even more reduced, confirming[4,11], which suggest the use of specific test instructions to reduce the test time and test program length.

Table 6 complements the test time information presented in Table 5. It presents the total test time for the mixed test approach for the benchmarks using the four partitions.

As expected, the software partition takes longer test time to complete the test. This difference is due to the overhead induced by the data transfer from the processor to the core over the bus, and the test serialization. In the hardware approach, these steps are executed in parallel, as a pipeline structure.

Table 4. Instruction memory requirements, in bytes, for the hardware/software partitions.

Core	Software	LFSR in Software MISR in Hardware	Adder in Software MISR in Hardware
All	564	428	394

Table 5. Time in clock cycles to apply a new pattern to the UUT.

Pattern Type	Hardware	Software	LFSR software MISR hardware	Adder software MISR hardware
Pseudo-random	1	185	102	73
Deterministic	10	175	43	43

Table 6. Total test time, in clock cycles, for four test approach.

Core	Hardware	Software	LFSR software MISR hardware	Adder software MISR hardware
c1355	1400	40168	28034	18817
c1908	646	35313	23246	17552
c3540	3785	94232	61906	46790
s1196	4749	110046	53452	48723

The third and the fourth test partitions reduce the total test time. The MISR implemented in hardware reduces the traffic in the bus, since the responses are compacted in hardware, sending back to the processor only the signature (see the test program in Figure 5). The test time in the last partition was reduced by using a processor specific instruction to generate the pseudo-random patterns. This has an advantage over the LFSR emulation because a new pattern is created with only one instruction, reducing the test time and the test program length[4,11]. The discussion of the results is presented in the next Section.

6. CONCLUSIONS

Software-based test minimizes the two main drawbacks related to BIST: performance degradation and additional hardware area. According to the literature, software-based test increases test time and the test memory requirements. This work presents a quantitative evaluation of these drawbacks. Actually, the test time is an important drawback for manufacturing test, since the test time using only software-based test is much longer compared to the hardware-based test. As shown, the test time is minimized when the test modules are partitioned into hardware/software modules. The test time is also minimized using an adder to generate patterns. On the other hand, the instruction memory requirement for software-based test is low. A careful design of the test program can enable its reuse to test different cores. It is also possible to reduce the data memory requirement using compression algorithms in the deterministic patterns, an issue not explored in this work. Thus, software-based test can be applied to field test since the memory requirement, the main bottleneck to this kind of test, is minimal.

The added area by the wrapper insertion, Table 1, is used to integrate the cores to the bus, without test-related circuitry. The only area overhead induced by the software-based test is the memory space used to store the test program.

The second contribution of this work is the developed CAT environment, integrating software-based test to a generic and simple to use SoC design flow.

As ongoing work, the CAT environment is being validated with ITC02 benchmark. Other criterions, such as hardware overhead and power dissipation are also being evaluated. We are also evaluating software based-test in other communication architectures like network-on-chip[14].

ACKNOWLEDGEMENTS

Fernando Moraes gratefully acknowledges the support of the CNPq through research grant number 307665/2003-2.

7. REFERENCES

1. L. Chen, and S. Dey, Software-Based Self-Testing Methodology for Processor Cores. *IEEE Transactions on Computer-Aided Designs*. **20**(3), 369-380 (2001).
2. J.-R. Huang, M. K. Iyer, and K.-T. Cheng, A Self-Test Methodology for IP Cores in Bus-Based Programmable SoCs. In: *VLSI Test Symposium*. 198-203 (2001).
3. S. Hwang, and J. A. Abraham, Reuse of Addressable System Bus for SOC Testing. In: *ASIC/SOC Conference*. 215-219 (2001).
4. W. C. Lai, and K. T. Cheng, Instruction-Level DFT for Testing Processor and IP Cores in System-on-a-Chip. In: *Design Automation Conference*. 59–64 (2001).
5. A. Krstic, W. C. Lai, K. T. Cheng, L. Chen, and S. Dey, Embedded Software-Based Self-Test for Programmable Core-Based Designs. *IEEE Design and Test of Computers*. **19**(4), 18–27 (2002).
6. S. M. Thatte, and J. A. Abraham, A Methodology for Functional Level Testing of Microprocessors. In: *International Symposium on Fault-Tolerant Computing*. 90-95 (1978).
7. N. Kranitis, G. Xenoulis, D. Gizopoulos, A. Paschalis, and Y. Zorian, Low-Cost Software-Based Self-Test of RISC Processor Cores. In: *Design, Automation and Test in Europe Conference*. 164-168 (2003).
8. A. J. Van de Goor, Using March Tests to Test SRAMs. *IEEE Design and Test of Computers*. **10**(1), 8–14 (1993).
9. Y. Zorian, and A. Ivanov, An Effective BIST Scheme for ROM's. *IEEE Transaction on Computers*. 41(5), 646-653 (1992).

10. S. Hellebrand, H. J. Wunderlich, and A. Hertwig, Mixed-Mode BIST Using Embedded Processors. In: *International Test Conference*. 195-204 (1996).
11. J. Rajski, and J. Tyszer, *Arithmetic Built-In Self-Test for Embedded Systems* (Prentice Hall, Upper Saddle River, 1998).
12. Altera Inc. Nios Embedded Processor: 32 bits Programmer's Reference Manual. (version 2.1, 2002, 124 p).
13. C. V. Krishna, and N. A. Touba, Reducing Test Data Volume Using LFSR Reseeding with Seed Compression. In: *International Test Conference*. 321-330 (2002).
14. A. M. Amory, E. Cota, M. S. Lubaszewski, and F. G. Moraes, Reducing Test Time with Processor Reuse in Network-on-Chip Based Systems. In: *Symposium on Integrated Circuits and System*. 111-116 (2004).

OPTIMIZING SOC TEST RESOURCES USING DUAL SEQUENCES*

Wei Zou[1], Chris C.N. Chu[2], Sudhakar M. Reddy[1], Irith Pomeranz[3]
[1]*Electrical & Computer Engineering, University of Iowa, Iowa City, IA, 52242, USA*
[2]*Electrical & Computer Engineering, Iowa State University, Ames, IA, 50011, USA*
[3]*Electrical & Computer Engineering, Purdue University, West Lafayette, IN, 47907, USA*

Abstract: **In this paper, we propose a new data structure called dual sequences to represent SOC test schedules. Dual sequences are used together with a simulated annealing based procedure to optimize the SOC test application time and tester resources. The problems we consider are generation of optimal test schedules for SOCs and minimizing tester memory and test channels. Results of experiments conducted on ITC'02 benchmark SOCs show the effectiveness of the proposed method.**

1. INTRODUCTION

As SOC (system on a chip) design moves toward mainstream use, the problem of effectively testing the IP blocks (called cores) within the SOC needs to be addressed. SOC test requires considering the following issues: test access mechanism (TAM) design, core wrapper design, test scheduling, tester memory and tester channels.

TAM is the hardware infrastructure, which transports the test data between the SOC pins and the core wrappers. The core wrapper primarily consists of scan chains placed around the core to isolate the core from its

* The work of W. Zou and S. M. Reddy supported in part by NSF Grant CCR-0097005 and SRC Grant 001-TJ-949. Work of I. Pomeranz supported in part by NSF Grant CCR-0098091 and by SRC Grant 2001-TJ-950.

in IFIP International Federation for

Please use the following format when citing this chapter:

Zou, Wei, Chu, Chris, C.N., Reddy, Sudhakar, M., Pomeranz, Irith, 2006, in IFIP International Federation for Information Processing, Volume 200, VLSI-SOC: From Systems to Chips, eds. Glesner, M., Reis, R., Indrusiak, L., Mooney, V., Eveking, H., (Boston: Springer), pp. 181-196.

surrounding logic and serves as an interface between the TAM and the core. A number of approaches [1-6, 11] have been proposed for the core wrapper design.

SOC test scheduling is the procedure of deciding the test start time of every core so as to obtain a minimum test application time for the SOC under certain constraints, such as TAM width (i.e. the number of SOC pins), power dissipation during test, etc.. Since test scheduling depends on the SOC TAM design and the core wrapper design, SOC test requires co-optimization of the TAM, the core wrapper design and the test schedule. Recently a number of works proposed solutions to this problem. Marinissen et al.[3] presented several methods to design TAMs. Larsson and Peng [7, 8] considered co-optimization of SOC test time and the number of SOC pins under the assumption that a wrapper for each core is given. Chakrabarty [9] developed an integer linear programming model for minimizing test application time by co-optimization of bandwidth distribution and test bus assignment. Huang et al.[10] formulated the co-optimization problem as a rectangle packing problem and solved it by using a best-fit heuristic algorithm. To solve the co-optimization problem, a SOC test schedule representation called sequence pair was used together with simulated annealing or heuristics. [6, 16] Other works [11-15] have investigated the same problem using specialized heuristic procedures.

It was shown that test application time for benchmark SOCs using a simulated annealing algorithm was most often shorter than all earlier proposed heuristic solutions and also shorter than an ILP based procedure when the run time of the ILP procedure was limited (to several hours).[6] The SOC schedules were represented in the simulated annealing by what are called sequence pairs [19].

In this paper, we introduce a simple and effective data structure called Dual Sequences (DS) to represent SOC test schedules and use this to obtain optimal SOC test schedules using simulated annealing. Experimental results show that test schedules obtained using DS with simulated annealing are as good as or better than those obtained using sequence pairs while the run time of the simulated annealing procedure is greatly reduced. Another problem we consider is minimization of tester memory and tester channels, again using DS to represent test schedules together with simulated annealing.

2. THE FEATURES OF SOC TEST TIME

The core wrapper is the interface between the core and the SOC TAM. It provides several kinds of operation modes, such as normal function, interconnect test, bypass test, etc. The test time for a core is derived by the following formula.[1]

$$T = \{1+\max(Si, So)\} * P + \min(Si, So) \qquad (1)$$

where P is the number of test patterns, and Si (So) denotes the length of the longest wrapper scan input (output) chain for the core. The core test time T is decided by the length of the longest wrapper scan chain. So one goal of the core wrapper design is to shorten the longest wrapper scan chain. For this purpose, balanced wrapper design [1, 11] was proposed, which partitions the wrapper scan elements among the wrapper chains to make the length of the wrapper chains as equal as possible.

Next we consider the relationship between the core test time and the core wrapper width. It is known that the test time for a core is a staircase function, which means that there are only some wrapper width values where the core test time changes. These points are called pareto-optimal points.[11] If the core wrapper is represented by a rectangle with the width representing the wrapper width and the height representing the core test time, there is a set of candidate rectangles for every core corresponding to the pareto-optimal core wrapper widths. In co-optimizing wrapper design and SOC test time, one of these rectangles is chosen for each core.

The problem of SOC test scheduling we are considering is stated below.

Given are a SOC with N pins and N_c cores. Each core C_i ($1 \leq i \leq N_c$) has a set of N_i permissible wrapper configurations. Each wrapper configuration is represented by a pair (W_{ij}, $T(W_{ij})$), where W_{ij} stands for the width of the j-th wrapper configuration for core C_i and $T(W_{ij})$ stands for the test time of core C_i with wrapper width W_{ij}. The objective is to pick one wrapper design for each core, determine the mapping from the SOC pins to the core wrapper pins, and set the test start time for each core such that the SOC test application time is minimized.

This problem can be transformed into the well-known two-dimensional bin packing problem, in which the SOC is represented by a bin with width N and the set of N_i SOC wrappers for every core is represented by a set of U_i rectangles with width W_{ij} and height $T(W_{ij})$.[10] The objective is to choose a rectangle for every core C_i and pack all the rectangles in the bin, such that height of the bin is minimum.

3. DUAL SEQUENCES

In Fig. 1, we illustrate a test schedule for a SOC with six cores. The vertical axis is time and the horizontal axis represents SOC pins. In this schedule, testing of Core 1, Core 2 and Core 3 starts simultaneously at time $t = 0$. Testing of Core 4 is initiated at $t = t_3$. Testing of Cores 5 and 6 is initiated at t = t_4. The two parts of Core 5, denoted 5_1 and 5_2, indicate that Core 5 is tested through two non-consecutive subsets of TAM pins. Testing of the SOC is completed at t_6. As seen from Fig. 1, every test schedule corresponds to a rectangle placement in the bin representing the SOC.

In this section, a new representation called Dual Sequences (DS) is introduced to express the rectangle placement.

The DS for a placement of a set of n rectangles (cores) is a pair of sequences ($\mathcal{R}$, $\mathcal{W}$), in which $\mathcal{R}$ is a sequence of the names of the n rectangles and $\mathcal{W}$ is a sequence of the widths of the n rectangles listed in $\mathcal{R}$. For example, (< R3 R1 R2 R4 >, < 4 2 1 5 >) is a DS from which we can see that the placement is composed of four rectangles with widths 4, 2, 1 and 5, respectively. Next we discuss how to represent a rectangle placement by a DS and how to obtain the rectangle placement corresponding to a DS.

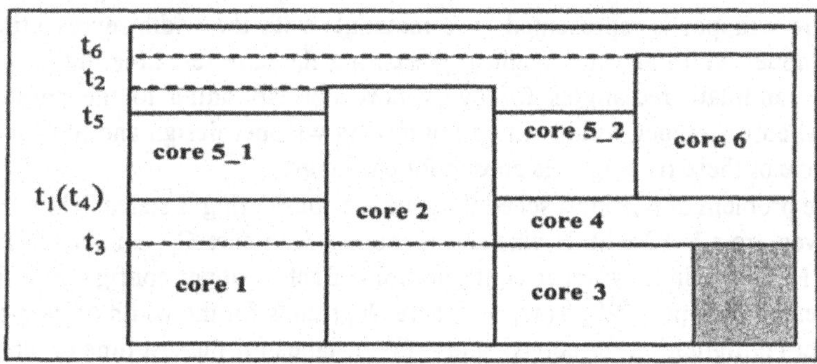

Figure 1. SOC test schedule

3.1 DS extraction from rectangle placement

Given a placement of rectangles, the corresponding DS can be obtained by visiting every rectangle in the placement from bottom to top and from left to right. During the visitation, the rectangles we encounter are recorded in $\mathcal{R}$ in the order of visiting them and the width corresponding to the discovered rectangles are recorded in $\mathcal{W}$. If a rectangle is split into several sub-rectangles in the placement, the sub-rectangles are merged into one rectangle for representation in ($\mathcal{R}$, $\mathcal{W}$) and its position in $\mathcal{R}$ is decided by the first sub rectangle and the width in $\mathcal{W}$ is the sum of the widths of the sub-rectangles. A rectangle placement corresponding to a SOC test schedule may have split a rectangle corresponding to a core since its wrapper pins are connected to non-consecutive SOC pins. For example consider the rectangle placement in Fig.1, which has six rectangles R1, R2, R3, R4, R5, R6 corresponding to the six cores with widths, say Φ1, Φ2, Φ3, Φ4, Φ5, and Φ6, respectively. The rectangle R5 is divided into two sub-rectangles R5_1 and R5_2. As explained next, by visiting each rectangle within the placement and merging the sub-rectangles, we obtain the Dual Sequences (<R1 R2 R3 R4 R5 R6>, < Φ1 Φ2 Φ3 Φ4 Φ5 Φ6 >). Since testing of R1, R2 and R3 are all scheduled at time zero, they are visited before R4 which is scheduled for testing at t_3. Within the set of rectangles R1, R2, and R3, R1 is visited first since it is left of R2, followed by R2 and then R3. Next R4 is visited. After visiting R4, R5 and R6 are visited but R5 is visited before R6.

3.2 Mapping from DS to placement of rectangles

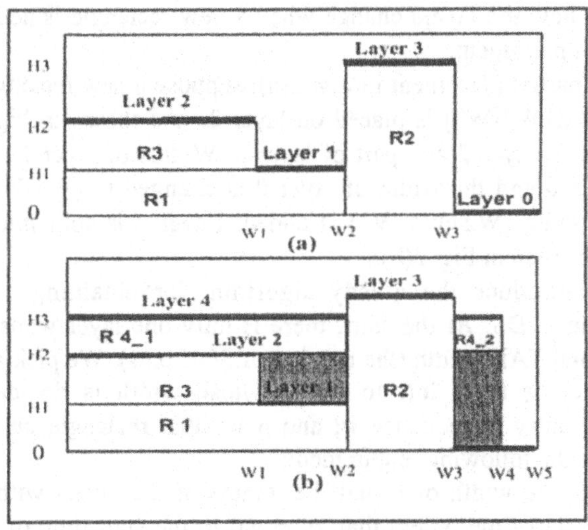

Figure 2. **The changes in the layers during bin packing**

To obtain a placement from a DS, a greedy algorithm based on two dimensional bin packing is used. The basic idea of this algorithm is that given the sequence $\mathcal{R}$ and the width sequence $\mathcal{W}$, we pick rectangles from $\mathcal{R}$ one at a time in the order of their appearance in $\mathcal{R}$ and pack a selected rectangle at a position which is as low as possible (i.e., we schedule the start of the test of the core corresponding to the rectangle as early as possible). It is important to point out the distinction between dual sequences and sequence pairs [6, 16] to represent a bin packing. Given a sequence pairs the corresponding bin packing is uniquely defined and is obtained by a longest path procedure run over two graphs derived from the sequence pairs. The bin packing corresponding to a given DS is not unique. The sequence $\mathcal{R}$

determines the order in which the rectangles are considered and $\mathcal{W}$ restricts the choice of the width (i.e. wrappers) of the rectangle corresponding to the core being packed. Any procedure to pack the rectangles in the order given by $\mathcal{R}$ can be used.

In the proposed method to obtain a rectangle placement from a given DS, a data structure called a layer is used, which corresponds to a position where a yet unplaced rectangle can be placed. A layer has two attributes: starting time and width. The starting time is the height of the layer in the bin and the width indicates the space available at this height. For example, in Fig. 2(a), we have four layers, layer 0 to layer 3, which are indicated by the thick dark lines. The start time and the width of the layer can be seen from Fig. 2(a). For example, the start time of layer1 is H1 and the width is (W2-W1). Before we describe the procedure to obtain the rectangle placement from a DS, we show how the layers change when a new rectangle is added into an existing partial placement.

Given the partial placement in Fig. 2(a), suppose a new rectangle, say R4 with width (W2+W4-W3) is placed on layer 2. As shown in Fig. 2(b), R4 occupies layer 1, layer 2 and part of layer 0. Widths of layer 1 and layer 2 are changed to 0 and the width of layer 0 is changed to (W5-W4). A new layer 4 with width (W2+W4-W3) is added. Layer 4 is split into two sub-layers as can be seen in Fig. 2(b).

Next we introduce the greedy algorithm for obtaining a rectangle placement from a DS. At the start, there is only one layer whose width is equal to the total TAM width (the number of SOC pins). We pick a rectangle from sequence $\mathcal{R}$ from left to right, whose width is decided by the corresponding entry in sequence $\mathcal{W}$, and place that rectangle on a layer L, which satisfies the following requirement:

The sum of the width of L and the widths of the layers with non-zero widths whose start time is less than or equal to the start time of layer L is greater than or equal to the width of the rectangle being placed.

The start time of layer L is the lowest among all layers that satisfy the above requirement.

If there is a power constraint, the core placed at that layer should not violate this constraint.

Following the placement of the rectangle, a new layer with width equal to the width of the just placed rectangle is added to the placement and the widths of the other layers are updated. This procedure is repeated until all the rectangles in sequence $\mathcal{R}$ are packed. It should be pointed out that the procedure proposed above is sub-optimal. One reason for this sub-optimality is that when the width of a layer is reduced it may preclude the use of some packing space. For example the shaded area in Fig. 2(b) is not available for future packing after rectangle R4 is placed.

Compared to test schedule representation using sequence pairs for obtaining SOC test schedules, the search space for DS is of size $(Nc! \times \prod^{Nc} Ki)$, while for the sequence pairs representation the search space is of size $((Nc!)^2 \times \prod^{Nc} Ki)$, where N_c is the number of cores in the SOC and K_i is the number of wrapper configurations for core i. The size of the search spaces given above is obtained by computing the number of distinct dual sequences and sequence pairs, respectively. Since the search space using DS representation of rectangle placement is smaller, it leads to a much lower run time for test schedule optimization using simulated annealing. As the experiments on ITC'02 benchmarks reported later show, the optimality of the obtained SOC test schedule is indeed not affected by using dual sequences instead of sequence pairs.

4. SIMULATED ANNEALING

Simulated annealing (SA) is a global stochastic optimization algorithm that was first introduced by Kirkpatric et al.[17]. The algorithm begins with an initial solution, and then a neighboring solution is created by perturbing the current solution. If the cost of the neighboring solution is less than that of the current solution, the neighboring solution is accepted; else it is accepted or rejected with some probability. The probability of accepting an inferior solution is a function of a parameter called the temperature. The probability function used is:

$$p = e^{-\frac{\Delta C}{T}},$$

where ΔC is the change in the cost between the neighboring solution and the current solution and T is the current temperature. The procedure we used to implement the simulated annealing algorithm for finding an optimal SOC test schedule that minimizes the expected test completion time is given below.

Objective: Find an optimal solution S_{opt}, which makes the cost function $C(S_{opt})$ minimum.

Procedure:

1. Construct an initial solution S_{init};
2. Let the current solution be S_{cur}: = S_{init};
3. Set the initial temperature to T: = T_{init};
4. Set Counter: = 1;
5. While the stopping criteria is not met do begin
6. While T >T_{final} do begin
7. For i: = 1 to Niter do begin
8. Generate a neighboring solution S_n from the current solution S_{cur};

9. Compute the change in the cost function
   $\Delta C = C(S_n) - C(S_{cur})$;
10. If $\Delta C \le 0$ then $S_{cur} := S_n$;
11. Else begin
12. q = random(0,1);
13. If $q < e^{\frac{-\Delta C}{T}}$ then $S_{cur} := S_n$;
14. End
15. End
16. Set new temperature T: = K * T;
17. End
18. Set T: = T_{new};
19. Counter: = Counter+1;
20. End

We use the SA algorithm described above to implement the SOC test scheduling based on dual sequences by specifying the parameters of the SA algorithm as follows.

Cost function C: The height of the bin where the rectangles are placed is defined as the cost function.

Neighboring solution S_n: A neighboring solution is defined by two types of moves over the dual sequences, given below.

M1: Exchange the position of two randomly chosen rectangles in the first sequence $\mathcal{R}$(note that $\mathcal{W}$ is also changed to reflect the exchange in $\mathcal{R}$).

M2: Change the width (and hence the height) of a rectangle in the sequence $\mathcal{W}$ to another allowed width of the rectangle.

During the process of optimization, the probabilities of moves M1 and M2 are set to 0.5 each.

Initial solution: The initial solution S_{init} can be set randomly. In order to accelerate the convergence of SA, the test schedule obtained by the heuristic procedure [14] is used as the initial solution in the experiment reported later.

Initial temperature: The initial temperature T_{init} is set to 4000. At the end of each outer loop, temperature T is reset to T_{new} = 4000 + 1000 * Counter.

Other parameters: These parameters include the final temperature T_{final}, the number of iterations Niter at every temperature, the stopping criteria and the temperature reduction multiplier K. In our implementation, these parameters are set as follows.

(1) $T_{final} = 10$;

(2)The number of iterations Niter at each temperature is set to 400*N_c where N_c is the number of rectangles.

(3)The stopping criteria can be decided by the user. In our experiment, if Counter is larger than 10, the procedure is stopped.

(4) The temperature reduction multiplier K is set to 0.98 when T < 10000; otherwise K = 0.93.

5. REDUCING ATE RESOURCES

Automatic Test Equipment (ATE) used in SOC test provides the ability to perform multi-site testing, which allows several copies of a SOC to be tested concurrently. When the number of ATE channels is given, to test a maximum number of SOCs at the same time requires minimization of the TAM width of the SOC while not violating the ATE memory depth constraint (decreasing the TAM width of the SOC will increase the test application time and hence the ATE memory depth requirement). In this section we discuss how the proposed method using DS representation of the test schedules can be used to minimize SOC TAM width as well as the ATE buffer memory depth for a given SOC TAM width.

When an SOC is tested by an ATE, the test channel memory depth required for the SOC is decided by the test data volume. The depth of the test channel memory can be approximated by the SOC test application time (the number of clock cycles). [20] Therefore, the problem of multi-site SOC test under ATE memory depth constraints can be considered as a problem of reducing the SOC TAM width while the total test application time is fixed. This allows testing of a maximum number of SOCs using a given number of test channels and their buffer memory depth. The SOC multi-site test problem can be solved using the two dimensional bin packing procedure with the width of the bin representing the memory depth constraint and the height of the bin representing the TAM width. We should point out that in the bin packing problem for multi-site testing, a rectangle cannot be divided into several sub-rectangles, which is different from the bin packing problem we discussed before. Dividing rectangles was permitted in the earlier problem since it is not necessary to connect the wrapper pins of a core to adjacent SOC pins. However, in multi-site testing, breaking a rectangle represents interruption of the test of a core, which may not be permitted. A simple way to accommodate the requirement that core tests cannot be interrupted is to require that the new rectangle to be packed must occupy contiguous layers only, thus avoiding division of rectangles.

Another issue that needs to be considered is illustrated by the rectangle packing shown in Fig. 3(a). Fig. 3(a) shows the case where the ATE test channels 1 and 2 are used to test core 2 and ATE test channel 3 is used to test core 1. Tests for core 1 occupy M1 bits of memory buffer for tester channel 3 and tests for core 2 occupy M2 bits of memory buffer for channels 1 and 2. Tests for core 3 use all three test channels and hence can only be started after completing the test of core 2. It should be pointed out that the tests are loaded into the buffer and shifted out to the inputs of the device under test. If the tester architecture is such that all test channel buffers are shifted at the same time and each channel has dedicated memory buffer then the buffer bits of channel 3 are don't cares from M1 to M2. However if the tester architecture is such that each test channel is individually controlled, then test channel 3 can be idled after testing core 1 until core 3 test is initiated. In this case the size of the buffer for test channel 3 need only be (M3-M2+M1). The packing shown in Fig. 3(b) for the same cores as in Fig. 3(a) illustrates the situation where the memory buffer contents for test channel 3 is such that don't cares occur only at the end. In this case after the testing of core 1 is complete test channel 3 can be idled. In general, if the packing is such that all the test channel buffers have don't cares only at the end the ATE memory management is simpler. [20, 21] Finally, in some ATE architectures the entire buffer memory can be configured as a single pool of memory that can be dynamically assigned to test channels. [21, 22] For such architectures the total memory requirements for a SOC test is important.

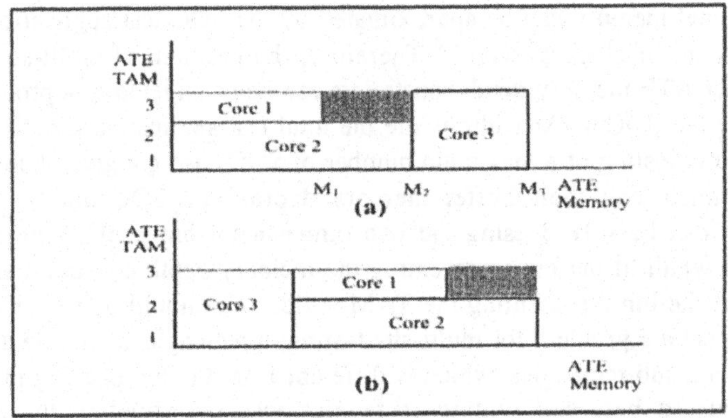

Figure 3. Rectangle packing to optimize ATE resources

For finding SOC test schedules to minimize the number of ATE test channels given the maximum depth of test channel buffers, we used two different procedures to obtain rectangle packings from dual sequences. The first one is a modified version of the procedure [20] to obtain rectangle packings such that all the don't cares in the memory buffers are at the end. The second procedure is the one described in the last section with the additional constraint that rectangles are not divided during packing.

6. EXPERIMENTAL RESULTS

The proposed simulated annealing based algorithm is implemented in C++ and executed on a PC with a Pentium IV 1.4GHZ processor and a 512

MB memory. The implemented procedure was applied to ITC'02 benchmark SOCs [18] under the assumption of no power constraint.

Table 1. Run times for the proposed method and method of ref. 6

Benchmark	Method	No.SOC pins						
		32	48	64	80	96	112	128
D695	DS	43.8	43.5	43.0	42.8	41.8	41.7	40.7
	6	52.3	54.2	56.1	56.9	58.9	61.3	67.7
P22810	DS	320.7	320.9	320.3	322.8	322.9	323.1	323.9
	6	559.1	580.1	616.3	642.0	642.4	686.3	678.2
P34392	DS	148.6	148.0	146.2	142.2	140.0	137.7	136.1
	6	233.1	261.2	279.2	305.1	324.6	335.5	342.0
P93791	DS	433.2	439.0	439.2	441.9	443.4	444.2	446.2
	6	686.3	732.4	756.2	782.6	774.4	781.7	783.9
G1023	DS	77.6	76.8	76.0	73.7	71.8	70.3	69.1
	6	122.9	128.0	135.2	143.1	149.0	153.6	156.2
U226	DS	26.6	27.5	26.9	26.8	26.4	26.2	26.0
	6	33.3	32.2	33.3	33.7	34.4	35.3	35.4
F2126	DS	12.6	11.8	11.4	11.4	11.4	11.4	11.4
	6	16.1	17.2	17.5	17.4	17.4	17.4	17.4
T512505	DS	330.2	357.6	354.7	359.0	362.2	356.7	348.2
	6	743.5	943.9	782.0	963.4	1023.2	1038.7	1043.4
A586710	DS	24.1	23.6	23.5	23.3	23.9	22.8	23.6
	6	24.4	25.2	26.4	26.1	26.2	25.7	27.4
Q12710	DS	12.3	12.3	12.3	12.3	12.3	12.3	12.3
	6	17.5	17.5	17.5	17.5	17.5	17.5	17.5
D281	DS	27.9	27.4	26.8	26.3	25.9	25.4	25.2
	6	37.0	39.5	41.3	42.8	44.3	45.1	45.4
H953	DS	28.5	27.1	26.3	25.9	25.9	25.8	25.8
	6	49.4	52.5	53.3	53.6	53.5	53.2	53.3

The results of applying the proposed method to SOC test scheduling together with the results reported by earlier methods are reported in Table 2. The proposed simulated annealing based procedure was run for ten iterations and the best schedule obtained is reported. The method used is indicated in column 2, where DS indicates the proposed method and the other methods are indicated by the number of the corresponding reference. The remaining columns give the SOC test application time for the number of SOC pins shown as the heading for the column. The entry for the method(s) achieving the best test application time is shown in bold. The method in ref. 6 also used a simulated annealing algorithm with test schedules represented by sequence pairs and had achieved better schedules than a heuristic method [16] that also used sequence pairs. For this method also we report the schedule obtained from ten iterations of the procedure. It can be seen that for all the benchmark SOCs, the proposed method achieves better or equal SOC test application time compared to ref. 6. It can also be observed that the proposed method achieves the same or better test application time than all other methods, except in the cases of P93791 with 80 SOC pins and A586710 with 32 SOC pins.

The run times for the proposed simulated annealing based procedure and the earlier procedure [6] using simulated annealing together with sequence pair are given in Table 1. The run times reported for both procedures are for a total of ten iterations of the procedures. From Table 1 it can be seen that using dual sequences instead of sequence pairs to represent rectangle placements improves the run time of simulated annealing based procedures. A 2X to 3X improvement in run time is obtained for most designs.

In Tables 3-6 we report the results on ATE tester channels and buffer memory for four circuits for which data of the earlier work [20] is available. In the first column we show the maximum memory allowed per test channel. In the next three columns we show the number of tester channels required to deliver the tests using the two procedures described in the last section and the method of ref. 20, respectively. Procedure DS1 is the proposed simulated annealing based procedure when the don't care bits in the buffer memory of the test channels are all at the end and DS is the procedure where the don't care bits are allowed to be anywhere in the buffer memory. In the next four columns we give the total ATE memory required to store the test input data. For method DS we report two entries. Under DSg we report total memory including the don't cares portion and under DS we report the total memory ignoring the don't care portion. For the other two procedures the don't care portions are not included in the totals reported.

From Tables 3-6 it can be seen that the simulated annealing based procedures require the same or smaller number of test channels compared to the heuristic procedure of ref. 20 for all the SOCs considered. It can also be seen that the total memory required is also smaller for the simulated annealing based procedures.

As an example of how reducing the number of ATE test channels helps reduce the cost of SOC test using multi-site testers, consider a tester with 128 test channels. Note that all SOCs under test can receive the test data simultaneously from the same tester channels. However the test responses from each SOC under test require separate test channels. From Table 4 for SOC p22810 with test channel buffer size limited to 640K, we note that using methods DS1, DS and of ref. 20, the number of test channels needed to apply tests to all SOCs under test is 12, 11 and 13, respectively. However each tested SOC needs the same number of separate test channels to obtain test responses. Thus in this case, the number of SOCs that can be simultaneously tested using a tester with 128 test channels will be 8 using the procedure of ref. 20, 9 using procedure DS1 and 10 if procedure DS is used. Thus by using procedure DS the number of SOCs tested per unit of time increases by 25% over the number tested if the procedure from ref. 20 is used.

7. CONCLUSIONS

A new data structure called dual sequences to represent rectangle packings is introduced. Using dual sequences together with simulated annealing procedures to obtain optimal SOC test schedules and to reduce ATE test resources were presented. Experimental results on ITC '02 SOC benchmarks showed that the proposed procedures yield better results than procedures proposed earlier.

References

1. E. J. Marinissen, S. K. Goel and M. Lousberg, Wrapper Design for Embedded Core test, pp.911-920, ITC ,2000.
2. P. Varma and S. Bhatia, A Structured Test Re-Use Methodology for Core-Based System Chips, pp. 294 –302, ITC, 1998.
3. E. J. Marinissen, R. Arendsen, G. Bos, H. Dingemanse, M. Lousberg, and C. Wouters, A Structured And Scalable Mechanism for Test Access to Embedded Reusable Cores, pp. 284-293, ITC, 1998.
4. P. T. Gonciari, B. M. Al-Hashimi and N.Nicolici , Addressing Useless Memory in Core-Based System-on-a-Chip Test, pp. 423-430, VTS, 2002.
5. E. J. Marinissen, R. Kapur, and Y. Zorian, On Using IEEE P1500 SECT for Test Plug-n-Play, pp. 770-777, ITC, 2000.
6. W. Zou, S. M. Reddy, I. Pomeranz and Y. Huang, SOC Test Scheduling Using Simulated Annealing, pp. 325-330 VTS, 2003.
7. E. Larsson and Z. Peng, An Integrated System-On-Chip Test Framework, pp. 138-144, DATE, 2001.
8. E. Larsson, Z. Peng and G. Carlsson, The Design and Optimization of SOC Test Solutions, pp. 523-530, ICCAD, 2001.
9. K. Chakrabarty, Design of System-on-Chip Test Access Architectures using Integer Linear Programming, pp. 127-134, VTS, 2000.
10. Y. Huang et. al., Resource Allocation and Test Scheduling for Concurrent Test of Core - Based SOC Design, pp. 265-270, ATS, 2001.
11. V. Iyenger, K. Chakrabarty and E. J. Marinssen, Test Wrapper and Test Access Mechanism Co-Optimization for System-on-Chip, pp. 1023-1032, ITC, 2001.
12. V. Iyenger, K. Chakrabarty and E. J. Marinssen, On Using Rectangle Packing for SOC Wrapper/TAM Co-Optimization, pp. 253-258, VTS, 2002.
13. V. Iyengar and K. Chakrabarty and E. J. Marinssen, Integrated Wrapper/TAM Co-Optimization, Constraint-Driven Test Scheduling, and Tester Data Volume reduction for SOCs, pp. 685-690, DAC, 2002.
14. Y. Huang et al., Optimal Core Wrapper Width Selection and SOC Test Scheduling Based on3-D Bin Packing Algorithm, pp. 74-82, ITC, 2002.
15. S. K. Goel and E. J. Marinissen, Effective and Efficient Test Architecture Design for SOCs, pp. 529-538, ITC, 2002 .

16. S. Koranne and V. Iyengar, On the Use of k-tuples for SoC Test Schedule Representation, pp. 539-548, ITC, 2002.
17. S. Kirkpatrick et al., Optimization by Simulated Annealing, pp.671-680, Science, Vol.220, No.4598, 1983.
18. E. J. Marinissen, V. Iyengar and K. Chakrabarty. ITC2002 SOC Benchmarking initiative, http://www.extra.research.philips.com/itc02socbenchm.
19. H. Murata, et. al., VLSI Module Placement Based on Rectangle-Packing by the Sequence-Pair, pp. 1518-1524, IEEE, TCAD, 1996.
20. V. Iyengar et. al., Test resource Optimization for multi-Site testing of SOCs under ATE memory Depth constrains, pp. 1159-1168, ITC, 2002.
21. P. T. Gonciari and B. M. Al-Hashimi, Useless Memory Allocation in System-on-Chip test: Problems and Solutions, pp. 423-429, VTS, 2002.
22. J. Bedsole, R. Raina, A. Crouch and M. S. Abadir, Very Low Cost Tester: Opportunities and Challenges, pp.738-747, ITC, 2001.

Table 2. Test Application Times for ITC'02 SOC Benchmarks

Benchmark	Method	No. SOC pins						
		32	48	64	80	96	112	128
D695	DS	41654	28161	21025	16962	14310	12134	10723
	6	41899	28165	21258	17101	14310	12134	10760
	11	41949	28327	21423	17210	16403	13023	12327
	12	43723	30317	23021	18459	15698	13415	11604
	15	44307	28576	21518	17617	14608	12462	11033
P22810	DS	433403	289332	219019	178402	147944	128887	110940
	6	438619	293019	219923	180004	151886	132812	112515
	12	452639	307780	246150	197293	167256	145417	136941
	15	458068	299718	222471	190995	160221	145417	133405
P34392	DS	960230	655607	544579	544579	544579	544579	544579
	6	965252	657561	544579	544579	544579	544579	544579
	12	1023820	759427	544579	544579	544579	544579	544579
	15	1010821	680411	551778	544579	544579	544579	544579
P93791	DS	1763528	1175756	887619	710211	594054	509845	445270
	6	1765797	1178397	893892	718005	597182	510516	451472
	11	1775099	1192980	899807	705164	602613	521806	463707
	12	1851135	1248795	975016	794020	627934	568436	511286
	15	1791638	1185434	912233	718005	601450	528925	455738
G1023	DS	30958	21233	16048	14794	14794	14794	14794
	6	31398	21365	16067	14794	14794	14794	14794
	15	34459	22821	16855	14794	14794	14794	14794
U226	DS	13416	10750	6746	5332	5332	4080	4080
	6	13416	10750	6746	5332	5332	4080	4080
	15	18663	13331	10665	8084	7999	7999	7999
F2126	DS	357088	335334	335334	335334	335334	335334	335334
	6	357088	335334	335334	335334	335334	335334	335334
	15	372125	335334	335334	335334	335334	335334	335334
T512505	DS	10530995	10453470	5268868	5228420	5228420	5228420	5228420
	6	10530995	10453470	5268868	5228420	5228420	5228420	5228420
	15	10530995	10453470	5268868	5228420	5228420	5228420	5228420
A586710	DS	42198943	27785885	21343768	19041307	15031300	13401034	11486601
	6	42198943	27785885	21735555	19041307	15071700	14709449	12754585
	15	41523868	28716501	22475033	19048835	15315476	13401034	12700205
Q12710	DS	2222349	2222349	2222349	2222349	2222349	2222349	2222349
	6	2222349	2222349	2222349	2222349	2222349	2222349	2222349
	15	2222349	2222349	2222349	2222349	2222349	2222349	2222349
D281	DS	7881	5329	4070	3926	3926	3926	3926
	6	7946	5485	4070	3926	3926	3926	3926
	15	8444	6408	5084	3964	3926	3926	3926
H953	DS	119357	119357	119357	119357	119357	119357	119357
	6	119357	119357	119357	119357	119357	119357	119357
	15	119357	119357	119357	119357	119357	119357	119357

Table 3. SOC g1023: TAM width and ATE memory

Memory	TAM width			Total ATE memory			
	DS1	DS	20	DS1	DSg	DS	20
32K	16	16	18	495679	496080	495639	511464
40K	13	13	15	495736	493047	492948	505107
48K	11	11	13	493358	494391	493717	507696
56K	9	9	11	492936	493713	492087	515956
64K	8	8	10	493475	493404	493124	514380
72K	7	7	9	490096	490086	488965	516732
80K	7	7	8	488759	488922	488910	514538
88K	6	6	7	489263	490291	490020	506660
96K	6	6	6	488639	490363	489495	507849
104K	5	5	5	490060	489592	489142	501840
112K	5	5	5	488244	490029	488886	500262
120K	5	4	5	487714	489018	487870	500209
128K	4	4	4	488271	488740	488740	497167

Table 4. SOC p22810: TAM width and ATE memory

Memory	TAM width			Total ATE memory			
	DS1	DS	20	DS1	DSg	DS	20
256K	30	28	30	7099492	7095609	6993356	7404961
320K	23	22	25	7027732	7003552	6967561	7134483
384K	19	19	21	6996904	7023324	6955179	7255983
448K	17	16	18	7086139	6994720	6913817	7326446
512K	14	14	16	6926695	7096052	7008133	7350768
576K	13	12	14	6862197	6955836	6935482	7390568
640K	12	11	13	6971280	6915493	6833044	7441084
704K	11	10	11	6804771	6886110	6820335	7234211
768K	10	9	11	6865817	6892560	6832015	7564350
832K	9	9	10	6839742	6917345	6835019	7601117
896K	9	8	10	6797761	6932763	6823407	7784192
960K	8	8	9	6837474	6940258	6843692	7642819
1M	7	7	8	6935563	6897983	6810674	7245774

Table 5. SOC p93791: TAM width and ATE memory

Memory	TAM width			Total ATE memory			
	DS1	DS	20	DS1	DSg	DS	20
1.00M	29	29	30	28801632	29165376	28701704	30569666
1.256M	23	23	23	28365553	28977417	28673994	28853177
1.512M	19	19	20	28416635	28749065	28485192	29587103
1.768M	16	16	17	28350143	28881393	28598386	30209460
2.000M	14	14	15	28448141	28576775	28221882	30570183
2.256M	13	13	13	28235620	28732954	28267254	29108758
2.512M	12	12	12	28304322	28902346	28223225	30385045
2.768M	11	11	11	28184680	28196995	28026911	29499548
3.000M	10	10	10	28157723	28665927	28160766	29635431
3.256M	9	9	9	28227717	28570811	28100678	29121214
3.512M	8	8	8	28056204	28417779	28194724	28853489
3.768M	8	8	8	28211308	28699383	28096145	29038354
4.000M	7	7	7	28301285	28420266	28189282	29096196

Table 6. SOC p34392: TAM width and ATE memory

Memory	TAM width			Total ATE memory			
	DS1	DS	20	DS1	DSg	DS	20
768K	21	21	23	15499573	15572458	15509657	15975513
896K	18	18	20	16037390	15381530	15348449	16676762
1.00M	16	15	16	15506240	15221266	15194202	15645989
1.128M	14	14	15	15399516	15267172	15213829	16227655
1.256M	13	12	14	15286752	15224186	15152249	15961051
1.384M	11	11	13	15336206	15345128	15268571	16713779
1.512M	11	10	12	15239519	15228804	15179270	15910317
1.640M	10	10	11	15177894	15304504	15165987	15474763
1.768M	9	9	10	15132399	15179765	15097375	15890652
1.896M	8	8	10	15124570	15176594	15127484	16330357
2.000M	8	8	9	15114833	15180006	15080645	16588577

A NOVEL FULL AUTOMATIC LAYOUT GENERATION STRATEGY FOR STATIC CMOS CIRCUITS

Cristiano Lazzari[1], Cristiano Domingues[1], José Güntzel[2], Ricardo Reis[1]
[1]UFRGS - Universidade Federal do Rio Grande do Sul
PPGC – Instituto de Informática
Porto Alegre – RS, Brazil
{clazz, cdviana, reis}@inf.ufrgs.br

[2] UFPEL – Universidade Federal de Pelotas
Departamento de Informática
Pelotas – RS, Brazil
guntzel@ufpel.edu.br

Abstract: The physical design of ASICs still relies on the standard cells because the design is well known and uses to produce good quality layouts. In addition, there are many choices of EDA tools that generate layout based on standard cells. However, in current CMOS technologies the standard cell approach is not able anymore to provide good performance predictability. Moreover, cell libraries have limited number of cells what imposes restrictions to layout synthesis. Automatic full-custom generators, on the other hand, do not use cell libraries and thus are more flexible to create optimized layouts. This chapter presents an automatic layout generator called PARROT PUNCH. Thank to a careful set of layout generation strategies and efficient algorithms, significant area and power optimization is achieved. Layouts generated by PARROT PUNCH are compared to those obtained by a similar automatic full-custom generator. Results show significant gain in area and delay.

Key words: Full automatic custom layout generation, Layout optimization, CMOS circuits.

Please use the following format when citing this chapter:

Lazzari, Cristiano, Domingues, Cristiano, Güntzel, José, Reis, Ricardo, 2006, in IFIP International Federation for Information Processing, Volume 200, VLSI-SOC: From Systems to Chips, eds. Glesner, M., Reis, R., Indrusiak, L., Mooney, V., Eveking, H., (Boston: Springer), pp. 197-211.

1. INTRODUCTION

Traditional physical level ASIC design flow still relies on standard cells libraries because it was the first automatic layout generation strategy to provide compact layouts. In fact it is not a full automatic layout generation because the cells are already designed. As the standard cells strategy has been used for more than twenty years, it is natural that this technique is well known and widely used. Moreover, within the customary pragmatism of industry, standard cells based layouts was a safe design flow because electrical performance could be accurately predicted, since the cells were pre-characterized.

However, with the advent of the deep submicron (DSM) technologies, geometries got smaller, clock frequencies increased and on-chip interconnect gains increased importance (Cong and Sarrafzadeh; 2000). In addition, problems in physical design are getting more complex and EDA (Electronic Design Automation) tools are essential to solve current design problems (Sarrafzadeh et al; 2001). Hence, the claimed predictability of the standard cells approach has been lost due to the difficulty on predicting the delays introduced by the routing.

Moreover, standard cells libraries have limited number of cells what imposes restrictions to layout synthesis. Also, different versions of each cell are required in order to drive different capacitive loads, thus increasing the total number of elements in the library to hundreds of cells. The larger is a library the more expensive is its update to a new fabrication technology.

An alternative to the cell-based layout generation approach is the automatic full-custom generation approach. An automatic full-custom layout generator does not use cells from a library. Instead, it generates each element (transistors and connections) according to a layout pattern that is intrinsically programmed within its algorithms. In addition, automatic generation can be flexible to create optimized layouts that are well tuned to a particular portion of the layout.

The pioneer works on automatic cell generation are those of Lopez and Law (Lopes and Law; 1980) and Uehara and Cleemput (Uehara and Cleemput; 1981). The former work introduced a layout style known as gate matrix, while the latter presented the so-called linear matrix layout style. Both works were originally developed having in mind the automatic generation of cell libraries. Current layout generators are mainly based on the linear matrix style and are able to generate combinational modules with up to tens of thousands of transistors.

PARROT PUNCH is an automatic full-custom generator based on the linear matrix layout style. PARROT PUNCH generates static CMOS layouts on demand for technologies with 3 or more metal layers. The tool tries to generate layouts with full over-the-cell routing. When it is not possible, channels are created between adjacent rows aiming at a complete routed circuit.

This paper is organized as follows. Section 2 presents the layout generation strategy. Section 3 shows the topology and the layout style used in the layout generation by PARROT PUNCH. In Section 4 some optimization techniques associated to the PARROT PUNCH layout generation are presented. Some experiments are reported in section 5 and the obtained results are presented in section 6.

2. LAYOUT GENERATION STRATEGY OVERVIEW

PARROT PUNCH is an automatic full-custom layout generator able to deal with circuits containing any kind of static CMOS gates, including complex ones.

Figure *1* presents the main characteristics of a layout generated by PARROT PUNCH. In this layout it is possible to observe the transistor folding technique (used to obtain wider transistors), the internal connection positions, the body tie placement in the diffusion gaps, the use of minimal transistor spacing and input/output contacts between PMOS and NMOS transistors.

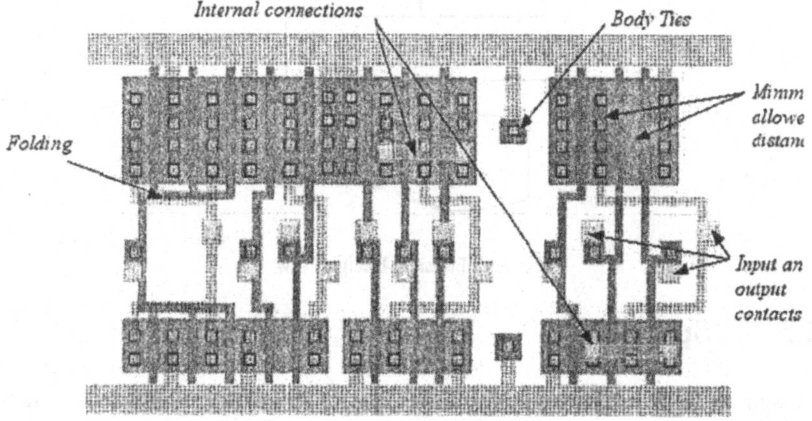

Figure 1. Main characteristics of a layout generated by Parrot Punch

In PARROT PUNCH generation follows the physical design flow shown in Figure 2. Each step is detailed in the next sub-sections.

2.1 Logic Gates Placement

Logic Gates placement consists on finding the position of each logic cell within the circuit surface. PARROT PUNCH generates layouts in which logic gates are placed in rows. Thus, the placement input is a list of logic gates partitioned in rows. Placement is performed as a separate step from the layout generation. Consequently, any placement tool can be used to realize this task.

The placement tool must be able to use an area estimate of each logic gate in the circuit because this information is not known in this step. Figure 2 shows the generation steps.

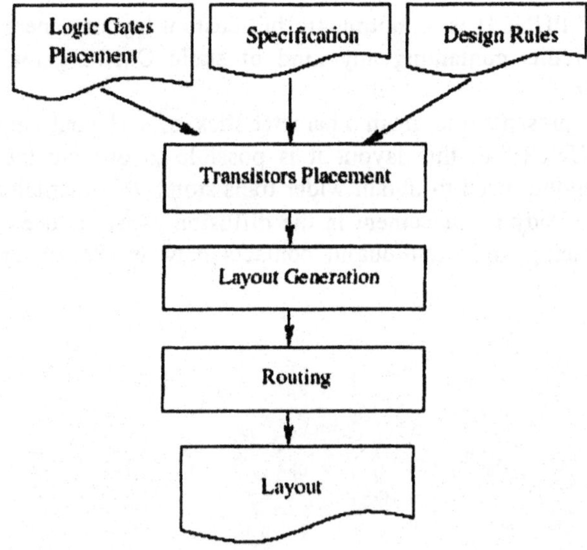

Figure 2. Physical Design Flow

2.2 Specification

Design specification is the set of desired characteristics for a circuit that is to be automatically generated. They are furnished by:

- A SPICE-like netlist
- The user parameters

PARROT PUNCH receives a SPICE-like netlist as input. This netlist is a set of logic gates defined by sub-circuits. Each sub-circuit in the netlist represents a logic gate and its position in the layout is specified in the placement output file.

User parameters consist on some definitions used in the layout generation as supply line characteristics and transistors width. These user characteristics influence on each logic cell in the circuit.

2.3 Design Rules

Design rules are the technologic rules furnished by the silicon foundry to develop the layout of masks that will be used in the fabrication process. They are basically the minimal dimensions for the rectangles or polygons in a given layer, metal enclosures to contacts and vias and minimal spacing between identical layers. The layout generation step tries to use minimal values of the chosen technology whenever it is possible.

2.4 Transistor Placement

Once logic gates and their positions in the layout are estimated, the layout can be generated. The transistor placement consists on searching the best as possible position for each transistor in a row. In the current version PARROT PUNCH only static CMOS circuits are generated. The used algorithm is responsible for ordering transistors in such a way that PMOS and NMOS transistors with common gate signal are easily connected together using only the polysilicon layer.

The Euler search algorithm is applied to a set of transistors (a logic gate) in which each transistor has an associated weight. The value of the weight is assigned to a transistor concerning the estimated position of the nearest point where it is connected to in the circuit. For example, a transistor that must be connected to another gate at its left side has more probability to be placed on the left. This is performed to reduce the wire congestion over the transistors of a gate.

2.5 Layout Generation

Layout generation consists on generating the geometries of each material (layer) that will be used to fabricate the circuit. In the proposed strategy, the layout generation can be divided into four main steps, being performed row by row. The layout generation steps are expressed by the pseudocode of Figure 3.

```
proc generateLayout( Row ) {
  contactsPlacement( Row );
  diffusionDesign( Row );
  polyRouting( Row );
  outputAndInternalConnect( Row );
}
```

Figure 3. Layout Generation Algorithm

The placement of contacts is performed first in the layout generation because of the grid router and the layout topology. Contacts must be placed regarding spaces defined by technology rules as the minimal spacing between metal layers and taking into account the enclosures of these metal layers on contacts and vias. Figure 4 shows the grid spacing definition as the function `contactsPlacement()`. In the grid spacing definition, all metal layers are evaluated to fit the real grid spacing without violating design rules.

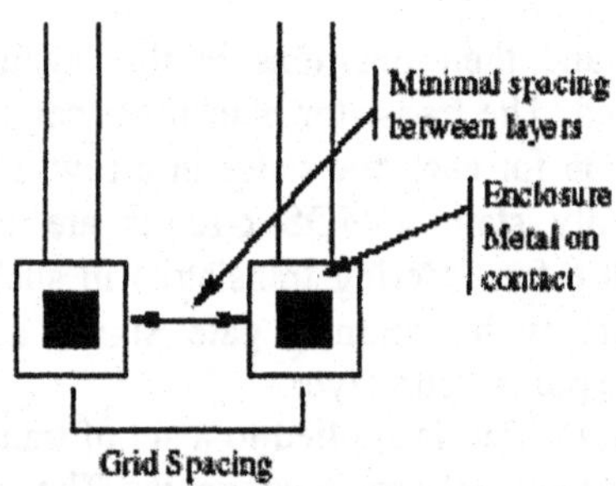

Figure 4. Grid router spacing

The used layout topology demands contacts placement as first step because contacts are inserted between PMOS and NMOS diffusion strips. Thus, the layout generation can be performed only when the positions of the contacts are already known.

The function `polyRouting()` performs polysilicon connection among contacts and gates in rows and function `diffusionDesign()` generates diffusion blocks. Once contact positions are known, diffusion blocks are performed with the aim of reducing source and drain areas of transistors. This is explained in more detail in section 3.

The layout of logic cells generated by PARROT PUNCH is performed using one metal layer (routing layers not included). Function `outputAndInternalConnect()` generates all internal connections and the output connection of each logic gate in the row.

The internal routing is divided in two parts: first, the polysilicon connections are generated and after, source/drains transistor nets are implemented.

2.5.1 Polysilicon Routing

Polysilicon routing consists on connecting transistors and input/output contacts that shares the same signal. Figure 5 shows four possible situations according to the transistor position and the contact inside the row. These situations are evaluated in the layout generation execution and the polysilicon routing is performed. They are the following:

- If the contact and the transistor are aligned with respect to the X-axis, they can be connected together by a straight line;
- If the contact and the transistor are not aligned with respect to the X-axis but there are no obstacles between the gate and the contact;
- When adjacent transistors are connected to the contact (folding technique);
- If the contact and the transistor are not aligned with respect to the X-axis and there are obstacles between the gate and the contact, the river routing algorithm is used to connect the gate and the contact together.

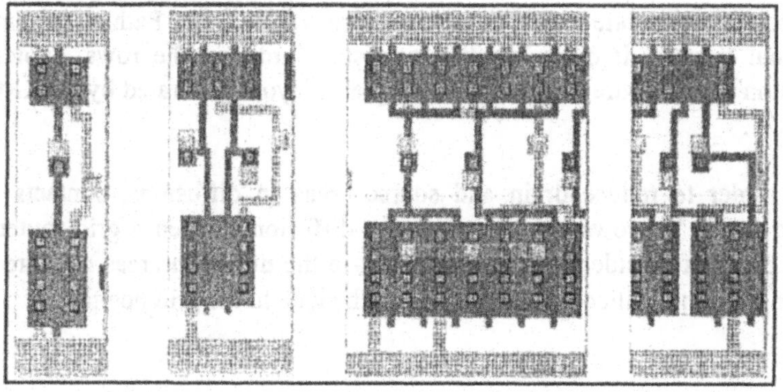

Figure 5. Four situations considered by the poly routing algorithm

2.5.2 Source/drain connections

These connections are implemented using the first and the second metal layers. The first metal layer is always used for wires between P and N diffusion strips. The second metal layer is used when the logic function demands connections over the transistors (as in the case of complex gates).

2.6 Routing

After the layout of each row is generated, connections among logic gates must be performed. A two-layer router integrated into the layout generation step realizes the routing. The first layer is used only for vertical wires while the second layer is used only for horizontal wires. The router starts at the top (north) of the layout, scanning and connecting related nets in direction to the bottom (south) of the circuit.

Once the routing is completed, the supply lines are generated. VDD and GND signals are connected to every row of the circuit and the contacts to substrate are placed in the layout.

3. PARROT PUNCH LAYOUT STYLE

The layout style generated by PARROT PUNCH is based on the *linear matrix* in which p- and n-diffusions are placed in parallel composing rows. Transistors are laid out in the diffusions strips in such a way they are easily connected to generate the layout of logic gates. The Euler Path algorithm is essential because it optimizes the transistors order in the rows, avoiding diffusion gaps. Figure 6 shows an example of layout generated by PARROT PUNCH.

In order to reduce drain and source areas in diffusions, contacts are placed inside the rows between p- and n-diffusions. When a grid router is used, contacts outside the rows can increase the diffusion areas or increase the length of polysilicon wires to connect the gate to the grid position.

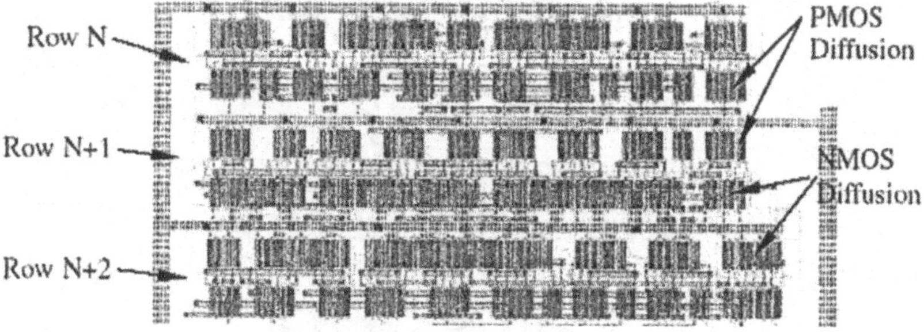

Figure 6. Layout generated by PARROT PUNCH

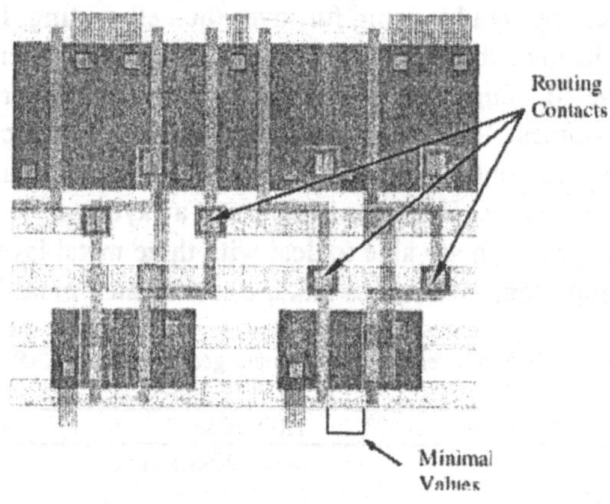

Figure 7. Contacts inside row reducing diffusion areas

Figure 7 shows a detail of a layout generated with the Parrot Punch. Due to the placement of contacts between p- and n-diffusion and according to the design rules, it is possible in the best case, to obtain minimal values of spacing between gates.

The whole layout (except routing) is performed using only one metal layer and one polysilicon layer. Supply lines are placed between adjacent rows in metal1. Rows are mirrored aiming at sharing supply lines and contacts to substrate (body ties) between each pair of adjacent rows. Figure shows an example of supply line between adjacent rows and contacts to substrate.

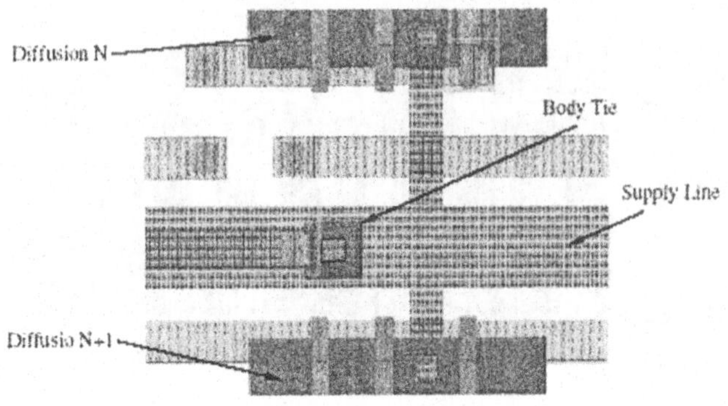

Figure 8. Supply lines and body ties are shared by adjacent rows

An important characteristic in the layout style of PARROT PUNCH is the attempt to reduce area by using full over-the-cell routing. In this routing style, routing channels between rows are only created when the routing cannot be completed using the minimal spacing between these rows.

TROPIC3 (Moraes, Reis and Lima;1997) is an automatic CMOS full-custom generator. Layouts generated by PARROT PUNCH and TROPIC3 contain some similarities. Both generators use a layout style based on the linear matrix one and both are able to deal with three metal layers. However, there are some important differences that are enumerated in table 1.

Table 1. Similarities and differences between layouts generated by PARROT PUNCH and TROPIC3

PARROT PUNCH	TROPIC3
Linear Matrix style	Linear Matrix style
3 or more metal layers	3 metal layers
Supply lines between rows (metal 1)	Supply lines inside rows (metal2)
Contacts inside the rows	Contacts outside the rows
No channel routing	Channel routing

Main differences between layouts generated by these tools rely on supply lines position, position of the contacts to routing and the attempt to use full over-the-cell routing by PARROT PUNCH.

4. OPTIMIZATION TECHNIQUES

4.1 Complex Gates and Power Optimization

Dynamic power consumption is still the dominant source of power consumption. However, high leakage current in DSM technologies is becoming a significant contributor to power consumption in DSM CMOS circuits when threshold voltage, channel length and gate oxide thickness are reduced (Roy, Mukhopadhyay, and H. Meimand; 2003).

Reis presents in (Reis, et.al; 1997) a method to map a set of Boolean equations into a set of static CMOS complex gates (SCCG) under a constraint in the number of serial transistors. In this work, a tool called TABA was developed to optimize a circuit under a set of complex gates in which the number of transistors is reduced. This kind of technology mapping is known as "library free mapping" and is particularly suited to be used along with a full automatic layout generator. Using the library free mapping together with a full automatic layout generator it is possible to generate an optimized set of cells with a reduced number of transistors in comparison to a standard cell mapping. In a logic mapping where the number of serial P and N transistors is constrained to 4, the number of possible logic functions is 3503 (Detjens and Rudell; 1987). However, this number is much greater than the possibilities offered by a regular library of cells. And in some cases it is possible to use more than 4 serial transistors, enlarging the possible number of logic cells.

The library free mapping is very important because it allows a reduction on the number of transistors of a circuit in comparison to a solution constrained by the use of library that has a restricted number of logic functions. Library free mapping can provide a 20% to 30% reduction in the number of transistors. Thus, the static power consumption is also reduced due to the reduction on the number of transistors.

4.2 The Gate Sizing Tool Integration

The accuracy of timing verification is completely dependent on the effectiveness of the used circuit model. By circuit models it is meant not only the physical delay model used to quantify the delay of each component, but also the models for computing circuit component delay and the circuit delay itself. Timing analysis associated with layout generation can improve accurate timing optimization characteristics to the circuit design. A tool called *TICTAC::Sizing* (Santos, et.al; 2003) is able to analyze a circuit and realize gate sizing based on timing constraints. The integration between *TICTAC::Sizing* and PARROT PUNCH is done in such a way that transistors' size information, given by the sizing tool, can be easily used in the layout generation.

In timing-driven layout synthesis, transistor widths tend to vary significantly and the use of conventional layout approaches may cause inefficient area utilization (Kim and Kang; 1997). The folding technique consists on breaking a large transistor into smaller ones, connecting them in parallel and placing them continuously with diffusion sharing. The folding technique is especially important in the case of row-based layouts because different transistor sizes in a row could cause non-uniform cell heights leading to significant waste of area. The used folding algorithm (Bastian, et.al; 2004) tries to look at every folded transistor as a unique transistor, applying the folding over a given transistor placement neither modifying their placement nor inserting new diffusion breaks in the row. The algorithm introduces a new approach in which transistors are classified according to even or odd multiplicity. Then, the algorithm is able to verify when a diffusion gap can be avoided, thus reducing gate capacitances.

5. PRACTICAL EXPERIMENTS

In order to validate PARROT PUNCH some experiments were performed. Circuits were generated with PARROT PUNCH and TROPIC3 (Moraes, Reis and Lima;1997). In the set of benchmarks, circuits with the number of transistors between 294 and 1356 were chosen.

To provide a fair comparison, the quadrature placement algorithm presented in (Moraes and Velasco; 2000) was used to perform the placement for both PARROT PUNCH and TROPIC3 cases. The algorithm consists on dividing the circuit into horizontal and vertical directions, minimizing the cut-size to each direction. These quadrants are processed line-by-line in the case of horizontal partition and column-by-column when vertical partition is performed.

The quadrature placement is also more efficient because of pin propagation. Thus, logic cells with common signals are placed in adjacent quadrants to reduce the wire length.

Area, delay and power were reported and compared in section 6. After the layout generation, layouts were inserted into CADENCE Virtuoso™ physical design environment where they were extracted using DIVA™ extractor and simulated with Spectre™ electrical simulator.

The used technology in theses examples was the AMS 0.35μm. The transistors width used in the layout generation were 4.00μm to PMOS transistors (W_P=4μm) and 2.00μm to NMOS transistors (W_N=2μm).

6. RESULTS

Table 2 shows the area occupation by circuits generated with Parrot Punch and TROPIC3. It is possible to note that there is a difference between the occupied areas of the layouts in a same technology. The main reason for these values is due to the routing strategy. In TROPIC3, routing is performed basically in routing channels between p and n diffusions while in PARRROT PUNCH routing is generated over the rows whenever it is possible. Results present between 33.2% and 47.5% of area reduction for the layouts generated with PARRROT PUNCH when compared with layouts generated with TROPIC3. The average gain in the area occupation is around 38%.

Table 2. Area occupation in TROPIC3 and Parrot Punch layouts (um2)

Bench	No. trans	P. PUNCH	TROPIC3	Gain (%)
C17	24	495.6	756.6	34.4
C432	804	20782.2	35443.2	41.3
C499	1556	54000.8	86224.3	37.3
C880	1802	59925.6	96014.4	37.5
C1355	2308	74613.9	111784.4	33.2
C1908	3482	116397.0	181575.6	35.8
C6288	10112	303376.8	578082.0	47.5

Table 3. Delay and power consumption in TROPIC3 and PARROT PUNCH layouts

Bench	No.	PARROT PUNCH		TROPIC3		Gain (%)	
	Trans	Delay	Power	Delay	Power	Delay	Power
C17	24	0.47	0.42	0.55	0.45	15.1	12.0
C432	804	0.78	15.31	1.07	16.73	26.9	9.3
C499	1556	0.31	18.61	0.42	17.8	26.0	-4.0
C880	1802	1.73	17.67	2.22	17.2	22.0	-2.0
C1355	2308	0.50	26.38	0.64	29.0	21.1	9.0
C1908	3482	2.67	35.1	3.66	45.56	27.0	22.0
C6288	10112	4.28	205.1	5.45	201.9	21.4	-1.5

Table 3 presents results given by electric simulation. Delay and power consumption of layouts are considered. Results related to delay of circuits show an average gain of around 22% when compared to TROPIC3 layouts. These results are obtained due to the difference of routing strategy and the attempt for optimizing drain/source areas on PARROT PUNCH strategy.

As reported in previously sections, the effort to optimize the layout is applied in all steps of the layout generation strategy of PARROT PUNCH. In addition, the reduction on the wire lengths obtained by the FOTC routing reduces the delay of the wires.

Power consumption in PARROT PUNCH layouts has an average gain of 4% over TROPIC3 layouts. In these circuits, it was applied the same simulation patterns and the switching activity is the same. Thus, the power consumption must be almost the same in layouts generated by both tools. The power consumption analysis for circuit C1908 shows a great difference between the layouts generated by PARROT PUNCH and by TROPIC3. This difference does not appear in other layouts. Verifying the simulation waveforms, glitches were found with more frequency in the layout generated by TROPIC3. It is believed that these glitches are responsible for the greater power consumption in the TROPIC3 layout.

7. CONCLUSION AND ON GOING WORK

This work presented an automatic full-custom generator called PARROT PUNCH. Layouts generated with PARROT PUNCH are characterized mainly by linear matrix style, full over-the-cell routing and 3 metal layers to implement all circuit connections. The previously mentioned features have led to an area reduction of 18.7% up to 44%, when compared to similar linear matrix layouts (TROPIC3). Also, delay and power were reduced.

The area occupied by layouts can still be reduced taking advantage on several metal layers available in current CMOS technologies. When full over-the-cell routing is target, the use of technologies with six or up to nine metal layers increases the success of this routing technique.

As long as most of commercial tools use standard cells based layout generation, we intend to perform practical comparisons between this style and PARROT PUNCH linear matrix style.

Some improvements are planned to PARROT PUNCH. Notably:

- Layout generation with more than three metal layers;
- Take advantage on the automatic layout generation to implement techniques of layout optimization such as gate sizing and buffer insertion.

REFERENCES

Bastian F., Lazzari, C., Güntzel, J. L., Reis, R. (2004). A New Transistor Folding Algorithm Applied to an Automatic Full-Custom Layout Generation Tool, PATMOS2004, *14th*

International Workshop on Power and Timing Modeling, Optimization and Simulation, Santorini, September 15-17, 2004. LNCS 3254 Springer. p. 732-741.

Cong, J. and Sarrafzadeh, M. (2000). Incremental Physical Design. In Proceedings of the 2000 International Symposium on Physical Design, pages 84-92. ACM Press.

Detjens, E. Rudell, R. Sangiovanni-Vinccentelli, A. and Wang, A. (1987). Technology Mapping in MIS. In *ICCAD*, pages 116–119.

Kim, J. and Kang, S. M. (1997). An Efficient Transistor Folding Algorithm for Row-based CMOS Layout Design. *DAC'97 – Design Automation Conference*, pages 456–459.

Lopez, A. and Law, H. S. (1980). A Dense Gate Matrix Layout Method for MOS VLSI. IEEE Transactions on Electron Devices, ED-27(8):1671-1675.

Moraes, F., Reis, R., and Lima, F. (1997). An Efficient Layout Style for Three-Metal CMOS Macrocells. In VLSI'97, pages 415-426.

Moraes, F. and Velasco, (2002). A. J. Deterministic Versus Non-Deterministic Placement Algorithms for Automatic Layout Synthesis Tools. In DCIS'02.

Reis, A., Reis, R. , Auvergne, D. and Robert, M. Library Free Technology Mapping. (1997). VLSI: Integrated Systems on Silicon, IFIP TC10 WG10.5 International Conference in Very Large Scale Integration, pages 303–314.

Roy, K. Mukhopadhyay, S. and Meimand, H. (2003). Leakage Current Mechanisms And Leakage Reduction Techniques in Deep Submicrometer CMOS Circuits. In *Proceedings of the IEEE*, volume 91, pages 305–327.

Sarrafzadeh, M., Bozorgzadeh, E., Kastner, R., and Srivastava, A. (2001). Design And Analysis of Physical Design Algorithms. In Proceedings of the 2001 International Symposium on Physical Design, pages 82-89. ACM Press.

Santos, C. L., Wilke, G., Lazzari, C., Guntzel, J., Reis, R. A. (2003). A Transistor Sizing Method Applied to an Automatic Layout Generation Tool. SBCCI2003. *16th Symposium on Integrated Circuits and Systems Design.* São Paulo, Septembre 8-11, 2003. p.303-307.

Uehara, T. and Cleemput, W. (1981). Optimal Layout of CMOS Functional Arrays. IEEE Transactions on Computer, C-30(5):305-312.

LOW POWER JAVA PROCESSOR FOR EMBEDDED APPLICATIONS

Antonio Carlos S. Beck and Luigi Carro
Universidade Federal do Rio Grande do Sul - Instituto de Informática - Av. Bento Gonçalves, 9500 - Campus do Vale - Porto Alegre, Brasil

Abstract: This chapter presents a low power architecture of a Java processor. We show that the use of techniques like pipeline and the implementation of the stack in a register bank instead of using the main memory allow aggressive reduction of power dissipation, with a very small area overhead. Besides, thanks to the forwarding technique and to the specific stack machine organization, huge power savings can be obtained when applying this technique to a pipelined implementation of the architecture. Several examples of embedded applications are used to show the power savings obtained through the architecture optimization

Key words: Java, Power Consumption, Stack Machines

1. INTRODUCTION

The embedded system market grows day by day. The production of specific processors to be used inside microwaves, videogames, printers, mp3 players, digital cameras, cellular phones and others appliances, is following the same growing path[1]. Particularly in portable embedded systems, where the battery lifetime is a crucial factor, some special care is needed in terms of power consumption of the system.

Java is becoming increasingly popular in embedded environments. It is estimated that devices with embedded Java such as cellular phones, PDAs and pagers will grow from 176 million in 2001 to 721 million in 2005 [2]. Nevertheless, is predicted that at least 80 percent of mobile phones will support Java by 2006 [3]. As one can observe, the importance of Java in

---

Please use the following format when citing this chapter:
Beck, Antonio Carlos, S., Carro, Luigi, 2006, in IFIP International Federation for Information Processing, Volume 200, VLSI-SOC: From Systems to Chips, eds. Glesner, M., Reis, R., Indrusiak, L., Mooney, V., Eveking, H., (Boston: Springer), pp. 213-228.

embedded systems is growing. This means a careful look on embedded Java processors and their performance versus power tradeoffs must be taken into account.

In this chapter we show a study about three different architectures capable of executing Java bytecodes and discuss their area, performance and mainly power requirements, focusing on embedded systems applications. We demonstrate that by the use of inexpensive techniques, one can optimize the execution of instructions and obtain a drastic reduction in the energy consumption, by a factor of more than 10. Moreover, it is demonstrated that the use of the forwarding technique, besides increasing the performance, allows further benefits to stack-like architectures, since power consumption is reduced because of the small number of writes in the stack. Less writes in the stack means a smaller number of register or memory writes, hence reducing memory accesses, one of the major sources of power dissipation in embedded processors[4,5,6]. Our experiments are supported by simulation, using three different architectures of the Femtojava Processor[7,8], executing different algorithms used in embedded system domain. The area was computed in number of logic cells, after synthesis of different VHDL versions of the processors.

This chapter is organized as follows: Section 2 shows a brief review of the existing Java processors. In Section 3 we discuss the different architectures of Java machines that will be evaluated, and present the advantages of using the forwarding technique in stack machines. Section 4 presents the simulation environment: the power simulator and the test case algorithms executed in the processors. Section 5 shows the results regarding power consumption, performance and area. The last Section draws conclusions and introduces future work.

2. RELATED WORK

A large number of Java processors aimed at the embedded systems market have already been proposed. Sun's Picojava I [9], a four stage pipelined processor, and Picojava II [10], with a six stage pipeline, are probably the most studied ones. Even though the specification of such processors allows a variable size for the data and instruction caches, and the floating point unit is optional, there is no special care on the underlying microarchitecture in order to reduce the area and power consumption of the system.

The same occurs to others Java processors: Komodo[11], a multithreaded Java microcontroller concerned especially with real time applications; and Traja[12], a dual issue pipelined processor that makes use of instruction reordering to avoid data dependencies. All of these and other examples of native Java execution machines always focus on obtaining the maximum possible performance, in order to leverage Java execution with RISC and VLIW architectures. However, in the domain of embedded systems, not only plain throughput is the correct metric. Other issues like power dissipation and software compatibility play a major role. This way, the research on low power Java processors, able to maintain enough performance to execute the target application with the smallest possible power budget, is the goal of this work.

3. ARCHITECTURE OF THE JAVA PROCESSORS

The Femtojava processor is a stack-based microcontroller that executes Java bytecodes. General characteristics of the Femtojava processor are: reduced instruction set, Harvard architecture and small size. This processor was designed specifically for the embedded system market. The size of its control unit is directly proportional to the number of different instructions used by the application. From an available tool[7], the Java bytecodes of the application are analyzed, and the control unit is generated, supporting only the instructions used by that application.

The first architecture evaluated is a multicycle version of Femtojava[7] that takes three to fourteen cycles to execute an instruction. Its microarchitecture can be observed in Figure 1.

The pipelined version[8] has five stages: instruction fetch, instruction decoding, operand fetch, execution, and write back, as shown in Figure 2. One of the main characteristics of the pipelined Femtojava is the presence of registers playing the role of operand stack and local variable storage.

The first stage, instruction fetch, is composed by an instruction queue of 9 registers of one byte each. The first instruction in the queue is sent to the instruction decoder stage. The decoder has three functions: the generation of the control word for that instruction, to handle data dependencies and to inform to the instruction queue the size of the current instruction, in order to put the next instruction of the stream in the first place of the queue. This is necessary because of the use of variable length instructions: they can have one or two immediate operands, or none at all. When at least 4 registers in the instruction queue are empty, a word of 32 bits comes from the instruction memory, pointed by the program counter.

The operands fetch is done in a variable size register bank, defined a priori in earlier stages of the design. Stack and the local variable pool of the methods are available in the register bank. There are two registers: SP and VARS. They point to the top of the stack and to beginning of the local variable storage, respectively. Depending on the instruction, one of them is used as base for the operands fetch. Once the operands are fetched, they are sent to the fourth stage, where they will be executed. There is no branch prediction, in order to save area. All branches are supposed to be not taken. If the branch is taken, a penalty of three cycles is paid.

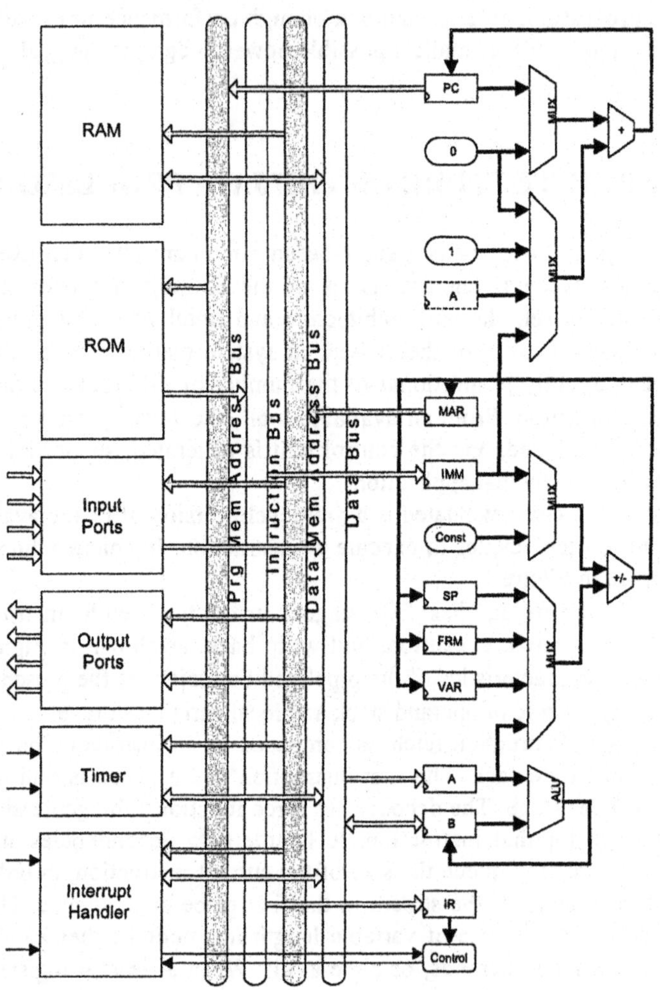

Figure 1. Multicycle Java Processor[7]

The write back stage saves, if necessary, the result of the execution stage back to the register bank, using the SP or VARS as base. As the register bank can not be simultaneously read and written, when an instruction in the fifth stage writes back its result and an instruction in the third stage wants to read an operand, a bubble is inserted in the pipeline. There is a unified register bank for the stack and local variable pool, because this facilitates the call and return of methods, taking advantage of the JVM specification, where each method is located by a frame pointer in the stack.

Finally, the third architecture evaluated is the same pipelined one described before, plus the increment of the forwarding technique: if there is an instruction in the execution stage that will write its result in the stack, and the following instruction accesses the stack in the operand fetch stage, a true dependency (RAW - read after write) is characterized. One of the solutions is to stall the pipeline, inserting bubbles on it, until the first instruction finishes the write back stage. Another solution is to make use of the forwarding technique[13], passing directly the result from the execution stage to the operand fetch stage.

In operand stack based processors, the use of such technique brings an advantage when comparing to register-like processors: in instructions that manipulate the stack, the operands forwarded to earlier stages will not be used anymore. As a consequence, there is no need to write back these operands to the stack. The result is the reduction on the power consumption, because the diminishing number of writes in the stack.

Two types of forwarding can occur: when the instruction in the execution stage consumes one operand from the top of the stack (like istore, which saves the top of stack in some place of the local variable pool); or when the instruction consumes two operands from the stack (like arithmetic operations: iadd, isub, ior). In the first case, the operand forwarded comes from the execution stage to the operand fetch. In the second case, the second operand comes from the write back stage.

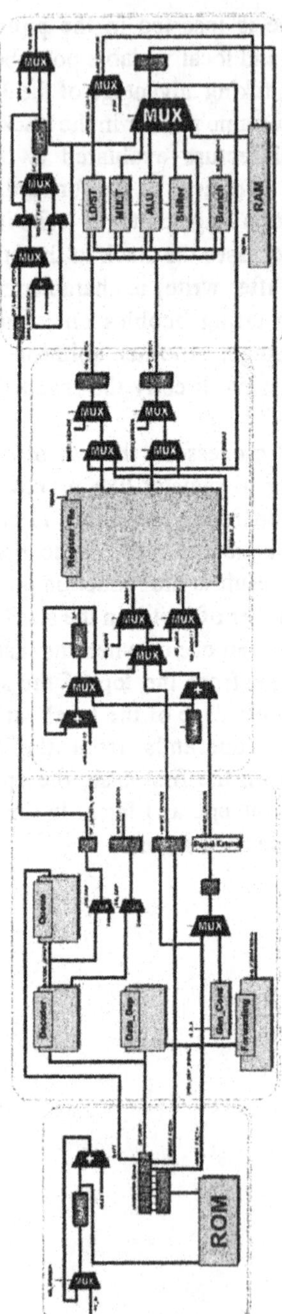

Figure 2. **Pipelined Java Processor**

4. SIMULATION ENVIRONMENT

4.1 Power Simulator

CACO-PS[8], a compiled-code cycle-accurate simulator, was used to provide data on the energy consumption, memory usage and performance. Power dissipation is evaluated in terms of switching capacitances, and as the processor has separated instruction and data memories, we also included an evaluation module concerning RAM and ROM memories, besides the register bank. This way, one can verify the relative power dissipation of the CPU, instruction memory, and data memory. It is important to measure the impact of each one of these blocks, so that one can better explore the design space.

We used the power simulator to collect the amount of capacitances that switch during the execution of a certain algorithm. This power estimation technique is comparable to the component-based approach [14,15,16].

4.2 Algorithms

Five different types of algorithms were implemented and simulated over the architectures described in Section 3. Sin Calculation, as a representative of arithmetic libraries; sort and search, used in schedulers; IMDCT (Inverse Modified Discrete Cosine Transformation), an important part of the MP3 decompression algorithm; and a library to emulate sums of floating numbers, since the Femtojava processors does not have a floating point unit in order to save area.

Sequential algorithms are used as representative of scalar computations. The first one is Sequential1 that performs the search in a sequential fashion in a non-ordered table. Even if the value has already been found, the algorithm stops just at the end of the table. The second algorithm (Sequential2) stops right after the required value is found in a table that has been previously ordered. The last search algorithm performs a binary search in the vector.

The sort algorithms arrange a set of ten numbers putting them in increasing order. Three different kinds of sort are performed: bubble sort, insert sort and select sort. The floating point sum algorithm makes 20 sums of two floating point numbers and puts its results in a vector in the memory. Finally, the sin algorithm uses the cordic method to calculate the result.

5. RESULTS

Table 1 shows the area occupied by the three different versions of the Femtojava processor. It is important to note that the register bank, used as stack and local variable pool, has 32 registers, the maximum required among all the applications. The computed area includes the control unit that supports all the Femtojava instruction set. The area was evaluated using the Leonardo Spectrum for Windows [17], and it is presented in logic cells. More details about the VHDL implementation can be found in [18].

Table 1. Area occupied by the VHDL version of the architectures

PROCESSOR	*Multicycle*	*Low-Power*	*Low-Power with forwarding*
AREA	1345 LCs	2916 LCs	3016 LCs

Table 2 shows the performance in number of cycles of the processors for each application.

Table 2. Performance of the different architectures

ALGORITHM	NUMBER OF CYCLES		
	Multicycle	*Low-Power*	*Low-Power with forwarding*
Sin	2447	1237	755
Sort – Bubble	6774	3234	2468
Sort – Select	5335	2703	1930
Sort – Insert	4093	2071	1571
Search - Binary	1162	602	403
Search - Sequencial 1	8497	3803	2765
Search - Sequencial 2	7586	2779	1997
IMDCT	140300	61841	40306
Floating Point Sums	30747	18735	14531

Operating at the same frequency, the core of the pipelined version with forwarding is the architecture which has the major power consumption per cycle, since this architecture is the more complex one, with several extra registers. This behavior can be observed in figure 3.

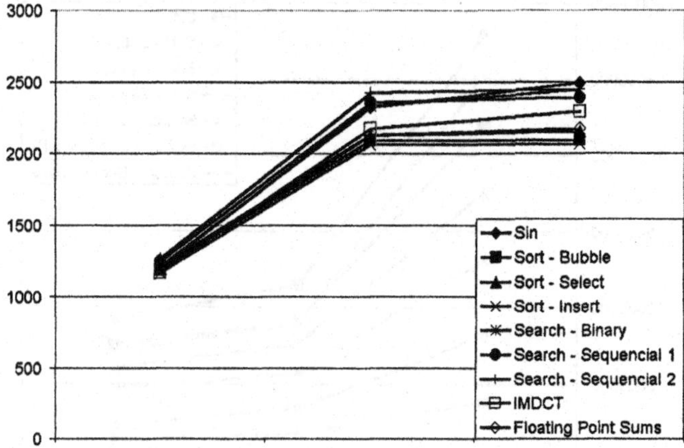

Figure 3. Power consumed per cycle

Even though the power consumed per cycle in the pipelined versions is greater then the multicycle one, the total energy consumption of the system is decreased, because of the high throughput reached by the pipelined versions and the decrease in the number of accesses in the memory.

As it can be observed in figures 4 and 5, the overall energy consumption is drastically reduced in some algorithms, when one changes the architecture of Femtojava. Others, like the algorithm of sums of floating point numbers, do not show such a great gain. The reason is that the implementation of this algorithm makes a lot of method calls, with a consequent increase in the main memory power consumption. Moreover, the algorithm makes an intensive use of the local variable pool of the methods, and although the forwarding technique is applied to the local variable pool as well, the forwarded operands must be written back in the registers, because there is no warranty that they will not be used in the future. The IMDCT and floating point sum algorithms are presented in a separated figure in order to ease the visualization.

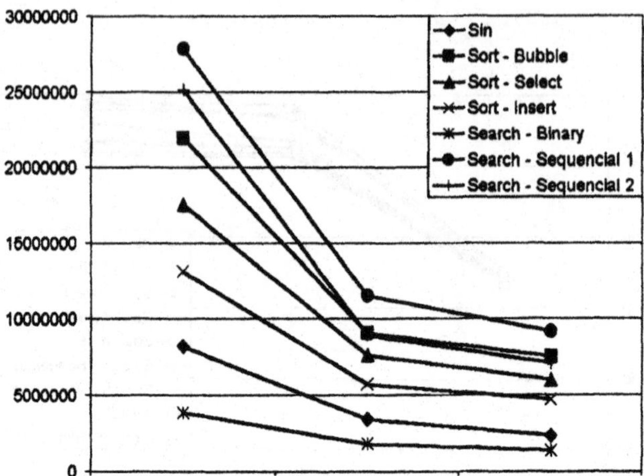

Figure 4. Energy consumption in the three architectures

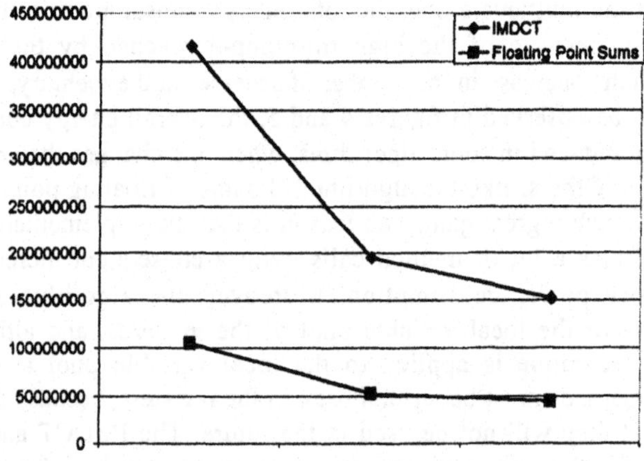

Figure 5. Continuation: Energy consumption in the three architectures

Figures 6 and 7 show the energy consumed because of RAM accesses. As Femtojava Multicycle uses the main memory as stack and variable pool, there is a big difference between this architecture and the pipelined ones. These last, in turn, just make accesses in the main memory in calls and returns of methods, and while executing specific instructions, such as *putstatic* and *getstatic*. Figures 6 and 7 show the advantage of implementing the stack and local variable pool in a register bank instead of using the main memory.

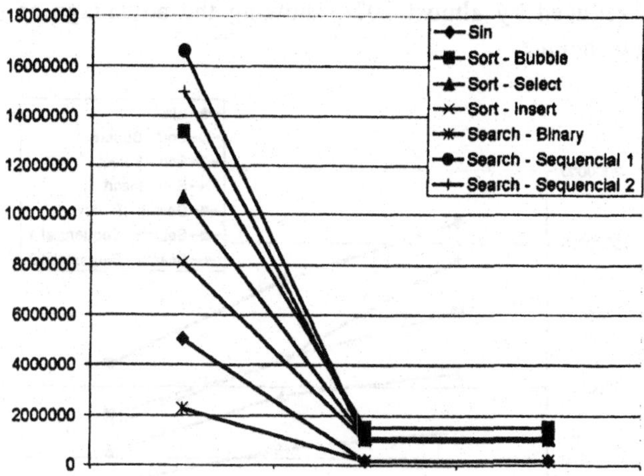

Figure 6. Energy caused by main memory accesses

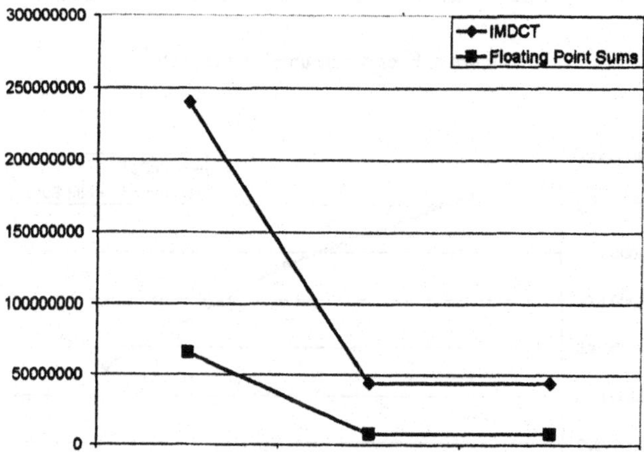

Figure 7. Continuation: Energy caused by main memory accesses

Figures 8 and 9 show the energy consumed by the core of the presented architectures. As one can observe, the power consumed in the core is reduced in the pipelined version with forwarding, in comparison to the version that does not use the technique. The difference on the energy consumption has two reasons: the better utilization of the functional units because of the reduction of pipeline stalls; and the decrease of the number of registers writes, since the average of power consumed in registers writes per cycle was reduced by almost 70%, showing the advantage of using the forwarding technique.

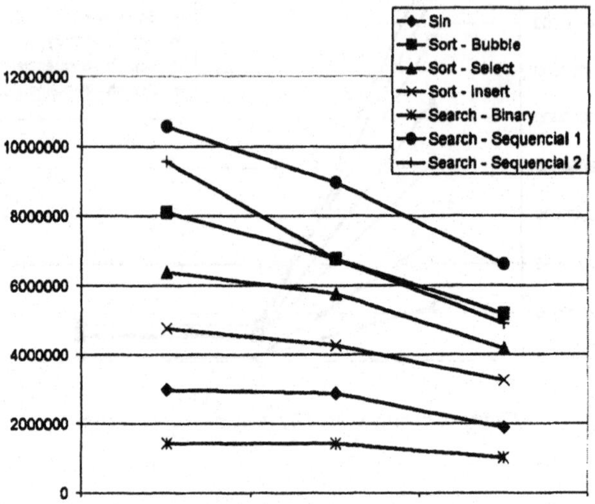

Figure 8. Energy consumed in the cores

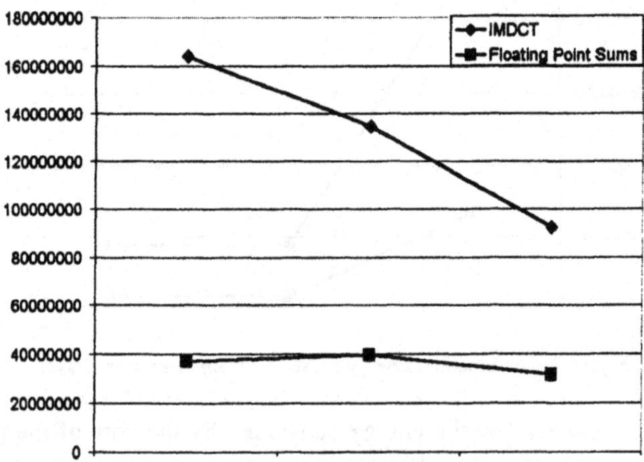

Figure 9. Continuation: Energy consumed in the cores

In embedded applications, many of them with real time requirements, a specific throughput must be warranted for the application. Assuming that this throughput is reached by the multicycle version, the frequency of operation of the pipelined versions can be decreased in order to save power, since these architectures can execute more instructions per cycle, as was showed in table 1.

Moreover, when assuming that the dynamic power is the dominant in the power consumed in the system, and all the gates of the microprocessor form a collective switching capacitance C with a common switching frequency f, one obtains:

$$P = C \cdot f \cdot Vdd^2 \quad (1)$$

As can be observed in [19], the voltage of the processor Transmeta TM5400 (known as Crusoe) [20], designed for embedded systems, can be decreased by a factor of 4,6% when the operation frequency is reduced by 10%.

Figure 10 shows the relative decrease in the energy consumption when the frequency of the pipelined version (without forwarding) is reduced to reach exactly the same throughput of the multicycle version, and when the voltage is reduced thanks to the decrease in the frequency, using as base the equation (1).

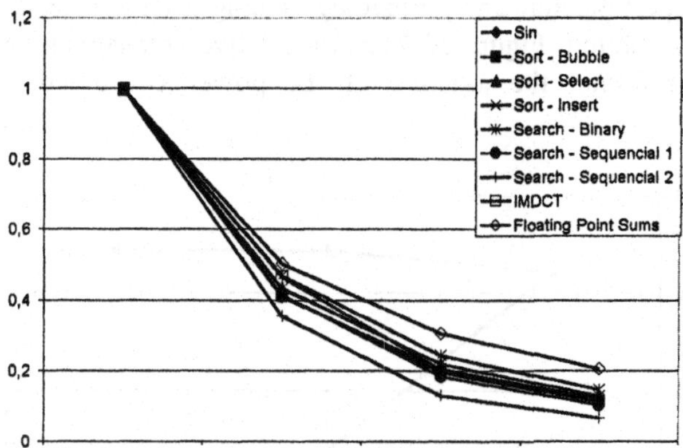

Figure 10. Energy consumed in the cores with the frequency and voltage reduced

Applying the forwarding technique, the energy consumption is reduced even more, as can be observed in figure 11, when it is shown the relative decrease in the power consumption comparing the pipelined version without forwarding with the one that uses the technique. Both had the frequency and voltage reduced to reach the same throughput of the multicycle version.

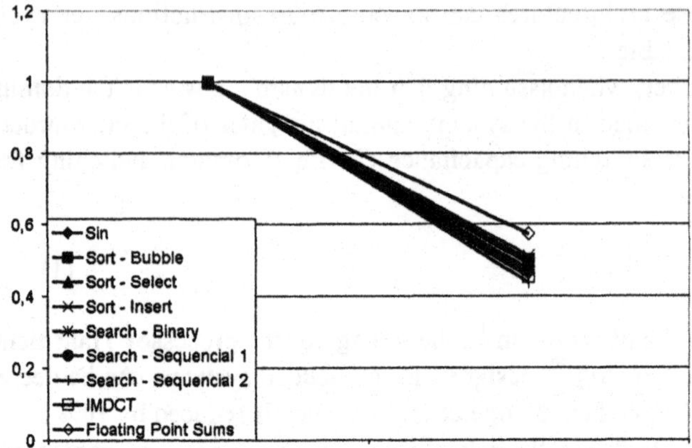

Figure 11. Comparison between the architectures without and with forwarding, both with the frequency and voltage reduced

Finally, we show the advantage of increasing the area of the processor in order to support pipeline and forwarding, trading it for a huge saving in the energy consumption. Figure 12 shows a relative comparison between the area overhead and the average of the power consumed by all the applications.

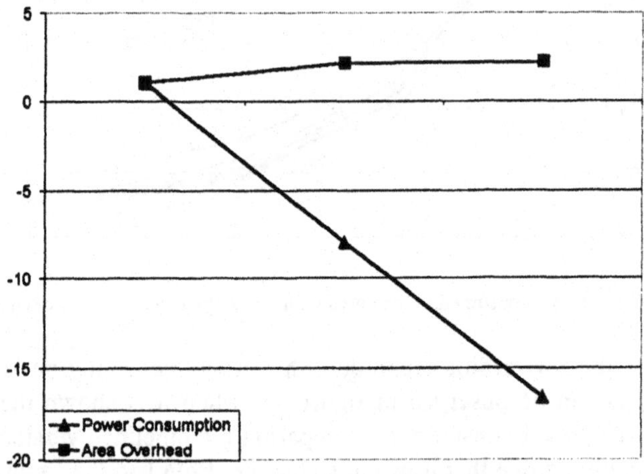

Figure 12. Area versus energy consumption

As can be observed, for twice the area overhead, one obtains a factor of 16 in the energy reduction.

6. CONCLUSION AND FUTURE WORK

We demonstrated that for embedded applications, which need a small stack for its operations, one can obtain a huge decrease in the power consumption with a proportionally small increase in the overall area of the system. Particularly for these machines, the use of the forwarding technique brings a large reduction in the power consumption of the core, taking advantage of the stack architecture and of algorithms that uses the stack intensively, like the IMDCT one. We believe that this behavior is true for others stream-based algorithms, since these are algorithms that make intensive use of the stack.

For future work, we will evaluate more algorithms of the embedded system domain, like an MP3 player. Moreover, we are studying the impact of applying the VLIW technique in Java machines, concerning mainly the power consumption[21], as well as alternatives to increase the VLIW performance, such as the use of techniques to find the instruction parallelism inside the program, like software pipelining, superblocks, and static speculative execution. Other techniques such as the use of a reconfigurable array working together with binary translation to detect the sequence of instructions and reconfigure the array at runtime, and CMP, where a set of processors with the same ISA work together in the same die, will also be evaluated, always giving special concern to the advantages and particularities of stack processors in these techniques.

7. REFERENCES

1. M. Schlett, "Trends in Embedded-Microprocessor Design", *Computer*, vol. 31, n. 8, 1998, pp. 44–49
2. D. Takahashi, "Java Chips Make a Comeback", *Red Herring*, 2001
3. G. Lawton, "Moving Java into Mobile Phones", *Computer*, vol. 35, n. 6, 2002, pp. 17-20
4. V. Tiwari, S. Malik, A. Wolfe, "Power Analysis of Embedded Software: A First Step Towards Software Power Minimization", *IEEE Transactions on VLSI Systems*, vol. 2, n. 4, Dec. 1994, pp. 437–445
5. T. Simunic, G. Micheli, L. Benini, "Energy-Efficient Design of Battery-Powered Embedded Systems", *Proceedings of the International Symposium on Low Power Electronics and Design (ISLPED99)*, Aug. 1999
6. G. Chen, R. Shetty, M. Kandemir, N. Vijaykrishnan, M. Irwin, "Tuning garbage collection for reducing memory system energy in an embedded java environment", *ACM Transactions on Embedded Computing Systems*, vol. 1, n. 1, Nov. 2002, pp. 27-55
7. S.A. Ito, L. Carro, R.P. Jacobi, "Making Java Work for Microcontroller Applications", *IEEE Design & Test of Computers*, vol. 18, n. 5, 2001, pp. 100-110
8. A.C.S. Beck, J.C.B. Mattos, F.R. Wagner, L. Carro, "CACO-PS: A General Purpose Cycle-Accurate Configurable Power-Simulator", *16th Brazilian Symp. Integrated Circuit Design (SBCCI 2003)*, Sep. 2003

9. J. M. O'Connor, M. Tremblat, "Picojava-I: the Java Virtual Machine in Hardware", *IEEE Micro*, vol. 17, n. 2, Mar-Apr. 1997, pp. 45-53
10. Sun Microsystems, *PicoJava-II Microarchitecture Guide*, Mar. 1999
11. J. Kreuzinger, R. Marston, Th. Ungerer, U. Brinkschulte, C. Krakowski, "The Komodo Project: Thread-based Event Handling Supported by a Multithreaded Java Microcontroller", *25th Euromicro Conference (EUROMICRO)*, Sep. 1999, pp. 2122-2128
12. N. Shimizu, M. Naito, "A Dual Issue Queued Pipelined Java Processor TRAJA-Toward an Open Source Processor Project", *Proceedings of Asia Pacific Conference on ASIC (AP-ASIC)*, 1999, pp. 213-216
13. J. L. Hennessy, D. A. Patterson, *Computer Architecture: A Quantitative Approach*, Morgan Kaufmann Publishers, 3th edition, 2003
14. V. Dalal, C. P. Ravikumar, "Software Power Optimizations in an Embedded System". *VLSI Design Conference*, IEEE Computer Science Press, Jan. 2001, pp. 254-259
15. K. Choi, A. Chatterjee, "Efficient Instruction-Level Optimization Methodology for Low-Power Embedded Systems". *International Symposium on System Synthesis*. Montréal, ACM, Oct. 2001, pp 147-152
16. R. Chen, M. J. Irwin, R. Bajwa, "Architecture-Level Power Estimation and Design Experiments". *ACM Transactions on Design Automation of Electronic Systems*, vol. 6, n. 1, Jan. 2001, pp 50-66
17. Leonardo Spectrum, available at homepage: http://www.mentor.com/synthesis
18. V. Gomes, A.C.S. Beck; L. Carro, "A VHDL Implementation of a Low Power Pipelined Java Processor for Embedded Applications". *X Workshop Iberchip*. Cartagenas, mar. 2004.
19. J. Pouwelse, K. Langendown, H. Sips, "Dynamic Voltage Scaling on a Low-Power Microprocessor", *The Seventh Annual International Conference on Mobile Computing and Networking*, 2001, pp. 251-259
20. Transmeta Corporation, *Tm5400 processor specifications*, http://www.transmeta.com
21. A.C.S. Beck, L. Carro, "A VLIW Low Power Java Processor for Embedded Applications", *17th Brazilian Symp. Integrated Circuit Design (SBCCI 2004)*, Sep. 2004

IMPACT OF GATE LEAKAGE ON EFFICIENCY OF CIRCUIT BLOCK SWITCH-OFF SCHEMES

Stephan Henzler[1], Philip Teichmann[1], Markus Koban[1], Jörg Berthold[2], Georg Georgakos[2], Doris Schmitt-Landsiedel[1]

1 Technical University of Munich, 2 Infineon Technologies

Abstract: Two different schemes to switch-off unused circuit blocks (ZigZag-cut-off scheme[1] and n-/p-block MTCMOS cut-off scheme[2,3,4,5,6] are examined in deep-submicron technologies by analytical investigation and simulation. The theoretical basis of the ZigZag-scheme is given and particular design constraints are discussed. It is shown that the power-saving benefits of the ZigZag-scheme are critically dependent on the gate-leakage, whereas n- or p-block switching keep their effectiveness. Finally it is derived that n-block switching tends to cause severe glitch activity during power-up process degrading both power-up-time and energy loss. The ZigZag-scheme however does not suffer from this effect. The advantages and drawbacks of the two schemes are compared depending on the available technology generation. Finally recent extensions to ZigZag are discussed.

Key words: MTCMOS, ZigZag, ZZSCCMOS, Circuit Block Switch-Off, Sleep-Transistor Scheme, Gate-Tunneling, GSCMOS

1. INTRODUCTION

Ongoing technology scaling increases speed and area efficency of digital systems, but transistor parasitics increase rapidly and effects up to now neglected gain strong influence on circuit and system design. Even in low-power devices off-currents increase by a factor of ten per technology node[7]. Consequently for SoC efficient power management strategies on transistor as well as on system level are required both in mobile and in high-end

Please use the following format when citing this chapter:

Henzler, Stephan, Teichmann, Philip, Koban, Markus, Berthold, Jörg, Georgakos, Georg, Schmitt-Landsiedel, Doris, 2006, in IFIP International Federation for Information Processing, Volume 200, VLSI-SOC: From Systems to Chips, eds. Glesner, M., Reis, R., Indrusiak, L., Mooney, V., Eveking, H., (Boston: Springer), pp. 229-245.

applications. Among others, a very promising concept is to turn-off unused circuit blocks by separating them from the supply (power-gating)[2,3,4,5,6].

Beside the task of finding the optimum switch dimension[8,9], there are a couple of possibilities to position the switch relatively to the logic[1,3]. A recently proposed strategy is the so called ZigZag-scheme that uses both an n-channel and a p-channel switch and assigns the different gates either to one switch or to the other one[1]. In this paper we describe the theoretical principles and technological suppositions of the ZigZag-scheme and derive the particular design rules for this scheme. Thereafter we discuss the different loss mechanisms for conventional power switching and ZigZag-scheme. The scaling of the different mechanisms is examined for state of the art deep-sub-micron technologies and extrapolated to future technologies.

Subsequently we discuss the behaviour of a circuit block during activation for conventional and ZigZag-schemes. Predictions about scaling limits and circuit behaviour are derived analytically and proven by simulation. This paper represents an example of technology dependent design: In deep sub-micron-CMOS not only transistor level but even system level design depends severely on technology properties. Therefore every technology generation will require its own adopted design strategy.

2. LEAKAGE-MECHANISMS IN DEEP-SUB-MICRON CMOS

At the beginning of the CMOS era standby power dissipation wasn't worth mentioning because of the high threshold voltage, long channels, thick gate oxides etc. During the last years reduction of the supply voltages required low threshold voltages in order to meet the performance requirements. The subthreshold current given by

$$I_S = I_0 \, exp\left(\frac{V_{GS} - V_{th}}{\eta V_T}\right)\left(1 - exp\left(-\frac{V_{DS}}{V_T}\right)\right); \qquad I_0 = \mu_{eff} C_{ox} V_T^2 (\eta - 1)\frac{W}{L} \qquad (1)$$

is exponentially dependent on the threshold voltage V_{th} and therefore becomes more and more important. Scenarios given by the ITRS-Roadmap[7] predict the subthreshold current, together with the gate-tunneling current covered next, to be the dominant mechanisms of power dissipation in future CMOS technologies. In a logic circuit with the effective gate width W_{eff}^n of all n-channel branches and W_{eff}^p of all p-channel branches, the average static power dissipation caused by subthreshold current can be estimated as:

$$P_{sub} = \frac{1}{2}\left(W^{n}_{eff} I'_n + W^{p}_{eff} I'_p\right) \cdot V_{DD}; \qquad I'_n = I'_{0_n} exp\left(-\frac{V^n_{th}}{\eta V_T}\right) \quad I'_p = I'_{0_p} exp\left(\frac{V^p_{th}}{\eta V_T}\right) \tag{2}$$

I_n' and I_p' describe the subthreshold currents of the n-channel and the p-channel transistors respectively normalized to the gate width. There are two characteristics of the subthreshold current that are important for this work: Although Eq. (1) indicates that for voltages $V_{DS} >> V_T$ the current is not a function of this voltage, the subthreshold current is exponentially dependent on V_{DS} because of the influence of Drain Induced Barrier Lowering (DIBL) on the threshold voltage in short channel devices:

$$V_{th} = V^0_{th} + V^1_{th} V_{DS}; \; V^1_{th} := \left.\frac{d}{dV_{DS}} V_{th}\right|_{V_{DS}=0} \tag{3}$$

The second property is the temperature dependence of the threshold voltage.

Another type of leakage which gains more and more importance, if gate oxide thickness is scaled further, is the leakage produced by the gate-tunneling current. In the general case it is very difficult to describe the flow of carriers over a tunneling barrier[10,11], but fortunately the tunneling in static CMOS logic is very simple, because apart from the switching process, there are only two states of transistor operation: Either the transistors are in linear region with a vanishing V_{DS} , or they are in subthreshold region. In the first case, there is a location-independent inversion channel. Thus the tunneling current is constant below the whole gate. The tunneling out of the source and drain overlap region respectively is different from the current density out of the channel, but because one single channel length is used the channel-to-overlap ratio is constant and the tunneling currents through the different transistor regions can be merged into an effective current density:

$$\begin{aligned} I_T &= \int_A j(x,y)\, dxdy \\ j_T &= \frac{\partial I_T}{\partial y} = \int j(x,y)\, dx \end{aligned}$$

A : Area under Gate

$j(x,y)$: local density of tunneling current

j_T : tunneling current per unit width (4)

In this context current density does not mean current per unit area but current per transistor width. The second important factor controlling the tunneling current is the potential drop across the gate oxide, which in static CMOS is given by the supply voltage V_{DD}-V_{SS}. For low voltages one effect of this potential drop is the impact on the shape of the potential barrier, but the more important one is the influence on the inversion charge density.

In the case of a transistor operating in subthreshold region, there is no inversion channel under the gate. The only leakage path is the drain-to-gate overlap region. Although there is a voltage drop of V_{DD}-V_{SS} across the barrier, the total current is smaller because of the reduced tunneling area.

3. CONVENTIONAL POWER-SWITCHING

With ongoing technology scaling, standby-power dissipation gets an important impact on the standby-time of mobile systems or the thermal power in high performance digital systems, respectively. Therefore different methods of switching-off unused circuits have been proposed. (MTCMOS = Multi Threshold CMOS combined with Circuit-Block-Switch-Off (CBSO)) Usually these schemes use one or more cut-off transistors that separate the unused logic block from the power-rails. Normally n-switches are preferable because of their lower on-resistance. The width of the switch transistors can usually be chosen much smaller than the total transistor width of the logic, therefore the total leakage is reduced. To reduce the leakage further, several techniques have been proposed in literature:

- high V_{th} switch-off-devices
- small switch width and boosting of switch device during active mode
- negative (positive) V_{GS} for switches in off-state (super-cut-off)[6]

If the leakage of the switch transistor is much lower than the total leakage of the logic, each node capacitance in the logic block will be charged (or discharged) by leakage to a potential near the unswitched power-rail. For instance switching-off a logic block by an n-channel device between V_{SS} and the circuit and waiting for a short time, each signal node and even the switched power-rail, which is referred to as virtual power-rail (V_{VSS}), reach the same potential slightly below V_{DD}. As a result all transistors operate in subthreshold region.

The upper bound of the remaining leakage current is determined by the subthreshold and gate-tunneling current of the switch device. For thick-oxide-switches the gate-tunneling current is negligible and the standby-power dissipation is given by

$$P_{standby}^{CBSO} = I'_{0S}\, exp\left(-\frac{V_{th}^{switch}}{\eta V_T}\right) W_{switch} \cdot V_{DD} \tag{5}$$

To describe the gain in standby power dissipation we introduce the leakage-reduction-ratio (LRR), which is the ratio between the total leakage in active and in inactive mode:

$$LRR = \frac{\frac{1}{2}\left(W^n_{eff}I'_n + W^p_{eff}I'_p\right) + I^{logic}_{gate}}{I'_S W_{switch}}$$
$$I'_S = I'_{0S}\, exp\left(-\frac{V^{switch}_{th}}{\eta V_T}\right) \qquad (6)$$

It is important to notice that there is no significant gate-tunneling current in a cut-off logic block because a few moments after switching-off the circuit, each node within the block has the same potential.

Summarizing: All schemes cutting-off the V_{DD} or V_{SS} power-rail (n-block or p-block-switching respectively) reduce the standby power significantly and suppress additional gate leakage paths completely. Therefore these schemes can be called gate-leakage tolerant strategies and will therefore also be applicable in future technologies with significantly higher gate leakage.

4. PRINCIPLE OF ZIGZAG-SCHEME

The main disadvantage of n- and p-block switching respectively is that all nodes are charged or rather discharged after switching-off the circuit. After turning on the system again, approximately half of the nodes have to be discharged to a logical zero and the rest of the nodes have to be recharged to a logical high state. This process requires not only energy, but also a significant power-on-time which contributes directly to the minimum power-down time, i.e. the minimum period of time for which it is useful to suspend a block by cutting-off the power supply. However, to minimize total leakage losses the power-on time should be as short as possible. The idea finally leading to the ZigZag-scheme[1], is to reduce the voltage swing of the virtual power rail, to maintain the system state at each logical node and not to increase total leakage over the switch. All this is possible by using both n-block and p-block switches and assigning each gate to one of these switches in a sophisticated way.

Assume a one-stage static CMOS gate with a particular logic level at its output. If this gate is assigned to the switch that cuts-off the power rail complementary to the output level, the logic level of the circuit will be conserved and the output capacitance does not have to be recharged. Furthermore the transistors, which operate in subthreshold region, are subject to the stack effect as they are in series with the switch transistor, i.e.

they get a negative V_{GS} and a reduced V_{DS} compared to the active-mode. In contrast to n- or p-block switching, even in off-state there is a non-vanishing V_{DS} voltage of the transistors in the switched block. Thus the voltage swing of the virtual power-rails is reduced. The strategy of assigning each gate to the proper cut-off switch can be verbalized as follows:

1. Assure fixed valid logic levels at all inputs during the power-down phase
2. Calculate the level of each node as if the circuit was not switched-off
3. Analyze each gate: If the output is high, assign the corresponding gate to the n-switch. If the output is low assign the gate to the p-switch.

The benefits of the ZigZag-Scheme depend on the strong monotonicity of subthreshold current around V_{GS}:

$$\left.\frac{\partial I_D}{\partial V_{gs}}\right|_V > 0 \quad \text{for each} \quad V \in K_\epsilon(0) \tag{7}$$

5. INPUT CIRCUITRY FOR ZIGZAG-SCHEME

The theory of the ZigZag-Scheme states that the cut-off switch should be in series to the non-conducting transistor group (n-block or complementary p-block). Indeed this includes that complex gates are allowed, but only those that have a single stage structure. Multi-stage gates would require that the first stage is assigned to one and the following stage to the other switch. This is not practical for library based design flows.

The main difficulty is that for the switch assignment the logical state has to be known for the whole block, especially for the input vector. Thus in most cases it will be necessary to set this condition explicitely before the system can be switched off. Unfortunately this state-forcing procedure requires additional energy and therefore the minimum power-down time increases. We propose four different methods to assure the system state:

1. Driving each input with a resetable flipflop the output stage of which is not switched-off[1].
2. Input multiplexers: Inserting additional gates (NAND or NOR gates for instance) that in dependence of a control signal at the first input transmit either the second input or a fixed level to the output. This method means some area overhead and a slightly longer critical path. Another drawback

is the high fan-out control signal that has to be valid during power-down period.

3. Special Input gates: These gates disconnect the input and assure the required logic level by an additional state-keeper transistor. Slight area and delay penalties result from this method.
4. Additional power rails: The idea of this method is to use additional power rails for the first logic stage. The gates of this stage do not operate in ZigZag-mode. By selecting a n- or p-switch their output level can be ensured without knowledge of the input vector. Anyway, this level does not correspond exactly to any supply potential as the output node is floating. Because of timing reasons this strategy means a serious area overhead.

It should be mentioned that the inner state of a circuit block before cut-off has to be known during design. Hence ZigZag scheme cannot be applied if the inner state has to be conserved by state retention flip-flops. Special state retention flip-flops forcing a certain output level during sleep mode and switching to the conserved level after activation can be used but result in additional switching activity. The additional energy overhead of this switching process increases the minimum sleep time.

6. POWER CONSUMPTION IN ZIGZAG-CBSO

As mentioned above, in a purely n-block or p-block switched system, respectively, the residual leakage is given by the properties of the switch transistor only:

$$I_{standby}^{n-CBSO} = W_s I_s' \tag{8}$$

In a ZigZag environment all logic nodes retain their respective logic levels, and therefore about half of all transistors operate in triode region, where there is a channel under the total gate area. The potential drop over their gate oxides is equal to the supply voltage V_{DD}-V_{SS}. Thus, conditions for maximum gate-leakage currents are given, according to section 2.

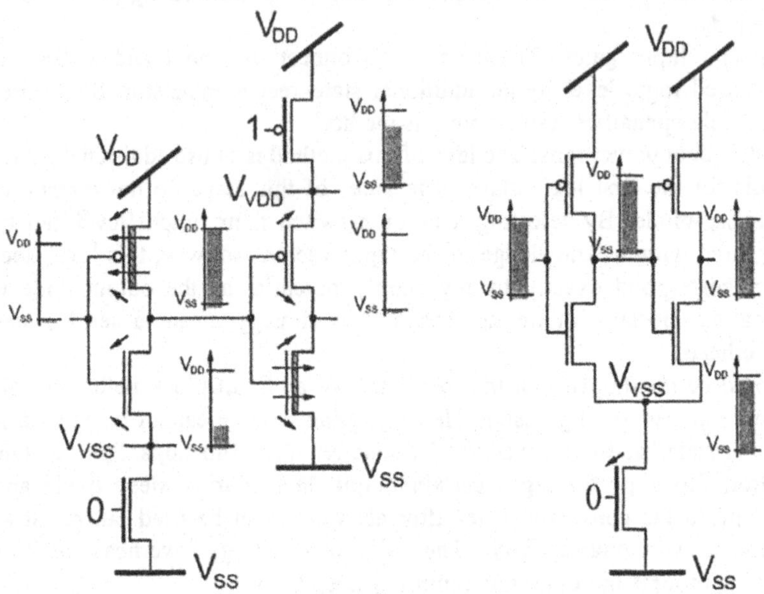

Figure 1. Principle of ZigZag-scheme (left) and n-block sleep transistor scheme (right) with node potentials and gate-tunneling paths.

The different tunneling paths are depicted in Figure 1. Consequently, in a ZigZag-environment the non-vanishing gate currents represent an additional leakage mechanism and in spite of the stack effect one has to expect an increased total leakage compared to an n- or p-block switching scheme:

$$\begin{aligned} I_{standby}^{ZigZag} &= W_S I'_{Sn} r + \frac{1}{2}\alpha \left(j_T^n W_{eff}^n + j_T^p W_{eff}^p \right) \\ r &:= \omega r_n + (1-\omega) \frac{I'_{Sp}}{I'_{Sn}} r_p \\ r_n &:= exp\left(\frac{V_{th,Sn}^1}{\eta V_T} (V_{DD} - V_{VSS}) \right) \\ r_p &:= exp\left(-\frac{V_{th,Sp}^1}{\eta V_T} V_{VDD} \right) \end{aligned} \tag{9}$$

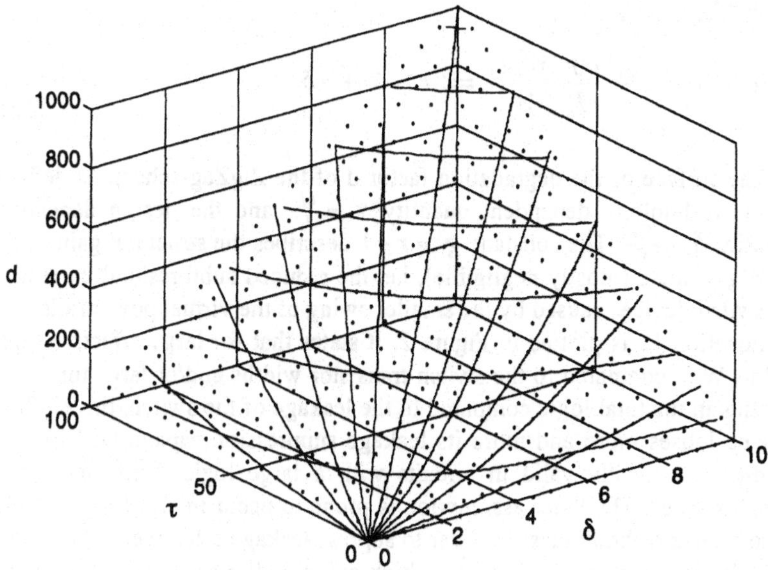

Figure 2. Dependence of degradation factor d on the design parameter δ and the technology parameter τ.

r is a correction because of the usage of two different types of switch transistors with width ratio

$$\frac{W_s^n}{W_s^p} =: \frac{\omega}{1-\omega} \tag{10}$$

and r_n and r_p respectively describe the reduced subthreshold current of the switches caused by the reduced swing of the virtual power-rails. α is an empirical factor describing the fact that the tunneling current is determined not by the effective gate width but by the total gate area of the active transistors.

The relative difference between the two schemes is given by

$$\begin{aligned} d &:= \frac{\Delta I_{standby}}{I_{standby}^{n-CBSO}} \\ &= r - 1 + \frac{\alpha}{2W_s I_s'} \left(j_T^n W_{eff}^n - j_T^p W_{eff}^p \right) \end{aligned} \tag{11}$$

Using the approximation $W_{eff} := W_{eff}^{n} = W_{eff}^{p}$ and substituting $j_T := \frac{1}{2}\left(j_T^{n} - j_T^{p}\right)$:

$$d = r - 1 + \frac{\alpha j_T}{I_s'} \frac{W_{eff}}{W_s} =: r - 1 + \tau\delta \tag{12}$$

The surface of the degradation factor d of the ZigZag-scheme is defined by the technology dependent quantity $\tau := \frac{\alpha j_T}{I_s'}$ and the design dependent quantity $\delta = \frac{W_{eff}}{W_s}$. The constant $g := r - 1$ describes the structural gain of the ZigZag-scheme (mostly negligible), i.e. the reduced subthreshold current of the switch devices caused by the smaller swing of the virtual power rails.

Equation 12 is plotted in Figure 2. It states that for large effective logic widths W_{eff} compared to the switch transistor widths and/or large tunneling currents in the total logic compared to the leakage of the switch, the ZigZag-scheme looses more and more its leakage suppression capability. The first condition is mostly valid in circuits with a large logic depth and many complex gates. The latter assumption is prone to occur in sub 100 nm high-performance technologies: In order to supress leakage effectively, the switch normally is made up of a high-V_{th}, thick oxide and/or super-cut-off[6] device. The logic however consists of leaky low or ultra-low-V_{th} devices with thin oxides. The following table shows a scenario for τ depending on transistor parameters specified by the ITRS-Roadmap:

Table 1. Efficiency of ZigZag scheme in future technologies predicted by ITRS-Roadmap

YEAR	2001	2002	2003	2004	2005	2006
L [nm]	90	75	65	53	45	40
I_S' @ 25C [pA/µm]	1	1	1	1	1	1
EOT [nm]	1.45	1.35	1.3	1.15	1.05	0.95
j_T [10^3 pA/µm]	4.64	14.4	25.6	144	464	1200
$\tau/\alpha = j_T / I_S'$ [10^3]	4.64	14.4	25.6	144	464	1200

7. POWER-ON PROCESS

The behaviour of the ZigZag-scheme during the power-on process differs significantly from the n- or p-block switching. As mentioned above in an n-block switching scheme each node is charged to a high potential. If the cut-off switch turns on, the V_{VSS} potential will break down quickly. Therefore both the gate-to-source and the drain-to-source voltages of all n-channel devices increase, subthreshold current increases exponentially and all logic nodes start discharging. Consequently the corresponding voltages of the p-channel-devices decrease causing an increased short-circuit current. When the V_{VSS} potential drops further the gate-to-source voltages of some n-devices reach the threshold voltage and the node capacitances connected to the corresponding gates discharge even faster. If their voltage is low enough the n-devices of the following stage turn off and these stages charge their outputs to V_{DD}. Although this process means additional energy loss, the main problem is the following: At the beginning each node and each gate have the same state. The size of the different capacitances as well as the drivability of the connected gates determine how fast the different nodes are discharged and therefore whether a particular node is discharged completely or recharged. However the final state is determined by the input vector and the logical topology of the circuit. Consequently there exists a significant glitch probability because of the two competing mechanisms described above. These glitches cause additional dynamic losses and enlarge the power-on time significantly. Both phenomena increase the minimum power-down time and debase the applicability of the scheme.

In the ZigZag-scheme however, each logic node keeps a valid level compatible to the topology of the logic. Therefore, if the system is switched on while the same input vector is applied that was forced during the power-down phase by the input circuitry, no glitches will appear. Furthermore, no logic nodes need to be charged or discharged. Consequently switching on a circuit block according to the ZigZag-scheme ensures significantly reduced power-on time.

8. SIMULATION RESULTS AND EXAMPLES

In order to verify the analytical considerations we chose two test circuits commonly used in SoC: A 5-bit carry-look-ahead adder and a 8x8-bit booth(2) multiplier. The circuits are given in Figure 3 and Figure 4, respectively.

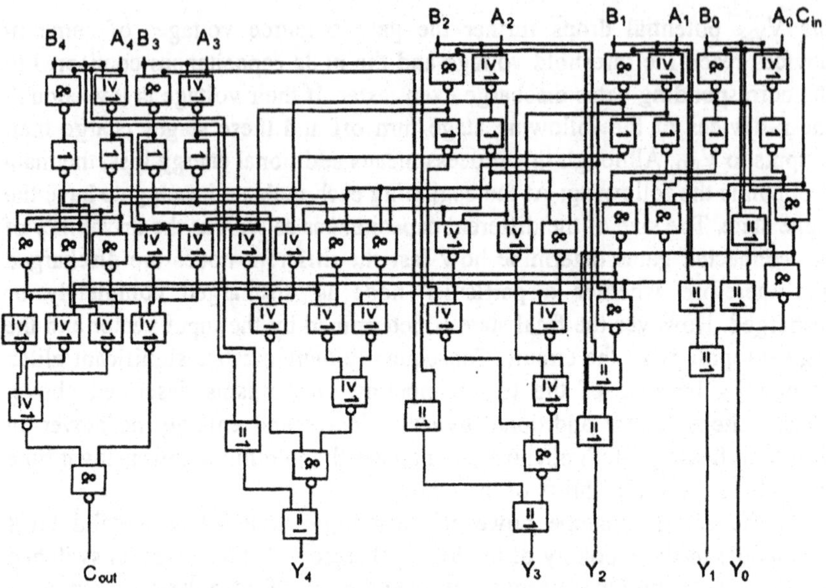

Figure 3. 5 Bit Carry Look Ahead Adder as first testbench

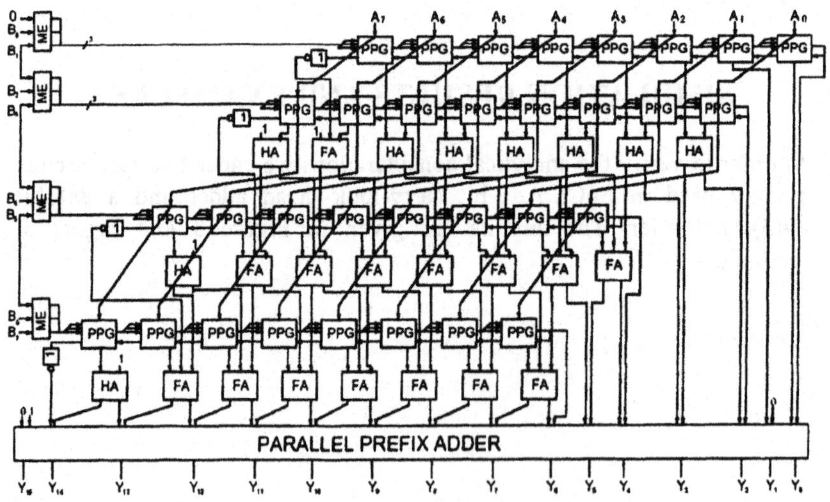

Figure 4. 8x8 Bit Booth(2) Multiplier as second testbench

We simulated both circuits assuming a gate tunneling current density of $j_T = 1.7 \cdot 10^3 \frac{pA}{\mu m}$ and a subthreshold current of $I_S' = 8.3 \frac{pA}{\mu m}$. Thus we get $\tau = 205.8$. The ratio of the logic width to the width of the switch is set to 5 in the case of the adder and 10 for the multiplier. Consequently we have to expect a degradation factor $d_{adder} = 1029$ or $d_{multiplier} = 2058$, respectively. The simulations are carried out using SPICE and BSIM 4.2 MOSFET-Model. To examine the influence of the tunneling currents the appropriate switches in the model (igcMod, igbMod) are either turned on or off.

Figure 5a shows the simulated leakage current of the adder over temperature for n-block-switching, p-block-switching and ZigZag-scheme. As expected the subthreshold current shows an exponential temperature dependence, whereas the gate tunneling current has little temperature dependence. Nevertheless in ZigZag-scheme the gate tunneling current is significantly larger than the subtreshold current even at high temperatures. In the n-block or p-block environment, leakage current varies only slightly if gate-tunneling is turned on or off. The slight difference is caused by the small tunneling current over the drain-to-gate overlap region of the switch transistor (thick gate oxide). The ZigZag implementation of the adder uses extra virtual power-rails for the first logic stage. Therefore a slight difference exists between the results in equation 11 and the simulation. For large circuits however or by using another strategy for the first stage this difference is negligible.

Fig 5b shows the simulation results for the multiplier. In ZigZag-scheme the first stage is driven by resetable flip-flops, so no extra power-rails are necessary. The subthreshold current varies exponentially over temperature and differs only slightly from total leakage in the n- or p-block switching scheme. In the ZigZag case there is a huge difference between subthreshold current of the switch and the total leakage. The strong dependence of the degradation factor d on the ratio τ of the two leakage components states that even if the gate current in BSIM was modelled inaccurately the prediction about the decreasing ZigZag benefits in future technologies would hold. Therefore, analytical examination and simulation are sufficient to derive this dependency, without a testchip manufactured in silicon.

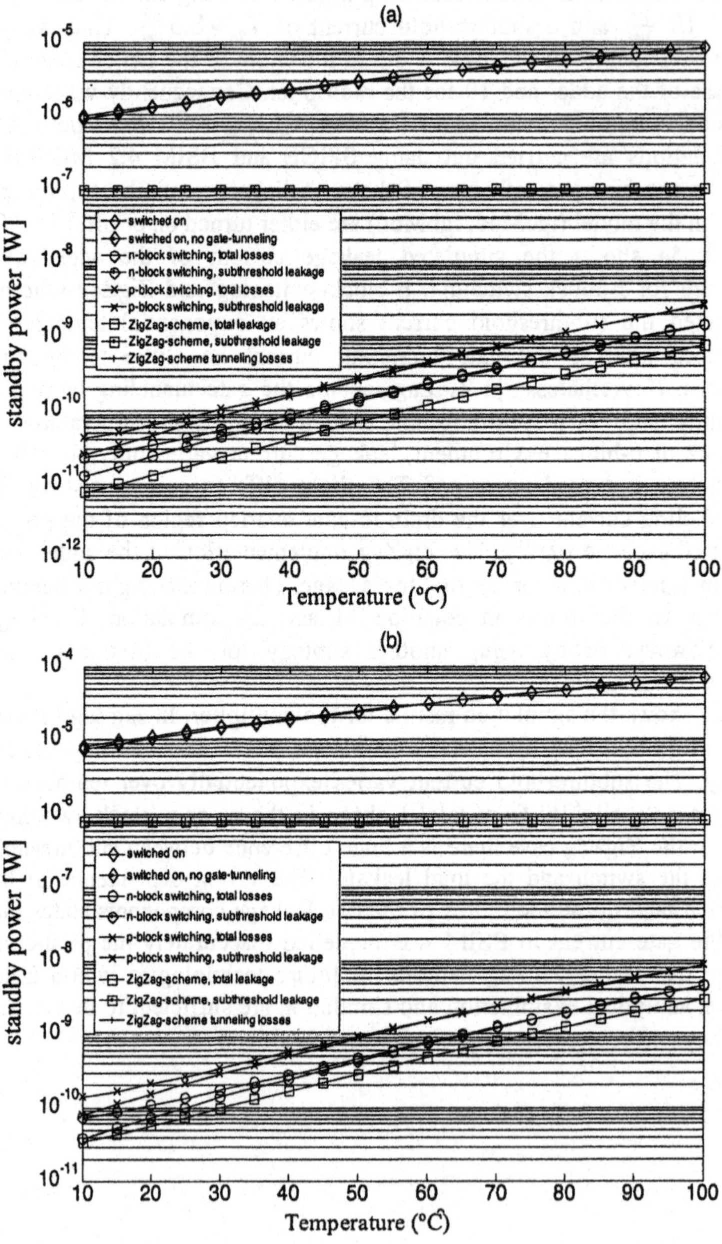

Figure 5. Leakage losses over temperature of the different cut-off schemes for the adder (upper) and the booth(2)-multiplier (lower)

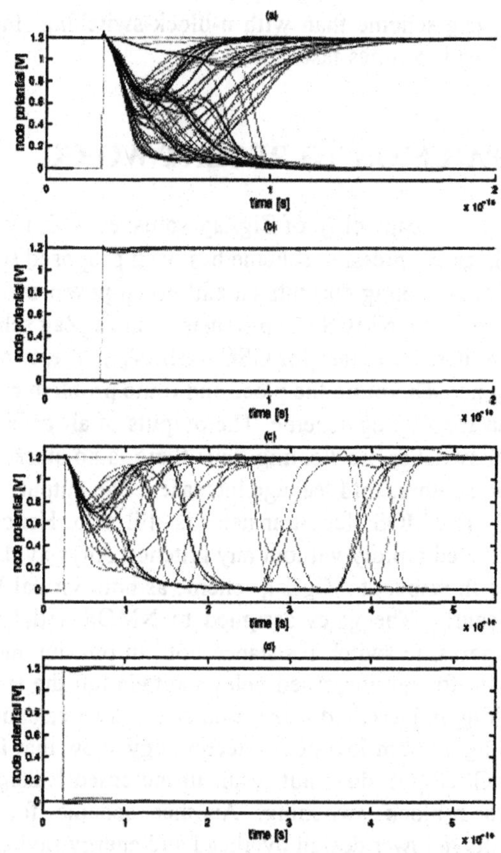

Figure 6. **Behaviour of inner nodes during power-on process for ZigZag and n-block circuit block switch-off scheme: (a) adder with n-block sleep-transistor, (b) adder according to ZigZag scheme, (c) n-block switched booth multiplier, (d) multiplier according to ZigZag scheme.**

Next we examined the behaviour of all signal nodes during the turn-on phase. Figure 6 a,b show the inner node potentials of the adder in the case of n-block-switching and ZigZag. As predicted, in the ZigZag case there are absolutely no glitches or switching processes, whereas in the n-block switched adder there exist many power- and time-consuming glitch activities. The multiplier has a considerably deeper logic structure. Therefore even more severe glitches have to be expected. Figure 6 c, d show some node potentials for the two switching strategies. Again many glitches can be observed in n-block scheme and none in ZigZag-scheme. Comparing the power-on times these simulations show that the adder can be reactivated 4 times faster in ZigZag-scheme than with n-block-switching. In the case of the multiplier, ZigZag is 5 times faster.

9. COMPARISION TO RECENT WORK

As the power saving capability of ZigZag vanishes with increasing gate leakage a gate leakage suppression scheme has been proposed (GSCMOS)[12]. To suppress the gate tunneling currents an additional power switch is added to the gates assigned to the NMOS sleep-transistor in ZigZag scheme. Hence there are five power nets necessary for GSCMOS: V_{DD}, V_{SS}, two virtual V_{DD} and one virtual V_{SS} net. Therefore the place and route problem becomes even more complex than for ZigZag scheme. The outputs of all gates assigned to PMOS and NMOS switches are floating, thus there is no stack effect in the following stage and subthreshold leakage increases. The output voltage of all gates assigned only to PMOS sleep-transistor is still zero. Hence the circuit block can be reactivated rapidly without any glitches. Anyway the activation time is increased with respect to ZigZag scheme as both virtual V_{DD} nets are discharged significantly. The gates assigned to NMOS and PMOS sleep-transistors have a parasitic switch resistance both in pull-up and pull-down path. To compensate for the increased delay degradation the switches have to be upsized resulting in increased leakage currents and area consumption.

Simulations using a 90nm low-power technology show that for the same delay degradation GSCMOS does not result in increased leakage reduction ratio compared to n-block switching. Another scheme for fast block activation that results in lower design overhead and energy savings similar to n-block switching has been proposed in [13]. Glitches are avoided by turning on two distinct groups of gates with slight latency.

10. CONCLUSIONS

Two methodologies to switch-off unused circuit blocks have been examined and compared. It has been shown that ZigZag-scheme will loose more and more its leakage suppression capability if technologies with significant gate leakage are used. n- or p-block switching keeps its leakage suppression property but suffers from considerable glitches during power-up phase. Thus it depends on the application and the average power-down time whether ZigZag or n-block switching is preferable: If there are only very short periods during which the circuit block is not needed, ZigZag will be preferable. In modern high performance technologies however the leakage suppression of this scheme is weak. If the power-down periods are long and the times of usage of the block can be estimated well, the reactivation time will be of inferior significance. Thus n- or p-block switching are the preferable schemes because of their excellent leakage supression property.

REFERENCES

1. K-S. Min, H. Kawaguchi, T. Sakurai et al., Zigzag Super Cut-off CMOS Block Activation with Self-Adaptive Voltage Level Controller: An Alternative to Clock-Gating Scheme in Leakage Dominant Era, *ISSCC 2003*
2. R. K. Krishnamurthy, A. Alvandpour, S. Mathew, M. Anders, V. De, S. Borkar et al., High-performance, Low-power, and Leakage-tolerance Challenges for Sub-70nm Microprocessor Circuits", *ESSCIRC* (2002)
3. S. Mutoh, T. Douseki, Y. Matsuya, T. Aoki, S. Shigematsu, J. Yamada et al., 1-V Power Supply High-Speed Digital Circuit Technology with Multithreshold-Voltage CMOS, *JSSC*, **30**(8) (1995)
4. B. Calhoun, F. Honoré, A. Chandrakasan, A Leakage Reduction Methodology for Distributed MTCMOS, *JSSC* **39** (5) (2004)
5. J. Tschanz, S. Narendra et al., Dynamic Sleep Transistor and Body Bias for Active Leakage Power Control of Microprocessors, *JSSC* **38** (11) 1838-1845 (2003)
6. H. Kawaguchi, K. Nose, T. Sakurai, A Super Cut-Off (SCCMOS) Scheme for 0.5-V Supply Voltage with Picoampere Stand-By Current, *JSSC* **35** (10) 1498-1501 (2000)
7. S.I.A., ITRS, International Technology Roadmap for Semiconductors, 2002
8. J. Kao, A. Chandrakasan, D. Antoniadis, Transistor Sizing Issues and Tool for Multi-Threshold CMOS Technology, *DAC* (1997)
9. St. Henzler, G. Georgakos, J. Berthold, D. Schmitt-Landsiedel, Two Level Compact Simulation Methodology For Timing Analysis Of Power-Switched Circuits, PATMOS 789-798, (2004).
10. W-C. Lee, C. Hu et al., Modeling CMOS Tunneling Currents Through Ultrathin Gate Oxide Due to Conduction- and Valence-Band Electron and Hole Tunneling, *IEEE Transactions on Electron Devices*, **48**(7), (2001)
11. J.W. Fattaruso, Buss et al., Current Trends in analog Design, *Short Course on System-on-a-Chip, ISSCC 2003*
12. M. Drazdziulis, P. Larsson-Edefors et al., A Power Cut-Off Technique for Gate Leakage Suppression, *ESSCIRC* (2004)
13. St. Henzler, G. Georgakos, J. Berthold, D. Schmitt-Landsiedel, A Fast Power-Efficient Circuit-Block Switch-Off Scheme, *IEE Electronics Letters*, **40**(2) 103-104, (2004)

EVALUATION METHODOLOGY FOR SINGLE ELECTRON ENCODED THRESHOLD LOGIC GATES

Casper Lageweg, Sorin Cotofana and Stamatis Vassiliadis
Computer Engineering Laboratory
Faculty of Electrical Engineering, Mathematics and Computer Science
Delft University of Technology, Delft, The Netherlands
{Casper,Sorin,Stamatis}@CE.ET.TUDelft.NL

Abstract Single Electron Tunneling (SET) is an emerging technology, with a switching behavior which is completely different from MOS technology. The ability to control the transport of individual electrons within SET circuits creates the conditions for Single Electron Encoded Logic (SEEL). Although it is expected that, when compared with other approaches, SEEL circuits have both reduced delay and reduced energy consumptions, a method for evaluation is required. This paper proposes a methodology to evaluate delay, power consumption, maximum fanin, and maximum fanout for buffered SEEL linear threshold gates. Furthermore, we discuss the implications of the proposed methodology on practical networks of such gates. We estimate that buffered threshold gates operating at room temperature can potentially switch with a delay of 6 ps and have a packing density of 10^9 gates per cm^2.

Keywords: SET, single electron technology, single electron encoded logic, threshold gates.

1. Introduction

During the last six decades we have witnessed spectacular increases in the processing power of logic and arithmetic circuits. Since the seventies the microelectronics industry has followed Moore's "law" [Moore, 1965], doubling processing power every 18 months. This increase in processing power can to a large extend be contributed to advances in algorithms and device technology [Hennessy and Patterson, 1996]. Focussing on device technology, one can observe that the feature size reduction in microelectronic circuits, and the corresponding increase in the number of transistors per cm^2, has been one of the main contributing factors to this dramatic increase.

Currently, circuit technology is primarily based on (C)MOS devices and it is anticipated that CMOS feature sizes will continue to be scaled down in the near future. The 2003 edition of the International Technology Roadmap for

---

Please use the following format when citing this chapter:
Lageweg, Casper, Cotofana, Sorin, Vassiliadis, Stamatis, 2006, in IFIP International Federation for Information Processing, Volume 200, VLSI-SOC: From Systems to Chips, eds. Glesner, M., Reis, R., Indrusiak, L., Mooney, V., Eveking, H., (Boston: Springer), pp. 247-262.

Semiconductors [TRS03, 2003] predicts that in 15 years ultra-thin body (UTB) MOSFETs will reach gate lengths of 10 nm. However, it also predicts that such devices will increasingly operate in a quasi-ballistic mode, i.e., with electrical currents dwindling down to individual electrons. Note that this is already occurring in today's circuits in the form of undesired tunnel currents through thin oxides [Brennan and Brown, 2002, Iwai and Momose, 1998, Momose et al., 1998]. Consequently, the switching behavior of future MOSFET devices will greatly differ from MOSFET behavior in the traditional sense. On the other hand, it is generally accepted that sooner or later MOS based circuits cannot be reduced further in (feature) size due to fundamental physical restrictions [Taur et al., 1997].

Given the anticipated end of the CMOS era, several candidate technologies based on alternative operating principles have been under investigation for the last two decades. These candidate technologies include, amongst others, Single Electron Tunneling (SET), Carbon Nanotubes, Rapid Single Flux Quantum (RSFQ), Resonant Tunneling Diodes (RTD), and Magnetic Spin devices (see [TRN99, 1999] for an overview of some of these emerging technologies). As of yet it is undecided which of these technologies, if any, will supplant (C)MOS. If a candidate technology is to replace (C)MOS in at least some application areas, it must satisfy as many of the following criteria as possible. First, the candidate technology should have greater feature size scaling potential then CMOS, i.e., larger device density. Second, given the anticipated device density the candidate technology should have extremely low power consumption. Note that power consumption is one of the main problems anticipated for (C)MOS in the near future [TRS03, 2003]. Third, it should be able to operate at room temperature as liquid cooling would encumber potential systems with added cost and bulk. Fourth, device switching should occur sufficiently fast, such that when the device is used in conjunction with the appropriate design techniques and architectures to build circuits and systems it can compete or outperform "equivalent" (C)MOS based designs.

As an emerging technology the SET technology [Korotkov, 1999] [Likharev, 1999] offers a number of advantages as follows. First, it offers a greater scaling potential than (C)MOS as the device structure is less complex. Second, SET has the potential to realize circuits that consume much less power than (C)MOS circuits. Third, recent advances in silicon based fabrication technology (see for example [Ono et al., 2000]) indicate that SET has the potential to operate at room temperature. An additional benefit of the SET technology is that SET quantum tunnel junctions, the basic SET circuit element, can be fabricated in many different ways. In order to illustrate the variety in possible implementation technologies, we present in Figure 1 two possible implementations of the SET inverter [Tucker, 1992]. Figure 1(a) depicts an SET inverter fabricated in a conventional lithographic technology on silicon [Heij

(a) Conventional process.

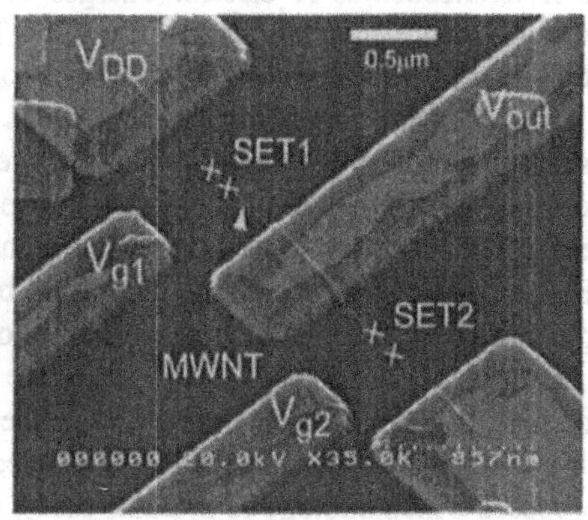

(b) Carbon nanotube.

Figure 1. SET inverter implementations.

et al., 2001]. In this case the tunnel junctions resemble conventional capacitors and consist of small gaps between conducting plates. Figure 1(b) on the other hand depicts a carbon nanotube based implementation [Ishibashi et al., 2001] in which case the tunnel junctions consist of small gaps in a multiwall carbon nanotube.

Similar to other future technology candidates, SET devices display a switching behavior that differs from traditional MOS devices. This provides new possibilities and challenges for implementing digital circuits. Currently, there are three design styles in creating SET-based logic circuits. The first approach is to implement the SET equivalent of the MOS transistor (see for example [Tucker, 1992]) an mimic the (C)MOS design style. The main disadvantage of this approach is that the current transport through an "open" transistor still comprises a large number of individual electrons "dripping" through the tunnel junctions. This results in increased delay and power consumption. The second approach is to design SET Boolean logic gates that operate according to the Single Electron Encoded Logic (SEEL) paradigm, i.e., charge transport due to switching activity is limited to a single electron. Given that this results in the transport of fewer electrons, it is expected that SEEL circuits have both reduced delay and reduced energy consumption. Earlier investigations revealed that (buffered) SEEL linear threshold logic gates can be constructed based on the Coulomb blockade of SET tunnel junctions [Lageweg et al., 2001a] and that such gates operate correctly in larger networks [Lageweg et al., 2002]. A third approach is to design high radix logic circuits in which a non-Boolean value X is encoded as X electron charges and implement electron counting based logic [Cotofana et al., 2003]. Currently electron counting based implementations still require additional amplification, i.e., OP-AMPs, for which a SET tunnel junction based implementation has not yet been found.

Comparing the above three design styles, we conclude that the SEEL based approach appears to be the most promising due to the limited amount of transported charge. Moreover, previous research has demonstrated that networks of SEEL threshold gates can be implemented by only utilizing SET circuit elements [Lageweg et al., 2002]. However, in order to compare this approach with others, a methodology for estimating the delay and power consumption is required. In order to evaluate the SEEL linear threshold gate, as well as to provide a means for evaluating other single electron logic approaches, this paper proposes a methodology to evaluate delay, power consumption, maximum fanin, and maximum fanout for buffered SEEL linear threshold gates.

The remainder of this paper is organized as follows. Section 2 introduces the SET theory. Section 3 describes the generic SET linear threshold gate and the SET buffer, which are combined as a generic SEEL buffered threshold gate. In Section 4 methods for analyzing the delay, the power consumption, the fanin and the fanout of buffered threshold gates operating in a network are proposed. Section 5 applies the proposed methods to evaluate parameter asymptotic bounds for practical SEEL threshold logic networks. Section 6 concludes the paper with some final remarks.

2. Background

The Single Electron Tunneling (SET) technology introduces the quantum tunnel junction as a new circuit element. A tunnel junction consist of two conductors separated by an extremely thin insulating layer. The insulating layer acts as an energy barrier which inhibits charge transport under normal (classical) physics laws. However, according to quantum physics theory, charge transport of individual electrons through this insulating layer can occur if this results in a reduction of the total energy present in the circuit. The transport of charge through a tunnel junction is referred to as *tunneling*, while the transport of a single electron is referred to as a *tunnel event*. Electrons are considered to tunnel through a tunnel junction strictly one after another.

Rather then calculating for each tunnel junction if a hypothetical tunnel event results in a reduction of the circuit's energy, we can calculate the critical voltage V_c, which is the voltage threshold needed across the tunnel junction to make a tunnel event through this tunnel junction possible. For calculating the critical voltage of a junction, we assume that the tunnel junction has a capacitance of C_j and that the remainder of the circuit, as viewed from the tunnel junction's perspective, has an equivalent capacitance of C_e. Given the approach presented in [Wasshuber, 1998], we calculate V_c for the junction as

$$V_c = \frac{q_e}{2(C_e + C_j)}. \tag{1}$$

In the above equation, as well as in the remainder of this discussion, we refer to the charge of the electron as $q_e = 1.602 \cdot 10^{-19}\ C$. Strictly speaking this is incorrect, as the charge of the electron is of course negative. However, it is more intuitive to consider the electron as a positive constant for the formulas which determine if a tunnel event will take place or not. We will of course correct for this when we discuss the direction in which the tunnel event takes place.

Generally speaking, if we define the voltage across a junction as V_j, a tunnel event will occur through this tunnel junction if and only if $|V_j| \geq V_c$. If tunnel events cannot occur in any of the circuit's tunnel junctions, i.e., $|V_j| < V_c$ for all junctions in the circuit, the circuit is in a *stable state*. For our research we focus on circuits where a limited number of tunnel events may occur, resulting in a stable state. Each stable state determines a new output value resulting from the distribution of charge throughout the circuit.

Given that electron tunneling as described by the orthodox tunneling theory (see for example [Wasshuber, 1998] for a more extensive introduction) is a quantum physical process, the transport of individual electrons can only be described by a stochastic process. This implies that we can at most calculate the chance that an electron has tunneled through a tunnel junction after a time interval t_d. In other words, there will always be a non-zero switching error probability P_{error}. In addition to the switching error probability there are two fundamental phenomena that may cause errors: thermal tunneling and cotunneling. Thermal tunneling events are tunnel events due to thermal agitation. Cotunneling events are two tunnel events that reduce the total amount energy of the system when they occur *simultaneously*, while the individual tunnel events each increase the total amount of energy present in the system. Given a maximum acceptable switching error probability, we must ensure that the thermal error probability as well as the cotunneling error probability are of the same order of magnitude or less.

The thermal error probability can be calculated as follows. For any temperature $T > 0K$ there exists a non-zero probability that a tunnel event will occur through a junction (even if $|V_j| < V_c$). The error probability P_{therm} due to *thermal tunneling* can be described by a simple formula as

$$P_{therm} = e^{-\Delta E/K_b T}, \tag{2}$$

where k_b is Boltzman's constant ($k_b = 1.38 \cdot 10^{-23} J/K$) and ΔE is the change in the energy as a results of the tunnel event. For practical purposes, this implies that, if we intend to add or remove charge to a circuit node by means of tunnel events, the total capacitance attached to such a circuit node must be less then $900aF$ for $1K$ temperature operation, or less then $3aF$ for $300K$ (room temperature) operation [Goossens, 1998]. This represents a major SET fabrication technology hurdle as even for cryostat temperature operation very small circuit features are required to implement such small capacitors.

For a multi-junction system in which a combination of tunnel events leads to a reduction of the energy present in the entire system there exists a non-zero probability that those tunnel events occur simultaneously (even if $|V_j| < V_c$ for all individual tunnel junction involved). This phenomenon is commonly referred to as *cotunneling* [Averin and Odintsov, 1989, Averin and Nazarov, 1990]. Although a detailed analysis of cotunneling is outside the scope of the present work, we remark that the best approach for reducing the cotunneling error probability appears to be the addition of strip resistors between the SET circuit and the supply voltage lines, as demonstrated in [Lotkhov et al., 1999, Zorin et al., 2000, Lotkhov et al., 2001]. This method can reduce the cotunneling rate without significantly increasing the switching delay. This is due to the fact that the delay added by a resistor is on the RC scale. Thus, assuming for example $R = O(10^6)\ \Omega$ and $C = O(10^{-18})\ F$, we find that the delay added by the resistor is $t_{RC} = O(10^{-12})\ s$. As these such RC values switching delay of tunnel events is typically at least one order of magnitude larger, the additional delay due to the cotunneling suppressing resistors would be negligible. Although the circuits discussed in the remainder of this paper do not contain such resistors, cotunneling suppressing resistors of appropriate value can be appended to the designs in order to reduce the cotunneling error to an acceptable error probability.

The next section introduces the SEEL linear theshold gate, which serves as an example circuit for the type of SEEL circuits which our proposed methodology evaluates in terms of delay, power consumption, fanin and fanout.

3. Single Electron Encoded Threshold Logic Gate

Threshold logic gates are devices which are able to compute any linearly separable Boolean function given by:

$$Y = sgn\{\mathcal{F}(X)\} = \begin{cases} 0 & \text{if } \mathcal{F}(X) < 0 \\ 1 & \text{if } \mathcal{F}(X) \geq 0 \end{cases} \tag{3}$$

$$\mathcal{F}(X) = \sum_{i=1}^{n} \omega_i x_i - \psi \tag{4}$$

where x_i are the n Boolean inputs and w_i are the corresponding n integer weights. The linear threshold gate performs a comparison between the weighted sum of the inputs $\Sigma_{i=1}^{n} \omega_i x_i$ and the threshold value ψ. If the weighted sum of inputs is *greater than or equal to* the threshold, the gate produces a logic

1. Otherwise the output is a logic 0. Note that threshold logic gate networks are fundamentally more powerful than networks of standard Boolean gates [Muroga, 1971], e.g., TL gate based implementations of Boolean functions potentially require a smaller number of gates and less gate levels.

A generic threshold gate scheme, which is displayed by Figure2(a), has been proposed in [Lageweg et al., 2001a]. The circuit operates as follows. The input voltages V^p (weighted by their input capacitors C^p) are added to V_j, while the input voltages V^n (weighted by their input capacitors C^n) are subtracted from V_j. The critical voltage V_c of the tunnel junction acts as the threshold value. The bias voltage V_b weighted by its input capacitor C_b adjusts the threshold value to the desired value. If the voltage V_j across the junction is less then V_c, no charge transport can occur and the circuit's output remains 'low'. If $|V_j| > V_c$, one electron tunnels through the junction (from node y to node x), resulting in a 'high' output. This scheme can therefore be used as a basis for implementing linear threshold gates with both positive and negative weights.

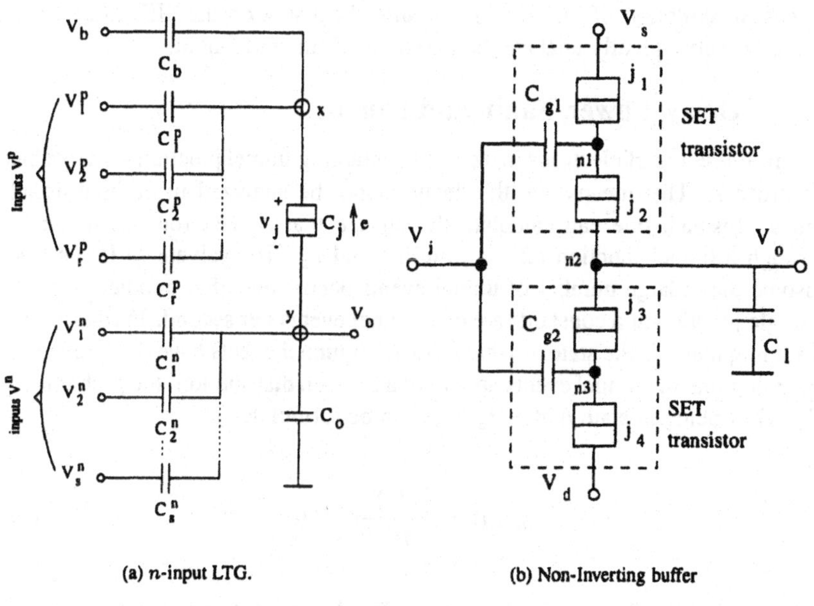

(a) n-input LTG. (b) Non-Inverting buffer

Figure 2. Linear Threshold Gates (LTG) and buffer.

Such threshold gates however do not operate correctly in networks due to the passive nature of the circuit. It was found in [Lageweg et al., 2001b] that augmenting the output of each threshold gate with a buffer results in correctly operating threshold gate networks. A buffer requires active components, for which SET transistors (see for example [Chen et al., 1996] for an introduction of the SET transistor) can be utilized. If two SET transistors share a single load capacitor, such that one transistor can remove a single electron from the load capacitor (resulting in high output) while the other can replace it, we arrive at the non-inverting static buffer as displayed in Figure 2(b). The circuit can also be modified, as described in [Lageweg et al., 2002], to become an inverting static buffer. The inverting static buffer can also function as a stand-alone inverter gate. Additionally, threshold gates can also be augmented with two cascaded buffers, such that both the normal and inverted output are available. Both the generic threshold gate and the buffer operate in accordance with the single electron encoded logic (SEEL) paradigm, i.e., charge transport due to switching activity is limited to one electron. Note that the buffer/inverter can be augmented with strip resistors in order to suppress cotunneling. Referring to Figure 2(b), this is achieved by adding strip resistors between junction 1 and V_s and between junction 4 and ground as suggested in [Lotkhov et al., 2001].

In the next section we discuss a methodology to calculate the delay and power consumption of SEEL logic circuits. Focusing on the SEEL threshold gates, we subsequently analyze the maximum fanin and fanout.

4. Delay, Power, Fanin and Fanout

The tunneling of electrons in a circuit containing tunnel junctions is a stochastic process. This means that the delay cannot be analyzed in the traditional sense. Instead, one can calculate the chance that an electron has tunneled through a tunnel junction after a time interval t_d. To evaluate t_d let us first assume that a large number of tunnel events occurs one after another through a single junction at a constant rate of Γ tunnel events per second. Furthermore, assuming that n is the state in which exactly n tunnel events have occurred, and assuming that the tunnel events adhere to a Poisson distribution, the probability $P_n(t_d)$ of being in state n after t_d time can be formulated as

$$P_n(t_d) = \frac{(\Gamma t_d)^n}{n!} e^{-\Gamma t_d}. \tag{5}$$

In the case of SEEL gates, such as the buffered threshold gate discussed in the previous section, in which the state transition diagram only consists of states $n = 0$ (before the tunnel event) and $n = 1$ (after the tunnel event), $P_0(t_d)$ is the probability that the tunnel event has not occurred after t_d seconds. If the tunnel event is the desired behavior of the circuit, then $P_0(t_d)$ or $P_{error} = e^{-\Gamma t_d}$ is the chance of an erroneous output after t_d seconds. Given that an acceptable error probability P_{error}, the time t_d needed to reduce the error probability to this value can be calculated as

$$t_d = \frac{-ln(P_{error})}{\Gamma}. \tag{6}$$

Assuming $|V_j| > V_c$, the rate at which electrons tunnel through a junction can be described as

$$\Gamma = \frac{-\Delta E}{q_e^2 R_t} \frac{1}{(e^{\Delta E/k_B T} - 1)}, \tag{7}$$

where R_t is the tunnel resistance (typically $R_t = 10^5 \Omega$), $k_B = 1.38 \cdot 10^{-23} J/K$ is Boltzman's constant and ΔE the reduction of the total amount of energy present in the circuit due to a single tunnel event, which can be expressed as

$$\Delta E = -q_e(|V_j| - V_c). \tag{8}$$

The above provide the basic framework for calculating the gate delay and power consumption of SEEL logic gates. Assuming $|\Delta E| >> k_B T$, one can combine Equations (6), (7) and (8) in order to describe the switching delay as a function of the error probability and the junction voltage as

$$t_d = \frac{-ln(P_{error}) q_e R_t}{|V_j| - V_c}. \tag{9}$$

The energy consumed by a single tunnel event occurring in a single tunnel junction can be calculated by taking the absolute value of ΔE. In order to calculate the power consumption of a gate, the energy consumption per tunnel event is multiplied by the switching frequency of the gate's output. The switching frequency in turn depends on the frequency at which the gate's inputs change and is input data dependent as a new combination of inputs may or may not results in a new output value. However, assuming that the gate's output switches at the maximum frequency $f_{max} = 1/t_d$, the theoretical maximum power consumption $P_{max} = \Delta E/t_d$ is

$$P_{max} = \frac{(|V_j| - V_c)^2}{-ln(P_{error}) q_e R_t}. \tag{10}$$

Assuming that gates operate in networks with a logic depth of 15 gates per pipeline stage, and that input data causes switching activity in half the time, we can estimate that the actual power consumption in SEEL circuits is about two orders of magnitude less then P_{max}.

Given that the SEEL threshold gate is intended to be utilized as a network component, in addition to gate delay and power consumption we also require a methodology to calculate the maximum fanin and fanout. When utilizing buffered threshold gates in a network, each gate influences the circuit node voltages inside neighboring gates (through capacitive division), causing feedback and feed forward effects. There are two fundamental contributors to these effects. First, the bias voltage(s) of a gate influences other gates through capacitive division. Given that bias voltages are DC signals, this results in a fixed contribution, which can be compensated for in the design phase. Therefore, the DC feedback and feed forward effects are ignored in this discussion. Second, the switching behavior of a gate also influences other gates. This effect can be considered as a 'random' disturbance of voltage levels, and therefore it cannot be compensated for through fixed biasing. The amount of 'random' disturbance at which the gates will cease to correctly perform their function therefore limits the fanin and fanout of gates operating in a network.

In the remainder of this paper it is assumed that Boolean logic values (input and output signals) are represented by predefined voltage levels, such that logic $'0' = 0$ Volt and logic $'1' = V_{high}$. If all inputs have the 'ideal' voltage levels of 0 and V_{high}, a buffered threshold gate will operate correctly for any number of inputs. If the inputs are not ideal then the fanin of a gate is limited by the quality of the input signals. The quality of an input signal V_{in} is defined by the parameter d, such that $|V_{in} - V_{ideal}| \leq d \cdot V_{high}$. For example, if $V_{high} = 100$ mV and $d = 0.02$, a logic 0 can be represented by any voltage in the range $[-2\,mV, +2\,mV]$.

In a buffered threshold gate an input is part of a weighted sum. Therefore, the quality of the input signal directly contributes to the sum which limits the discrimination of the gate. In practical threshold gate implementations the threshold is usually set in the middle of two integer values in order to maximize robustness for disturbances of the weighted sum. This implies that the maximum deviation of the weighted sum from its 'ideal' value should be less then 0.5. Consider for example a 3-input buffered threshold gate with $d = 0.2$ and all input weights $\omega_i = 1$. In this case the weighted sum can deviate from the intended value by as much as of $3d = 0.6$, which can result in a wrong threshold evaluation. For our discussion we define $fanin$ as the maximum sum of the absolute values of the input weights, which can be described as

$$fanin < \frac{0.5}{d}. \tag{11}$$

For the fanout calculations the following is assumed (see Figure 2): $C_\Sigma^p = C_b + \Sigma_{k=1}^r C_k^p$, $C_\Sigma^n = C_o + \Sigma_{l=1}^s C_l^n$, C_{max}^p is the largest input capacitance of a positively weighted input, C_{max}^n is the largest input capacitance of a negatively weighted input, C_{out} is the buffer's output capacitance, $C_{out} >> C_{max}^p$, $C_{out} >> C_{max}^n$. It is also assumed that $C_\Sigma^p >> C_j$ and $C_\Sigma^n >> C_j$ for the threshold gates. If the output of a gate controls another gate via a positively weighted input and the controlled gate switches to a high output value, the total charge on node x of the threshold gate becomes $q_x = -q_e$. This charge reduces the voltage of node x by $V_x = \frac{-q_e}{C_\Sigma^p}$. Given that $V_{high} = q_e/C_{out}$, V_x can also be described as $V_x = \frac{-C_{out}V_{high}}{C_\Sigma^p}$. The resulting feedback voltage V_{fb}^p on the output of the controlling gate is at most $V_{fb}^p = \frac{-C_{max}^p V_{high}}{C_\Sigma^p}$. Similarly, a negatively weighted input generates at most a feedback voltage $V_{fb}^n = \frac{C_{max}^n V_{high}}{C_{out}}$. If the maximum accepted deviation from ideal voltage levels is d, it is required that the sum of all feedback contributions due to fanout is less then $d \cdot V_{high}$. Therefore the following two constraints apply to fanout

$$p\_fanout < \frac{d \cdot C_\Sigma^p}{C_{max}^p} \; ; \; n\_fanout < \frac{d \cdot C_{out}}{C_{max}^n}, \tag{12}$$

where p_fanout is the fanout to positively weighted inputs and n_fanout is the fanout to negatively weighted inputs.

The next section discusses the implications of the proposed analysis methods on practical networks of buffered threshold gates. Given that we intend our discussion to be technology and application independent, we ignore the effects of interconnects on area, delay and power consumption. Also, given that it is probable that nanotechnology applications will be restricted to locally connected circuits it can be assumed that interconnects will not dominate in the area, delay and power calculations.

5. Discussion

The first practical implication of the proposed methodology, as discussed in the previous section, applies to the limits imposed on the operation temperature T. In order to ensure that the thermal energy $E_{therm} = k_B T$ does not mask the switching energy ΔE related to tunneling events, one must ensure that $\Delta E >> k_B T$. Additionally, Equation (7) implies the same in order to ensure a high tunneling rate. Assuming $\Delta E = 10 k_B T$ and utilizing Equation (8), the operating temperature as a function of $|V_j| - V_c$ is depicted in Figure 3. It can be observed that room temperature operation requires a difference of approximately 250 mV between the junction voltage V_j and the critical voltage V_c.

One of the key metrics for any novel technology is its performance it terms of gate delay. Given an error probability P_{error}, the gate delay solely depends on the tunnel resistance R_t and $|V_j| - V_c$ (see Equation (9)). The tunnel resistance depends on the physical implementation, but $R_t = 100k\Omega$ is commonly used in literature. For the switching error probability we assume $P_{error} = 10^{-12}$ as this is a reasonaly low value which can be further reduced by the application of error correction schemes if so desired. We note that smaller error probabilities can be achieved at the cost of increased delay, as suggested by Equation (9). Using the above parameter values for R_t and

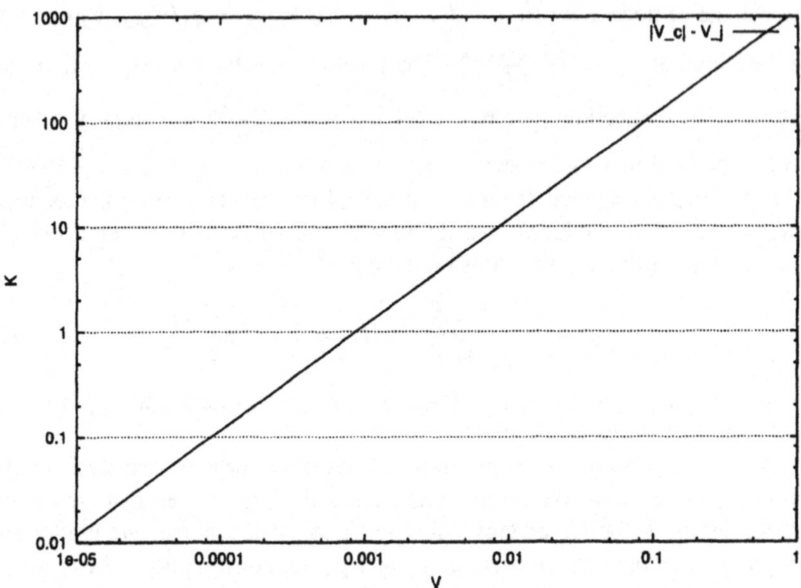

Figure 3. Maximum temperature as function of $|V_j| - V_c$.

P_{error}, the switching delay of a single junction as a function of $|V_j| - V_c$ is depicted in Figure 4. It can be observed that the switching delay is approximately 2 ps if we assume $|V_j| - V_c = 250$ mV, i.e., room temperature operation. Given that this is the delay per tunnel event, and that three tunnel events will sequentially occur in the buffered threshold gate when output switching occurs, we increase the delay estimate by a factor of 3, resulting in an estimated gate delay of 6 ps.

In order to estimate power consumption we assume $|V_j| - V_c = 250$ mV, i.e., room temperature operation, and the calculated gate delay of 6 ps for buffered threshold gates. Given these values, Equation (8) implies that the energy consumption per switching event is $\Delta E = 250$ eV, which implies $\Delta E = 750$ eV if a buffered threshold gate switches its output value (three switching events in total). Larger networks typically operate in pipelines stages, in which series of combinatorial gates are alternated with latches. If we assume 10 gate delays for a pipeline stage, the threshold gates can receive new input data every 60 ps. Assuming that data processing is synchronized by a clock, a system of system threshold gates can potentially operate on a $\frac{1}{60ps}Hz = 17$ GHz clock frequency. Assuming that new input data in 50% of the cases results in the gate switching output value, the buffered threshold gates can potentially switch output value with a frequency of 7.5 GHz. Combining the above, each

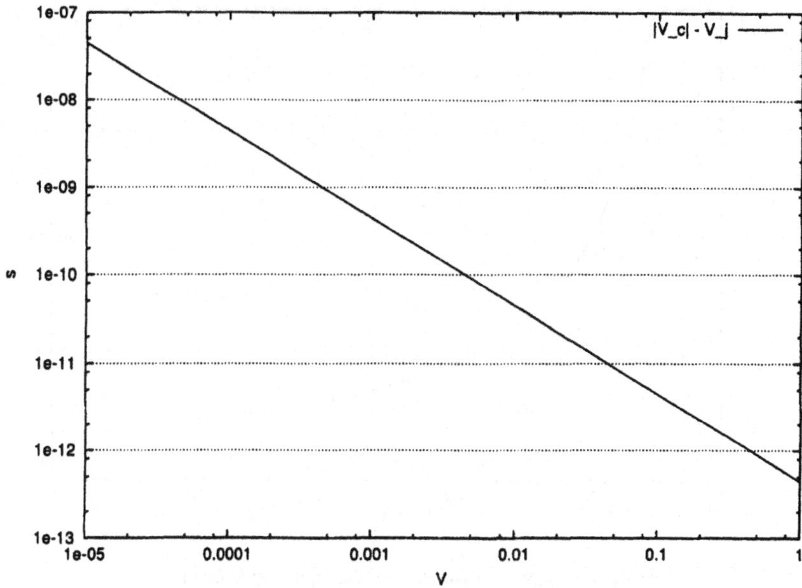

Figure 4. Switching delay as function of $|V_j| - V_c$.

gate consumes 56 nW. If we assume a heat cooling capability of 50 W/cm^2 while maintaining room temperature (100 W/cm^2 cooling capability will be required for mainstream applications by 2007 [TRS03, 2003]), we find a theoretical upper bound of 10^9 buffered threshold gates per cm^2. This is a huge improvement when compared with state-of-the-art CMOS technology that has around 10^6 standard Boolean gates per cm^2.

As discussed in the previous section, $fanin$ (see Equation (11)) is determined by the quality of the input signal d. Figure 5 depicts $fanin$ as a function of d. The maximum fanout to positively weighted inputs p_fanout (see Equation (12)) is determined by the quality d, and the ratio between the input capacitor and C_Σ^p. Figure 6 displays p_fanout as a function of d for various values of $\frac{C_\Sigma^p}{C_{max}^p}$. If for example we target a $fanin = p\_fanout = n\_fanout = 3$, this requires that $d < 0.2$, which results in the following constraints: $\frac{C_\Sigma^p}{C_{max}^p} > 9$ and $\frac{C_{out}}{C_{max}^n} > 9$. In general, increasing the ratios $\frac{C_\Sigma^p}{C_{max}^p}$ and $\frac{C_{out}}{C_{max}^n}$ results in larger fanin and fanout at the cost of larger delay, as increasing these ratios decreases the contribution of input voltages to the voltage V_j across the tunnel junction through which the charge transport occurs.

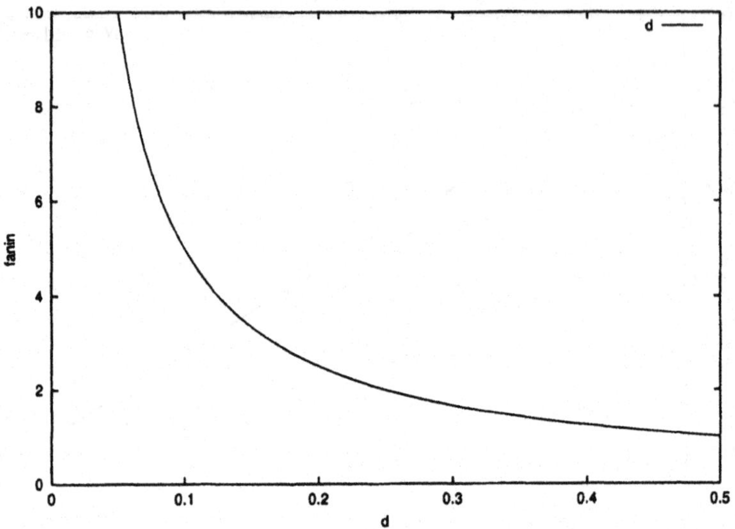

Figure 5. $fanin$ as function of the input signal quality d.

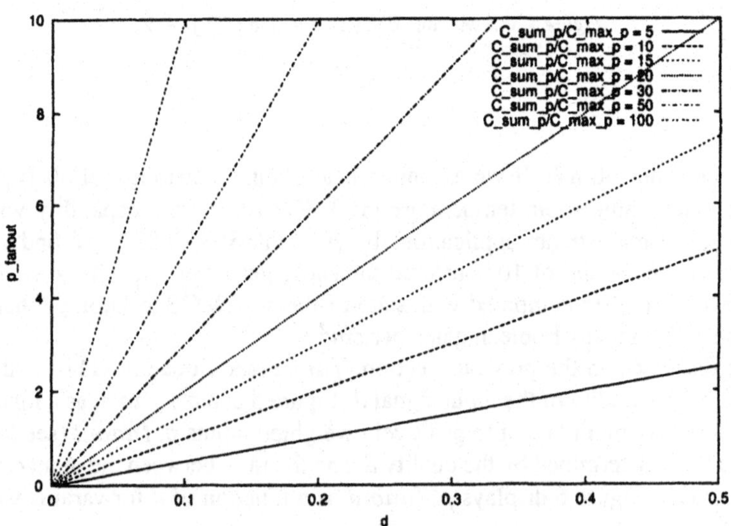

Figure 6. p_fanout as function of d for various $\frac{C_{\Sigma}^{p}}{C_{max}^{p}}$.

6. Conclusions

The SET technology allows for compact implementation of (buffered) SEEL threshold gates with both positive and negative weights. Although it is expected that, when compared with other approaches, SEEL circuits have both reduced delay and reduced energy consumptions, a method for evaluation is required. In this theoretical investigation we proposed a methodology to evaluate delay and power consumption associated with the transport of a single eletron through a tunnel junction. The proposed methodology is based on the orthodox theory and assumes that the stochatic behavior of electron tunneling adheres to a Poisson distribution. We also proposed methods to calculate the maximum fanin and fanout for such buffered SEEL linear threshold gates. Furthermore, we discussed the implications of the proposed methodology on practical networks of such gates. Our estimation indicates that buffered threshold gates operating at room temperature can potentially switch with a delay of $O(10^{-12})$ seconds and have a packing density of 10^9 gates per cm^2.

References

[Averin and Nazarov, 1990] Averin, D.V. and Nazarov, Yu.V. (1990). Virtual Electron Diffusion during Quantum Tunneling of the Electric Charge. *Physical Review Letters*, 65(19):2446–2449.

[Averin and Odintsov, 1989] Averin, D.V. and Odintsov, A.A. (1989). Macroscopic Quantum Tunneling of the Electric Charge in Small Tunnel Junctions. *Physics Letters A*, 140(5):251–257.

[Brennan and Brown, 2002] Brennan, K.F. and Brown, A.S. (2002). *Theory of Modern Electronic Semiconductor Devices*. John Wiley and Sons, Inc.

[Chen et al., 1996] Chen, R.H., Korotkov, A.N., and Likharev, K.K. (1996). Single-electron Transistor Logic. *Applied Physics Letters*, 68(14):1954–1956.

[Cotofana et al., 2003] Cotofana, S.D., Lageweg, C.R., and Vassiliadis, S. (2003). On Computing Addition Related Arithmetic Operations via Controlled Transport of Charge. In *proceedings of 16th IEEE Symposium on Computer Arithmetic*, pages 245–252.

[Goossens, 1998] Goossens, M. (1998). *Analog Neural Networks in Single-Electron Tunneling Technology*. PhD thesis, Delft University of Technology.

[Heij et al., 2001] Heij, C.P., Hadley, P., and Mooij, J.E. (2001). A Single-Electron Inverter. *Applied Physics Letters*, 78(8):1140–1142.

[Hennessy and Patterson, 1996] Hennessy, J. and Patterson, D. (1996). *Computer Architecture - A Quantitative Approach*. Morgan Kaufmann Publishers, San Franscisco, California, 2nd edition.

[Ishibashi et al., 2001] Ishibashi, K., Tsuya, D., Suzuki, M., and Aoyagi, Y. (2001). Fabrication of a Single-Electron Inverter in Multiwall Carbon Nanotubes. *Applied Physics Letters*, 82(19):3307–3309.

[Iwai and Momose, 1998] Iwai, H. and Momose, H.S. (1998). Ultra-Thin Gate Oxides - Performance and Reliability. In *International Electron Devices Meeting 1998*. IEEE.

[Korotkov, 1999] Korotkov, A.N. (1999). Single-Electron Logic and Memory Devices. *International Journal of Electronics*, 86(5):511–547.

[Lageweg et al., 2001a] Lageweg, C.R., Cotofana, S.D., and Vassiliadis, S. (2001a). A Linear Threshold Gate Implementation in Single Electron Technology. In *IEEE Computer Society Workshop on VLSI*, pages 93–98.

[Lageweg et al., 2001b] Lageweg, C.R., Cotofana, S.D., and Vassiliadis, S. (2001b). Achieving Fanout Capabilities in Single Electron Encoded Logic Networks. In *6th International Conference on Solid-State and IC Technology (ICSICT)*.

[Lageweg et al., 2002] Lageweg, C.R., Cotofana, S.D., and Vassiliadis, S. (2002). Static Buffered SET Based Logic Gates. In *2nd IEEE Conference on Nanotechnology (NANO)*, pages 491–494.

[Likharev, 1999] Likharev, K.K. (1999). Single-Electron Devices and Their Applications. *Proceeding of the IEEE*, 87(4):606–632.

[Lotkhov et al., 2001] Lotkhov, S.V., Bogoslovsky, S.A., Zorin, A.B., and Niemeyer, J. (2001). Operation of a three-junction single-electron pump with on-chip resistors. *Applied Physics Letters*, 78(7):946–948.

[Lotkhov et al., 1999] Lotkhov, S.V., Zangerle, H., Zorin, A.B., and Niemeyer, J. (1999). Storage Capabilities of a Four-Junction Single-Electron Trap with an On-Chip Resistor. *Applied Physics Letters*, 75(17):2665–2667.

[Momose et al., 1998] Momose, H.S., Nakamura, S., Katsumata, Y., and Iwai, H. (1998). Study of Direct-Tunneling Gate Oxides for CMOS Applications. In *3rd International Symposium on Plasma Process-Induced Damage*, pages 30–33. American Vacuum Society.

[Moore, 1965] Moore, G. (1965). Cramming More Components onto Integrated Circuits. *Electronics*, 38(8).

[Muroga, 1971] Muroga, S. (1971). *Threshold Logic and its Applications*. Wiley and Sons Inc.

[Ono et al., 2000] Ono, Y., Takahashi, Y., Yamazaki, K., Nagase, M., Namatsu, H., Kurihara, K., and Murase, K. (2000). Fabrication Method for IC-Oriented Si Single-Electron Transistors. *IEEE Transactions on Electron Devices*, 49(3):193 –207.

[Taur et al., 1997] Taur, Y., Buchanan, D.A., Chen, W., Frank, D., Ismail, K., Lo, S., Sai-Halasz, G., Viswanathan, R., Wann, H., Wind, S., and Wong, H. (1997). CMOS Scaling into the Nanometer Regime. *Proceeding of the IEEE*, 85(4):486–504.

[TRN99, 1999] TRN99 (1999). Technology Roadmap for Nanoelectronics. Downloadable from website http://www.cordis.lu/esprit/src/melna-rm.htm. Published on the internet by the Microelectronics Advanced Research Initiative (MELARI NANO), a European Commission (EC) Information Society Technologies (IST) program on Future and Emerging Technologies.

[TRS03, 2003] TRS03 (2003). International Technology Roadmap for Semiconductors, 2003 Edition, Executive Summary. Downloadable from website http://public.itrs.net/Home.htm. Available from SEMATECH, ITRS department, 2706 Montoppolis Drive, Austin TX 78741, USA.

[Tucker, 1992] Tucker, J.R. (1992). Complementary Digital Logic based on the "Coulomb Blockade". *Journal of Applied Physics*, 72(9):4399–4413.

[Wasshuber, 1998] Wasshuber, C. (1998). *About Single-Electron Devices and Circuits*. PhD thesis, TU Vienna.

[Zorin et al., 2000] Zorin, A.B., Lotkhov, S.V., Zangerle, H., and Niemeyer, J. (2000). Coulomb Blockade and Cotunneling in Single Electron Circuits with On-Chip Resistors: Towards the Implementation of the R Pump. *Journal of Applied Physics*, 88(5):2665–2670.

ASYNCHRONOUS INTEGRATION OF COARSE-GRAINED RECONFIGURABLE XPP-ARRAYS INTO PIPELINED RISC PROCESSOR DATAPATH

Jürgen Becker, Alexander Thomas and Maik Scheer
Institut für Technik der Informationsverarbeitung (ITIV), Fakultät für Elektrotechnik und Informationstechnik, Universität Karlsruhe (TH), Karlsruhe, Germany

Abstract: Nowadays, the datapaths of modern microprocessors reach their limits by using static instruction sets. A way out of these limitations is a dynamic reconfigurable processor datapath extension achieved by integrating traditional static datapaths with the coarse-grain dynamic reconfigurable XPP-architecture (eXtreme Processing Platform). Therefore, a loosely asynchronous coupling mechanism of the corresponding datapath units has been developed and integrated onto a CMOS 0.13 μm standard cell technology from UMC. Here the SPARC compatible LEON processor is used, whereas its static pipelined instruction datapath has been extended to be configured and personalized for specific applications. This allows a various and efficient use, e.g. in streaming application domains like MPEG-4, digital filters, mobile communication modulation, etc. The chosen coupling technique allows asynchronous concurrency of the additionally configured compound instructions, which are integrated into the programming and compilation environment of the LEON processor.

Key words: Reconfigurable Datapath, XPP Architecture, LEON Processor

1. INTRODUCTION

The limitations of conventional processors are becoming more and more evident. The growing importance of stream-based applications makes coarse-grained dynamically reconfigurable architectures an attractive alternative [3], [4], [6], [7]. They combine the performance of ASICs, which

Please use the following format when citing this chapter:
Becker, Jürgen, Thomas, Alexander, Scheer, Maik, 2006, in IFIP International Federation for Information Processing, Volume 200, VLSI-SOC: From Systems to Chips, eds. Glesner, M., Reis, R., Indrusiak, L., Mooney, V., Eveking, H., (Boston: Springer), pp. 263-279.

are very risky and expensive (development and mask costs), with the flexibility of traditional processors [5].

In spite of the possibilities we have today in VLSI development, the basic concepts of microprocessor architectures are the same as 20 years ago. The main processing unit of modern conventional microprocessors, the datapath, in its actual structure follows the same style guidelines as its predecessors. Although the development of pipelined architectures or superscalar concepts in combination with data and instruction caches increases the performance of a modern microprocessor and allows higher frequency rates, the main concept of a static datapath remains. Therefore, each operation is a composition of basic instructions that the used processor owns. The benefit of the processor concept lays in the ability of executing strong control dominant application. Data or stream oriented applications are not well suited for this environment. The sequential instruction execution isn't the right target for that kind of applications and needs high bandwidth because of permanent retransmitting of instruction/data from and to memory. This handicap is often eased by using of caches in various stages. A sequential interconnection of filters, which do the according data manipulating without writing back the intermediate results would get the right optimization and reduction of bandwidth. Practically, this kind of chain of filters should be constructed in a logical way and configured during runtime. Existing approach to extend instruction sets uses static modules, not modifiable during runtime.

Customized microprocessors or ASICs are optimized for one special application environment. It is nearly impossible to use the same microprocessor core for other applications without loosing the performance gain of this architecture.

A new approach of a flexible and high performance datapath concept is needed, which allows to reconfigure the functionality and make this core mainly application independent without losing the performance needed for stream-based applications.

This contribution introduces a new concept of loosely coupled implementation of the dynamic reconfigurable XPP architecture from PACT Corp. into a static datapath of the SPARC compatible LEON processor. Thus, this approach is different from those, where the XPP operates as a completely separate component within one Configurable System-on-Chip (CSoC), together with a processor core, global/local memory topologies and efficient multi-layer AMBA-bus interfaces [11]. Here, from the programmers' point of view the extended and adapted datapath seems like a dynamic configurable instruction set. It can be customized for a specific application and accelerate the execution enormously. Therefore, the programmer has to create a number of configurations, which can be uploaded to the XPP-Array at run time, e.g. this configuration can be used like a filter to calculate stream-oriented data. It is also possible, to configure more than one function at the same time and use them simultaneously. This concept promises an enormously performance boost and the needed flexibility and power reduction to perform a series of applications very effectively.

2. LEON RISC MICROPROCESSOR

For implementation of this concept we chose the 32-bit SPARC V8 compatible microprocessor [1][2], LEON. This microprocessor is a synthesizable, free available VHDL model which has a load/store architecture and has a five stages pipeline implementation with separated instruction and data caches.

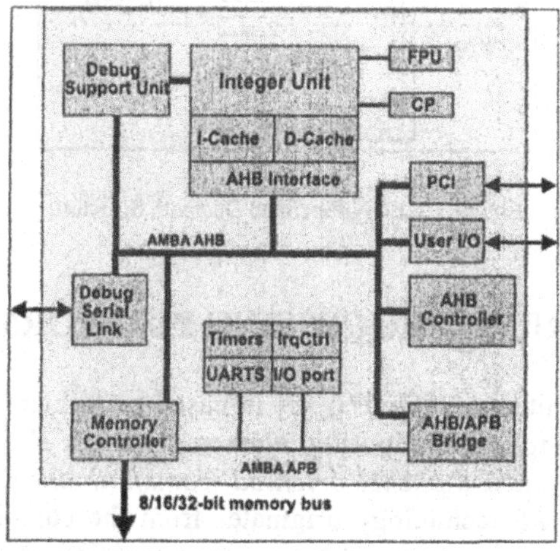

Figure 1. LEON Architecture Overview

As shown in Figure 1 the LEON is provided with a full implementation of AMBA 2.0 AHB and APB on-chip busses, a hardware multiplier and divider, programmable 8/16/32-bit memory controller for external PROM, static RAM and SDRAM and several on-chip peripherals such as timers, UARTs, interrupt controllers and a 16-bit I/O port. A simple power down mode is implemented as well.

LEON is developed by the European Space Agency (ESA) for future space missions. The performance of LEON is close to an ARM9 series but don't have a memory management unit (MMU) implementation, which limits the use to single memory space applications. In Figure 2 the datapath of the LEON integer unit is shown.

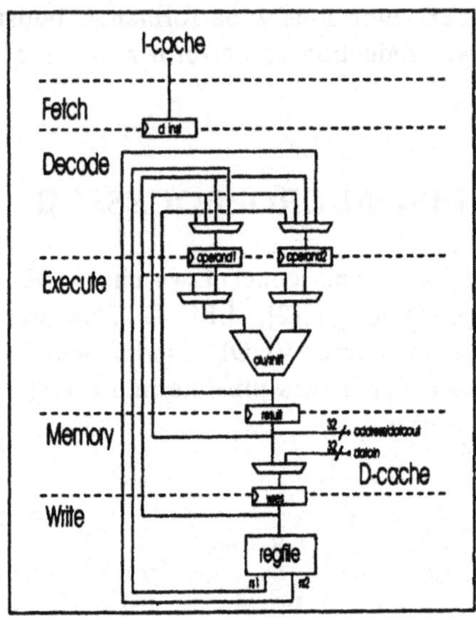

Figure 2. LEON Pipelined Datapath Structure

3. EXTREME PROCESSING PLATFORM – XPP

The XPP architecture [6], [7], [8] is based on a hierarchical array of coarse-grain, adaptive computing elements called *Processing Array Elements (PAEs)* and a *packet-oriented communication network*. The strength of the XPP technology originates from the combination of array processing with unique, powerful run-time reconfiguration mechanisms. Since configuration control is distributed over a *Configuration Manager (CM)* embedded in the array, PAEs can be configured rapidly in parallel while neighboring PAEs are processing data. Entire applications can be configured and run independently on different parts of the array. Reconfiguration is triggered externally or even by special event signals originating within the array, enabling self-reconfiguring designs. By utilizing protocols implemented in hardware, data and event packets are used to process, generate, decompose and merge streams of data.

The XPP has some similarities with other coarse-grain reconfigurable architectures like the Kress-Array [3] or Raw Machines [4], which are specifically designed for stream-based applications. XPP's main distinguishing features are its automatic packet-handling mechanisms and its sophisticated hierarchical configuration protocols for runtime- and self-reconfiguration.

3.1 Array Structure

A CM consists of a state machine and internal RAM for configuration caching. The PAC itself (see top right-hand side of Figure 3) contains a configuration bus which connects the CM with PAEs and other configurable objects. Horizontal busses carry data and events. They can be segmented by configurable switch-objects, and connected to PAEs and special I/O objects at the periphery of the device.

A PAE is a collection of PAE objects. The typical PAE shown in Figure 3 (bottom) contains a BREG object (back registers) and an FREG object (forward registers) which are used for vertical routing, as well as an ALU object which performs the actual computations. The ALU performs common fixed-point arithmetical and logical operations as well as several special three input opcodes like multiply-add, sort, and counters. Events generated by ALU objects depend on ALU results or exceptions, very similar to the state flags of a classical microprocessor. A counter, e.g., generates a special event only after it has terminated. The next section explains how these events are used. Another PAE object implemented in the XPP is a memory object which can be used in FIFO mode or as RAM for lookup tables, intermediate results etc. However, any PAE object functionality can be included in the XPP architecture.

3.2 Packet Handling and Synchronization

PAE objects as defined above communicate via a packet-oriented network. Two types of packets are sent through the array: data packets and event packets. Data packets have a uniform bit width specific to the device type. In normal operation mode, PAE objects are self synchronizing. An operation is performed as soon as all necessary data input packets are available. The results are forwarded as soon as they are available, provided the previous results have been consumed. Thus it is possible to map a signal-flow graph directly to ALU objects. Event packets are one bit wide. They transmit state information which controls ALU execution and packet generation.

3.3 Configuration

Every PAE stores locally its current configuration state, i.e. if it is part of a configuration or not (states „configured" or „free"). Once a PAE is configured, it changes its state to „configured". This prevents the CM from reconfiguring a PAE which is still used by another application. The CM caches the configuration data in its internal RAM until the required PAEs become available.

While loading a configuration, all PAEs start to compute their part of the application as soon as they are in state „configured". Partially configured applications are able to process data without loss of packets. This concurrency of configuration and computation hides configuration latency.

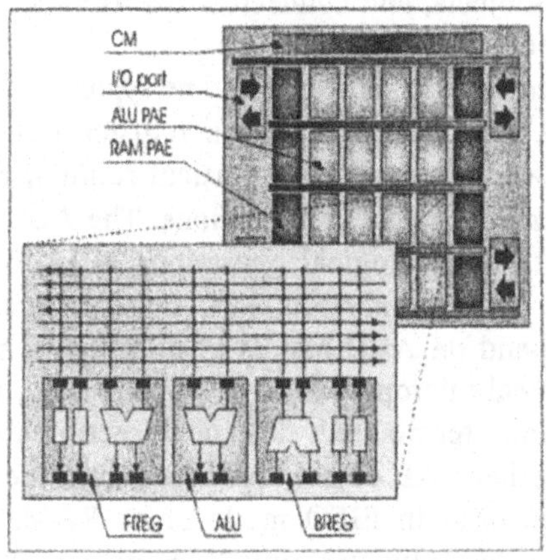

Figure 3. Structure of an XPP device

3.4 XPP Application Mapping

The Native Mapping Language (NML), a PACT proprietary structural language with reconfiguration primitives, was developed by PACT to map applications to the XPP array. It gives the programmer direct access to all hardware features.

In NML, configurations consist of modules which are specified as in a structural hardware description language, similar to, for instance, structural VHDL, PAE objects are explicitly allocated, optionally placed, and their connections specified. Hierarchical modules allow component reuse, especially for repetitive layouts. Additionally, NML includes statements to support configuration handling. A complete NML application program consists of one or more modules, a sequence of initially configured modules, differential changes, and statements which map event signals to configuration and prefetch requests. Thus configuration handling is an explicit part of the application program.

A complete XPP Development Suite (XDS) is available from PACT. For more details on XPP-based architectures and development tools see [6].

4. LEON INSTRUCTION DATAPATH EXTENSION

The system is designed to offer a maximum of performance. LEON and XPP should be able to communicate with each other in a simple and high performance manner. While the XPP is a dataflow orientated device, the LEON is a general purpose processor, suitable for handling control flow [1], [2]. Therefore, LEON is used for system control. To do this, the XPP is integrated into the datapath of the LEON integer unit, which is able to control the XPP.

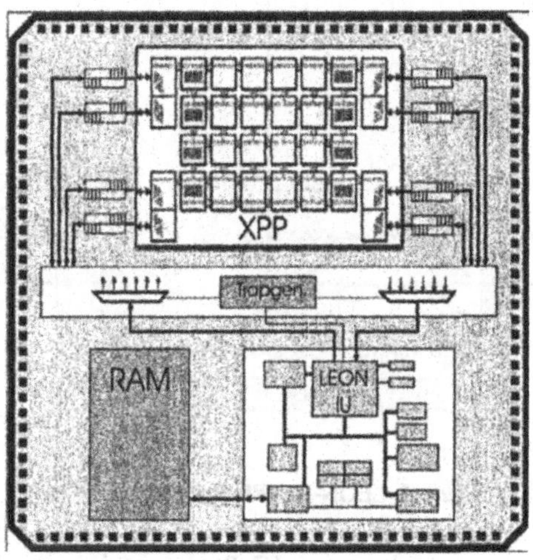

Figure 4. Extended Datapath Overview

Due to unpredictable operation time of the XPP algorithm, integration of XPP into LEON datapath is done in a loosely-coupled way (see Figure 4). Thus the XPP array can operate independent from the LEON processor, which is able to control and reconfigure the XPP during runtime. Since the configuration of XPP is handled by LEON, the CM of the XPP is not necessary and can be left out of the XPP array. The configuration codes are stored in the LEON RAM. LEON transfers the needed configuration from its system RAM into the XPP and creates the needed algorithm on the array.

To enable a maximum of independence of XPP from LEON, all ports of the XPP – input ports as well as output ports – are buffered using dual clock FIFOs. Dual-clocked FIFOs are implemented into the IO-Ports between LEON and XPP. To transmit data to the extended XPP-based datapath the data are passed through an IO-Port as shown in Figure 5. In addition to the FIFO the IO-Ports contain logic to generate handshake signals and an interrupt request signal. The IO-Port for receiving data from XPP is similar to Figure 5 except that the reversed direction of the data signals. This enables that XPP can work completely independent from LEON as long as there are input data available in the input port FIFOs and free space for result data in the output port FIFOs. There are a number of additionally features implemented in the LEON pipeline to control the data transfer between LEON and XPP.

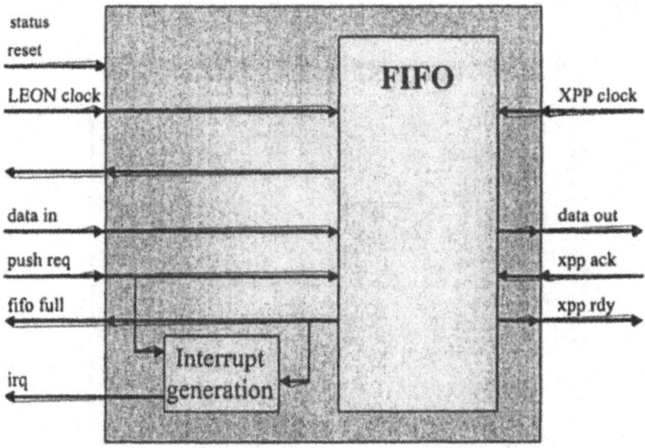

Figure 5. LEON-to-XPP dual-clock FIFO

When LEON tries to write to an IO-Port containing a full FIFO or read from an IO-Port containing an empty FIFO a trap is generated. This trap can be handled through a trap handler. There is a further mechanism - pipeline-holding - implemented, to allow LEON holding the pipeline and wait for free FIFO space during XPP write access respectively wait for a valid FIFO value during XPP read access. When using pipeline-holding the software developer has to avoid reading from an IO-Port with empty FIFO while the XPP, respectively the XPP input IO-Ports, contains no data to produce outputs. In this case a deadlock will occur and the complete system has to be reseted.

XPP can generate interrupts for the LEON when trying to read a value from an empty FIFO port or to write a value to a full FIFO port. The occurrence of interrupts indicates that the XPP array cannot process the next step because it has either no input values or it cannot output the result value. The interrupts generated by the XPP are maskable.

The interface provides information about the FIFOs. LEON can read the number of valid values the FIFO contains. The interface to the XPP appears to the LEON as a set of special registers (see Figure 6). These XPP registers can be categorized in communication registers and status registers.

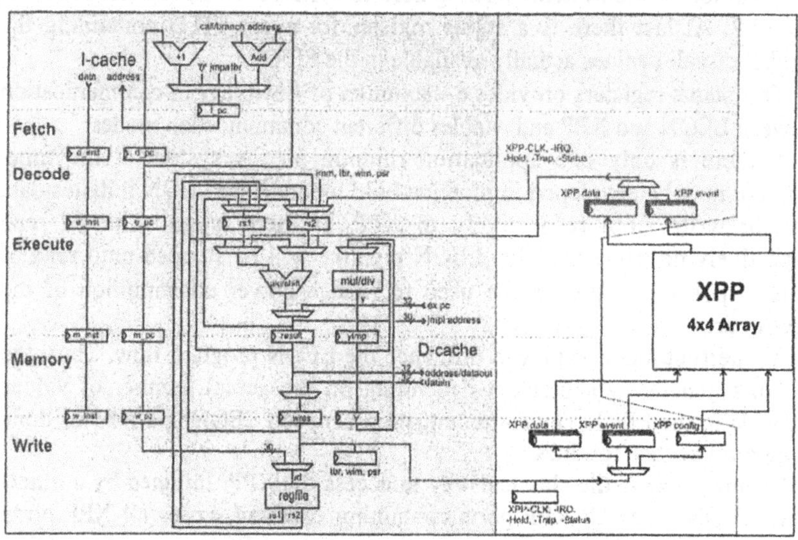

Figure 6. Extended LEON Instruction Pipeline

For data exchange the XPP communication registers are used. Since XPP provides three different types of communication ports, there are also three types of communication registers, whereas every type is splitted into an input part and an output part:

The data for the process are accessed through XPP data registers. The number of data input and data output ports as well as the data bit width depends on the implemented XPP array.

XPP can generate and consume events. Events are one bit signals. The number of input events and output events depends on the implemented XPP array again.

Configuration of the XPP is done through the XPP configuration register. LEON reads the required configuration value from a file – stored in his system RAM – and writes it to the XPP configuration register.

There are a number of XPP status register implemented to control the behavior and get status information of the interface. Switching between the usage of trap handling and pipeline holding can be done in the hold register. A XPP clock register contains a clock frequency ratio between LEON and XPP. By writing this register LEON software can set the XPP clock relative to LEON clock. This allows to adapt the XPP clock frequency to the required XPP performance and consequently to influence the power consumption of the system. Writing zero to the XPP clock register turns off the XPP. At last there is a status register for every FIFO containing the number of valid values actually available in the FIFO.

This status registers provides a maximum of flexibility in communication between LEON and XPP and enables different communication modes:

If there is only one application running on the system at the time, software may be developed in pipeline-hold mode. Here LEON initiates data read or write from respectively to XPP. If there is no value to read respectively no value to write, LEON pipeline will be stopped until read or write is possible. This can be used to reduce power consumption of the LEON part.

In interrupt mode, XPP can influence the LEON program flow. Thus, the IO-Ports generate an interrupt depending on the actual number of values available in the FIFO. The communication between LEON and XPP as done in interrupt service routines.

Polling mode is the classical way to access the XPP. Initiated by a timer-event LEON reads all XPP ports containing data and writes all XPP ports containing free FIFO space. Between those phases LEON can perform other calculations.

It is anytime possible to switch between those strategies within one application.

The XPP is delivered containing a configuration manager to handle configuration and reconfiguration of the array. In this concept the configuration manager is dispensable because the configuration as well as any reconfiguration is controlled by the LEON through the XPP configuration register. All XPP configurations used for an application are stored in the LEON's system RAM.

5. TOOL AND COMPILER INTEGRATION

The LEON's SPARC V8 instruction set [1] was extended by a new subset of instructions to make the new XPP registers accessible through software. These instructions based on the SPARC instruction format are not SPARC V8 standard conform. Corresponding to the SPARC V8 conventions of RISC (load/store) architectures the instruction subset can be split in two general types. Load/store instructions can exchange data between the LEON memory and the XPP communication registers. The number of cycles per instruction is similar to the standard load/store instructions of the LEON. Read/write instructions are used for transfers between LEON registers. Since the LEON register-set is extended by the XPP registers the read/write instructions are extended also to access XPP registers. Status registers can only be accessed with read/write instructions. Execution of arithmetic instructions or compare operations on XPP registers is not possible. Values have to be written to standard LEON registers before they can be used. The complete system is still SPARC V8 compatible. By doing this, the XPP part is completely unused.

The LEON is provided with the LECCS cross compiler system [9] standing under the terms of LGPL. This system consists of modified versions of the binary utilities 2.11 (binutils) and GNU cross compile 2.95.2 (GCC). To make the new instruction subset available to software developers, the assembler of the binutils has been extended by a number of instructions according to the implemented instruction subset. The new instructions have the same mnemonic as the regular SPARC V8 load, store, read and write instructions. Only the new XPP registers have to be used as source respectively target operands. Since the modifications of LECCS are straightforward extensions, the cross compiler system is backward compatible to the original version. The availability of the source code of LECCS has offered the option to extend the tools by the new XPP operations in the described manner.

The development of the XPP algorithms has to be done with separate tools, provided by PACT XPP Technologies.

5.1 Current Compiler Integration

Well, the actual status of the compiler integration allows the using of assembler macro calls within the C-code. Figure 7 shows the development flow in detail. The first step has to be done is the partitioning of your software tasks: Which part should be executed within the LEON datapath and which part should be executed by the XPP architecture. The resulted partition has a big influence on the communication strategy, which strongly affects the reachable performance. The next step is to select the communication strategy. Which transfer mode is the best, depends on several facts of your environment. The most important reason is the time point of the availability. Are my data words already in memory and can be accessed every time I want or there is no way to predict when my data is available. Those things decide which mode is the best for my situation. Like described above you can choose between three modes: Trap Mode, Hold Mode and Polling Mode. The best way to implement the communication is by using two or three modes in parallel. Therefore it is possible to use each register in its own mode. This feature is not implemented yet, but will be available shortly.

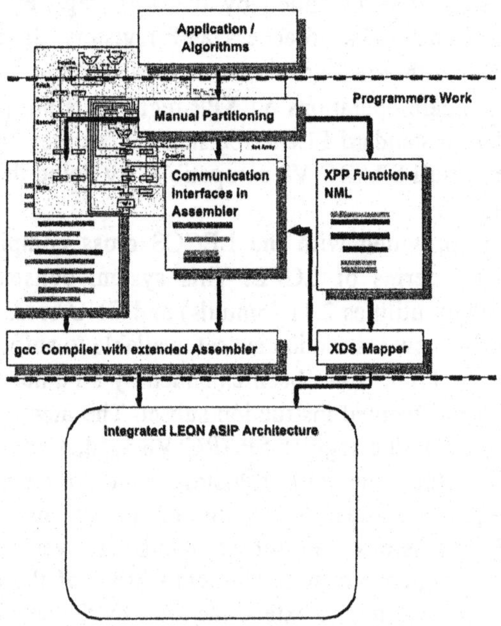

Figure 7. Application Development Flow

Now, the programmer has to take into account the exactly interface structure of his software which depends on his software specification demands. It is important to implement all needed synchronization mechanisms to guarantee full data consistency. The resulted communication structure has to be implemented in assembler and covered with macros. Of course, the assembler parts can be used directly within C-code by enclosing with assembler statement blocks.

The development of the XPP configuration has to be done with in the XDS, like already described above. The parameters of the resulted configuration, like the used IO-ports, duration of a calculation cycle, the size of the configuration stream, and so on must be taken into account within the communication interface realization. The size of a configuration stream e.g. affects the time which is needed to configure the XPP. If you have strong real-time demands, this size can lead to a violation of your real-time behavior and to prevent this mistake has to be taken into account. The other parameter like to IO port number implicates which XPP register should be used for the IO. Those parameters bind both sides of our application tightly where the assembler communication routines describe the way this binding is realized.

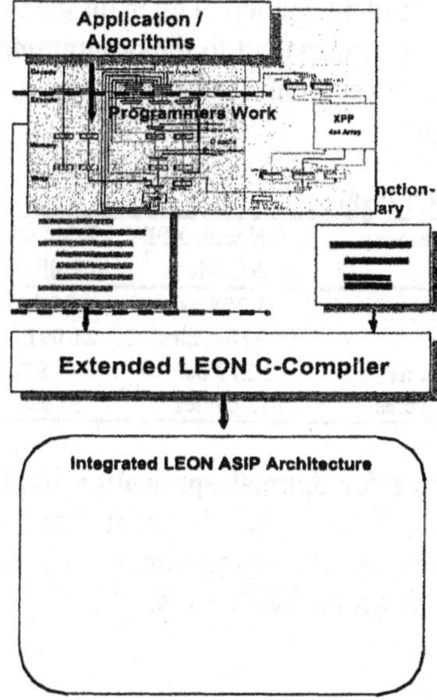

Figure 8. Future Application Development Flow

After the compilation of the resulted code we get an executable program with a data block which contains the implemented configuration for the XPP. The current way to develop an application for our architecture is surely not easy to use. Therefore we are working on the next generation compiler support which offers more transparency to users.

5.2 Future Compiler Integration

The future implementation of the software tools and compilers are shown in figure 8. The main advantage of this concept is the using of the XPP function library which contains a set of function templates with prepared interface descriptions. The programmer can decide at the development time which part of his application should be executed by LEON and which one by the XPP by using the XPP function library. This procedure is comparable to the partition in the current development flow with one difference: the user has just to decide the way how the XPP and LEON have to communicate but don't have to implement the communication routines. Within the C-Code he just inserts selected functions like he always do with standard C functions and the new compiler inserts the appropriate routines for the communication. Therefore the XPP function library contains beside the XPP function templates communication routine templates. The compiler uses those functional drafts for exactly environment parameterization and supports in this way the automatic implementation. The main gain of this concept is that the XPP unit seems nearly transparent to the programmer. He just has to take into account the exactly execution distribution and make for the XPP part appropriate function calls.

Table 1. Performance on IDCT (8x8)

	LEON alone	LEON with XPP in IRQ Mode	LEON with XPP in Polling Mode	LEON with XPP in Hold Mode
Configuration of XPP	———	71.308 ns 17.827 cycles	84.364 ns 21.091 cycles	77.976 ns 19.494 cycles
2D IDCT (8x8)	14.672 ns 3.668 cycles	3.272 ns 818 cycles	3.872 ns 968 cycles	3.568 ns 892 cycles

There is no guarantee for optimal application implementation by using the new development flow. This concept has no mechanisms for performance estimation so its programmers job to make an optimal application partitioning to get the best results.

6. APPLICATION RESULTS

As a first analysis application a inverse DCT applied to 8x8 pixel block was implemented. For all simulations we used 250 MHz clock frequency for LEON processor and 50 MHZ clock frequency for XPP. The usage of XPP accelerates the computation of the IDCT about factor four, depending on the communication mode. However XPP has to be configured before computing the IDCT on it. Table 1 also shows the configuration time for this algorithm.

As shown in figure 9, the benefit brought by XPP rises with the number of IDCT blocks computed by it before reconfiguration, so the number of reconfigurations during complex algorithms should be minimized.

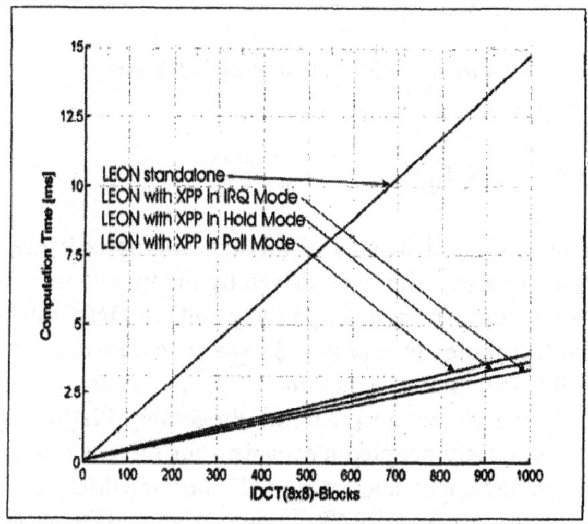

Figure 9. Computation Time of IDCT 8x8-Blocks

A first complex application implemented on the system is MPEG-4 decoding. The optimization of the algorithm partitioning on LEON and XPP is still under construction. In Figure 8 the block diagram of the MPEG-4 decoding algorithm is shown. Frames with 320 x 240 pixel was decoded. LEON by using SPARC V8 standard instructions decodes one frame in 23,46 seconds. In a first implementation of MPEG-4 using the XPP, only the IDCT is computed by XPP, the rest of the MPEG-4 decoding is still done with LEON. Now, with the help of XPP, one frame is decoded in 17,98 s. This is a performance boost of more then twenty percent. Since the XPP performance gain by accelerating the iDCT algorithm only is very low in the moment, we work on XPP implementations of Huffmann-decoding, dequantization and prediction-decoding. So the performance boost of this concept against the standalone LEON will be increased.

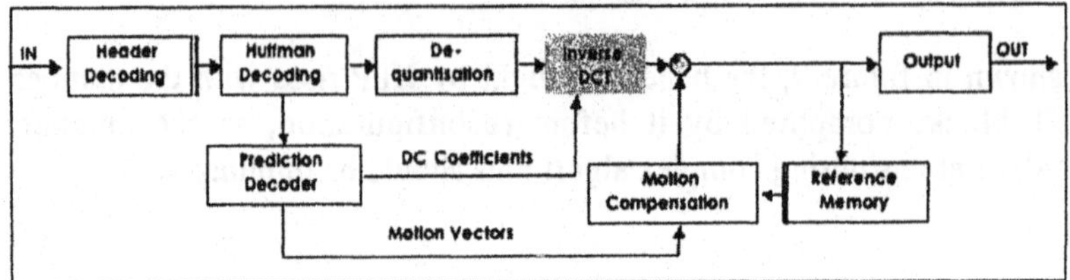

Figure 10. MPEG-4 Decoder Block Diagram

7. CONCLUSION

Today, the instruction datapaths of modern microprocessors reach their limits by using static instruction sets, driven by the traditional von Neumann or Harvard architectural principles. A way out of these limitations is a dynamic reconfigurable processor datapath extension achieved by integrating traditional static datapaths with the coarse-grain dynamic reconfigurable XPP-architecture (eXtreme Processing Platform). Therefore an asynchronously loosely-coupled mechanism for the given microprocessor data path has been developed and integrated onto the UMC CMOS 0.13 μm standard cell technology. The SPARC compatible LEON RISC processor has been used, whereas its static pipelined instruction data path has been extended to be configurable and personalize able for specific applications. This compiler-compatible instruction set extension allows a various and efficient use, e.g. in streaming application domains like MPEG-4, digital filters, mobile communication modulation, etc. The introduced coupling technique by flexible dual-clock FIFO interfaces allows asynchronous concurrency and adapting the frequency of the configured XPP datapath dependent on actual performance requirements, e.g. for avoiding unneeded cycles and reducing power consumption.

As presented above, the introduced concept combines the flexibility of a general purpose microprocessor with the performance and power consumption of coarse-grained reconfigurable datapath structures. Here, two programming and computing paradigms (control-driven von Neumann and transport-triggered XPP) are unified within one hybrid architecture with the advantages of two clock domains. The ability to reconfigure the transport-triggered XPP makes the system independent from standards or specific applications. This concept opens potential to develop multi-standard communication devices like software radios by using extended processor architectures with adapted programming and compilation tools. Thus, new standards can be easily implemented through software updates. The system is scalable during design time through the scalable array-structure of the used XPP extension. This extends the range of suitable applications from products with less multimedia functions to complex high performance systems.

In spite of all the introduced features of the resulted architecture the processor core (LEON processor) is still SPARC compliant. This advantage is necessary in the migration of the reconfigurable hardware extensions into the processor domain. The important advantage is the compatibility of the prior developed compilers and debugging tools which can still be used for developing purposes and older software version are still executable on this architecture. It is also important to keep the effort for compiler tool extension small just by using the available compiler technologies.

REFERENCES

1. The SPARC Architecture Manual, Version 8, SPARC international INC., http://www.sparc.com
2. Jiri Gaisler: The LEON Processor User's Manual, http://www.gaisler.com
3. R. Hartenstein, R. Kress, and H. Reinig. A new FPGA architecture for word-oriented datapaths. In Proc. FPL'94, Prague, Czech Republic, September 1994. Springer LNCS 849
4. E. Waingold, M. Taylor, D. Srikrishna, V. Sarkar, W. Lee, V. Lee, J. Kim, M. Frank, and P. Finch. Baring it all to software: Raw machines. IEEE Computer, pages 86-93, September 1997
5. J. Becker (Invited Tutorial): Configurable Systems-on-Chip (CSoC); in: Proc. of 9th Proc. of XV Brazilian Symposium on Integrated Circuit Design (SBCCI 2002), Porto Alegre, Brazil, September 5-9, 2002
6. PACT Corporation: http://www.pactcorp.com
7. The XPP Communication System, PACT Corporation, Technical Report 15, 2000
8. V. Baumgarte, F. Mayr, A. Nückel, M. Vorbach, M. Weinhardt: PACT XPP - A Self-Reconfigurable Data Processing Architecture; The 1st Int'l. Conference of Engineering of Reconfigurable Systems and Algorithms (ERSA'01), Las Vegas, NV, June 2001
9. LEON/ERC32 Cross Compilation System (LECCS), http://www.gaisler.com/leccs.html
10. M. Vorbach, J. Becker: Reconfigurable Processor Architectures for Mobile Phones; Reconfigurable Architectures Workshop (RAW 2003), Nice, France, April, 200
11. J. Becker, M. Vorbach: Architecture, Memory and Interface Technology Integration of an Industrial/Academic Configurable System-on-Chip (CSoC); IEEE Computer Society Annual Workshop on VLSI (WVLSI 2003), Tampa, Florida, USA, February, 2003

GRAY ENCODED ARITHMETIC OPERATORS APPLIED TO FFT AND FIR DEDICATED DATAPATHS

Eduardo A. C. da Costa[1], Jose C. Monteiro[2] and Sergio Bampi[3]

[1] Universidade Católica dePelotas (UCPel), Pelotas/RS-Brazil; [2]Instituto de Engenharia e Sistemas de Computadores (INESC-ID), Lisboa-Portugal; [3]Universidade Federal do Rio Grande do Sul (UFRGS), Porto Alegre/RS-Brazil

Abstract: **This paper addresses the use of low power techniques applied to FIR filter and FFT dedicated datapath architectures. New low power arithmetic operators are used as basic modules. In FIR filter and FFT algorithms, 2's complement is the most common encoding for signed operands. We use a new architecture for signed multiplication, which maintains the pure form of an array multiplier. This architecture uses radix-2^m encoding, which leads to a reduction of the number of partial lines. Each group of *m* bits uses the Gray code, thus potentially further reducing the switching activity both internally and at the inputs. The multiplier architecture is applied to the DSP architectures and compared with the state of the art. Due to the characteristics of the FIR filter and FFT algorithms, which involve multiplications of input data with appropriate coefficients, the best ordering of these operations in order to minimize the power consumption in the implemented architectures is also investigated. As will be shown, the use of the low power operators with an appropriate choice of coefficients can contribute for the reduction of power consumption of the FIR and FFT architectures. Additionally, a new algorithm for the partitioning and ordering of the coefficients is presented. This technique is experimented in a Semi-Parallel architecture which enables speed-up transformation techniques.**

Key words: **Hybrid encoding, DSP architectures, power consumption.**

Please use the following format when citing this chapter:

A. C. da Costa, Eduardo, Monteiro, Jose, C., Bampi, Sergio 2006, in IFIP International Federation for Information Processing, Volume 200, VLSI-SOC: From Systems to Chips, eds. Glesner, M., Reis, R., Indrusiak, L., Mooney, V., Eveking, H., (Boston: Springer), pp. 281-297.

1. INTRODUCTION

This work focuses on power optimization techniques at the architectural level applied to Digital Signal Processing (DSP) systems[1-4]. DSP applications require high computational speed and, at the same time, suffer from stringent power dissipation constraints[5]. Power consumption in VLSI DSP circuits has gained special attention mainly due to the proliferation of high-performance portable battery-powered electronic devices like cellular phones, laptop computers, etc. In DSP applications, Finite Impulse Response (FIR) and Fast Fourier Transform (FFT) are two of the most widely used algorithms.

In our work, FIR filter and FFT computations are addressed through the implementation of dedicated architectures, where the main goal is to reduce the power consumption by using transformation techniques.

Since multiplier modules are common to many DSP applications, one of the low power techniques used in this work is the use of efficient multiplier architectures in the dedicated DSP architectures[6]. These multiplier circuits, named Hybrid array multipliers, use coding as a method of decreasing the switching activity. As observed in this paper, DSP architectures that use the multiplier of[6] are more efficient than those that use the common Booth multiplier. Power savings above 35% are achievable in the FFT architecture using Hybrid array multiplier of[6]. This power reduction is mainly due to the lower logic depth in the multiplier circuit, which has a big impact on the reduction of the glitching activity in the FFT architectures.

In this work, the low power arithmetic modules are experimented in dedicated FIR filter and FFT architectures. In the FIR implementations, combinations of Fully-Sequential and Semi-Parallel architectures with pipelined version are explored. For the FFT algorithm, Fully-Sequential and Semi-Parallel architectures with pipelined version are also implemented.

Additionally, we propose an extension to the Coefficient Ordering technique[7] that aims at reducing the power dissipation by optimizing the ordering of the coefficient-data product computation. We have used this technique in the FIR and FFT implementations. As will be shown, the manipulation of a set of coefficients can contribute for reducing the power consumption in the dedicated architectures.

This work is organized as follows. In Section 2, we discuss the dedicated FIR filter and FFT realization. An overview of coding for low power is presented in Section 3. Section 4 describes the low power techniques use in this work. Performance comparisons between the architectures for the different low power techniques are presented in Section 5. Finally, in Section 6 we discuss the main conclusions of this work.

2. DEDICATED FIR AND FFT REALIZATION

We present Fully-Sequential and Semi-Parallel FIR filter architectures in the Pipelined form. The Pipelined version is also explored for the Fully-Sequential and Semi-Parallel FFT implementation. These different datapath architectures are compared with implementations that are 16-bit wide and use as examples: i) an 8-order FIR filter ii) a 16-point radix-2 common factor FFT with decimation in frequency. As should be emphasized, although we have presented FIR and FFT examples with a lower number of coefficients, the technique shown in this work could be applied to architectures with any coefficient order. However, the results of these more complex architectures are limited by the power estimation tool used in this work[8].

2.1 FIR Filter Datapaths

FIR filtering is achieved by convolving the input data samples with the desired unit impulse response of the filter. The output *Y[n]* of an *N*-tap FIR filter is given by the weighted sum of the latest *N* input data samples *X[n]* as shown in Eq. (1).

$$Y[n] = \sum_{i=0}^{N-1} H_i X[n-i] \qquad (1)$$

In the Direct Form FIR filter implementation, in each clock cycle a new data sample and the corresponding filter coefficients are simultaneously, producing considerable glitching at the primary outputs[2].

In our work, we address this problem by implementing an alternative Fully-Sequential architecture, called Pipelined form, as shown in Fig. 1.

The Fully-Sequential architecture is a manner to reduce hardware requirements for the FIR filter algorithm, shown in Fig. 1(a). In the sequential implementation the basic idea is to reduce hardware requirements by re-using as much of the hardware as possible.

In order to speed-up the FIR filter computations, we have experimented a Semi-Parallel architecture. In this architecture, shown in Fig. 1(b), hardware requirements are duplicated with respect to the Fully-Sequential, allowing two samples to be processed simultaneously. Again, we have constructed a Pipelined version of the Semi-Parallel architecture.

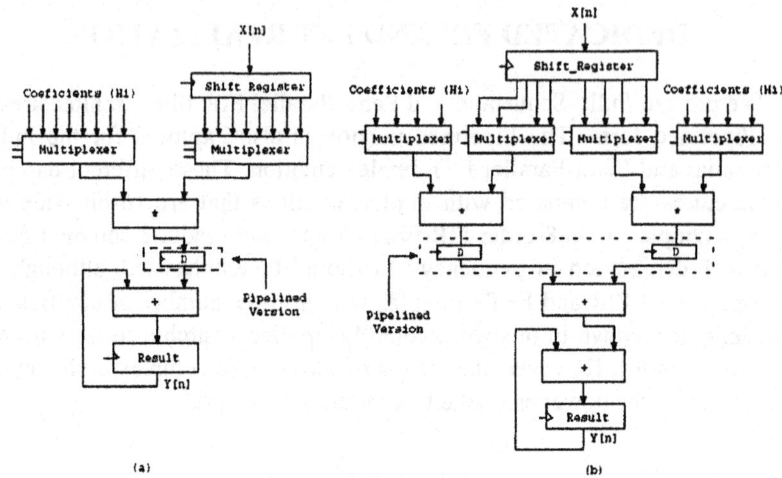

Figure 1. FIR Filter Fully-Sequential and Semi-Parallel Implementations

2.2 FFT Datapaths

The main goal of the FFT algorithms is to compute the Discrete Fourier Transform (DFT) efficiently[9]. The hierarchical computational blocks in the FFT structure are stages, groups, and butterflies. Each stage requires the computation of groups, and each group requires the computation of butterflies. The butterfly plays a central role in the FFT computation. For the common factor FFT algorithm with decimation in frequency, the butterfly allows the calculation of complex terms according to Eq. (2) and Eq. (3).

$$C_{complex} = A_{complex} + B_{complex} \tag{2}$$

$$D_{complex} = (A_{complex} - B_{complex}) * W_{complex} \tag{3}$$

As can be observed in the equations above, one complex addition, one complex subtraction and one complex multiplication are involved in the butterfly block. The arithmetic operators for the complex operation are shown in the Fig. 2 for a Fully-Sequential FFT implementation. In this figure, the arithmetic operators present in the butterfly block, enable the calculation of the real and imaginary parts. The results of these calculation are stored in appropriate register banks shown in the left side and right side of the Fig. 2 for the real and imaginary parts respectively. The set of multiplexers shown in this figure select the appropriate values to be stored in the register banks. Several modules of ROM are required for the storage of twiddle factors. We have omitted these modules to minimize the complexity of Fig. 2.

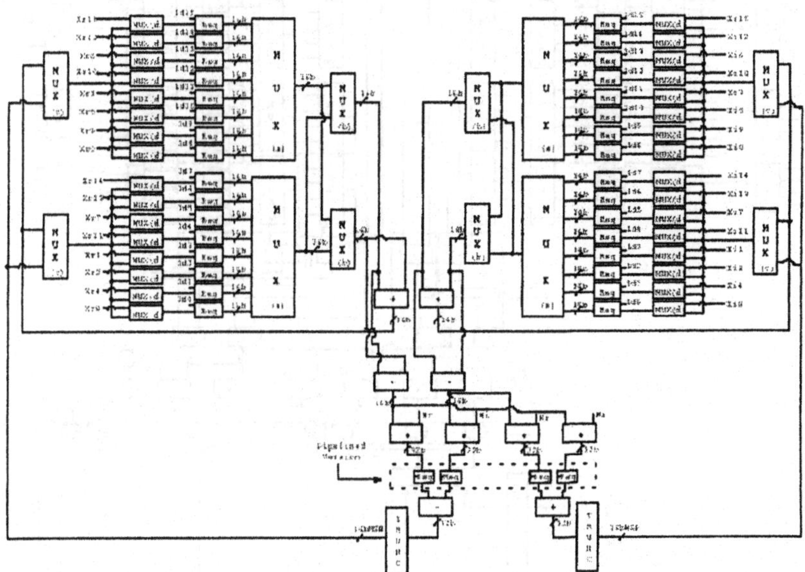

Figure 2. Datapath of FFT Fully-Sequential Implementations

The presence of a large number of multiplier operators in the FFT architecture leads to a significant amount of glitching in a transform computation. Thus, we have implemented a pipelined version with the insertion of registers at the multiplier outputs, as shown using the dotted lines in the Fig. 2.

In a 16-point Fully-Sequential FFT implementation, 32 real and 32 imaginary terms are performed in the butterfly (4 stages with 8 butterfly). Thus, 33 clock cycles are necessary for a full calculation in the FFT architecture (1 cycle for the 16 point load and 32 cycle for a transform computation in the butterfly). In order to speed-up the FFT calculation, we have implemented a Semi-Parallel architecture, presented in Fig. 3. In this architecture, hardware requirements in terms of arithmetic operators are duplicated with respect to the Fully-Sequential, because two butterfly are used and two transforms can be performed simultaneously. Thus, the full transform calculation is performed using half of the cycles used in the Fully-Sequential version. Again, we have implemented a Pipelined Semi-Parallel architecture, as shown in Fig. 3.

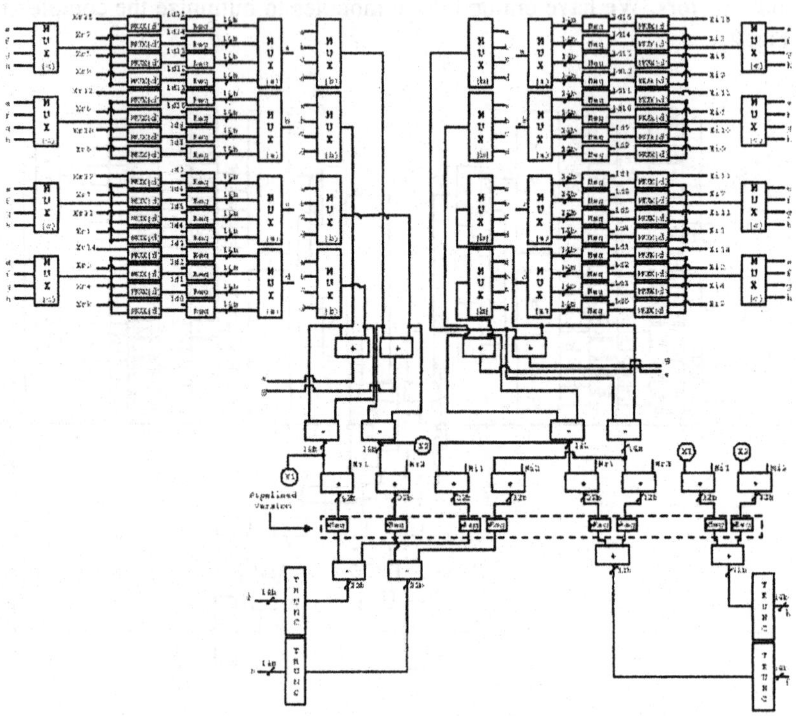

Figure 3. Datapath of FFT Semi-Parallel Implementations

2.3 Related Work on FIR and FFT Realization

Various architectures have been used in FIR filter and FFT realizations, where implementations in programmable DSP and hardwired architectures are addressed[1,3,10]. In case of applications where the flexibility of the programmable processor is not required, hardwired implementation is the preferred choice as such an implementation typically results in higher throughput and low power[7].

For the hardwired implementation, architectural transformations have targeted performance, power and computational complexity[7]. A very efficient technique when targeting low power consumption is to reduce the supply voltage, resulting in a power reduction proportional to the square of the reduction in the supply voltage. With this objective, parallel processing and pipelining have been applied to the implementation of FIR filters and FFT architectures[4,11-14] as a form of recovering the performance loss due to the lower supply voltages.

The work presented in this chapter will build on some of the transformation approaches mentioned, specially the techniques that target the increase in performance and switching activity reduction. In particular, similar transformations will be essayed on pipelined dedicated FIR filter and FFT architectures. In our work, we experiment the use of low power arithmetic operators in the dedicated architectures.

In the FIR filter operation, the output is performed by a summation of data-coefficient products. Thus, some techniques called Coefficient Ordering, Selective Coefficient Negation and Coefficient Scaling have addressed the use of coefficient manipulation in order to reduce the switching activity in the multipliers inputs[15,7]. The main goal of these techniques is to minimize the Hamming distance between consecutive coefficients in order to reduce power consumption in the multiplier input and data bus. The technique is only applied to a Fully-Sequential architecture. In our work an extension of the Coefficient Ordering technique is experimented in the FIR and FFT architectures. The proposed technique can be applied to both Fully-Sequential and Semi-Parallel architectures.

3. CODING FOR LOW POWER

Coding has long been used in communication systems to control the statistics of the transmitted data symbols, or in other words, to control the spectrum of the transmitted signal[16].

Low-Power techniques for global communication in CMOS VLSI using data encoding methods are overviewed in[17], where it is shown that such techniques can decrease the power consumed for transmitting information over heavy load communication paths (buses) by reducing the switching activity.

One technique that has been proposed in order to reduce the switching activity on buses is One-Hot Coding[16]. This technique is a redundant coding scheme with a one-to-one mapping between the n-bit data words to be sent and the m-bit data words that are transmitted. The main disadvantage of this technique is related to the wire quantity required, proportional to 2^n.

The Limited-Weight Codes is another technique proposed in order to obtain switching activity reduction on buses[18]. This technique requires transition signaling in order to reduce the switching activity, since with transition signaling only 1's generate transitions. According to[18], transition signaling is convenient for low-power as it offers a direct way of controlling the bus activity factor simply by reducing the number of logical 1's transmitted over the bus.

The Bus-Invert method as a means of encoding words for reducing I/O power, in which a word may be inverted and then transmitted if doing so reduces the number of transitions[19]. In this method an extra bus line, called *invert* is used.

The Transition Coding and Bit Prediction techniques were used in order to reducing the number of transitions observed in data and address buses[20]. The Transition Coding technique indicates that there is a transition on the bus every time the data to be transmitted is a 1 and there is no transition on the bus if the data to be transmitted is a 0. The Bit Prediction technique is used in address buses that exhibit a very high percentage of addresses that are sequential, so that a factor can be used to predict the value of the next data word with reasonably high accuracy.

One of the most promising encodings that can be used to reduce switching activity is the Gray code since only one bit changes between consecutive values. Therefore, for highly correlated signals the switching activity can be reduced significantly[16]. This code has been applied to code address lines for both instruction access and data access to reduce the number of transitions[16].

As presented above, there is a large number of techniques that resort to signal encoding in order to reduce switching activity on buses. These techniques have all been applied to address buses where data is highly sequential. In[6] similar techniques were applied to arithmetic operators that operate directly upon different coded inputs. In this work we have experimented the use of these operators in the dedicated FIR and FFT architectures.

4. LOW POWER TECHNIQUES

This section presents different low power techniques that will be experimented in the dedicated datapath architectures for DSP. The reduction of switching activity is addressed by using low power arithmetic operators and the manipulation of the filter and FFT coefficients.

4.1 Low Power Arithmetic Operators

In this section we summarize the methodology of[6] for the generation of regular structures for arithmetic operators using signed radix-2^m Hybrid representation.

4.1.1 2's Complement Radix-2^m Hybrid Multiplier Architecture

The idea of splitting the operands in groups of m-bits and encode each group using the Gray code can be used for operands that operate in 2's complement representation. Table 1 shows the 2's complement Hybrid encoding for 4-bit numbers and m=2.

Table 1. 2's Complement Hybrid Code Representation for m=2

Decimal	Hybrid	Decimal	Hybrid	Decimal	Hybrid	Decimal	Hybrid
0	0000	4	0100	-8	1100	-4	1000
1	0001	5	0101	-7	1101	-3	1001
2	0011	6	0111	-6	1111	-2	1011
3	0010	7	0110	-5	1110	-1	1010

For the operation of a radix-2^m multiplication, the operands are split into group of m bits. Each of these groups can be seen as representing a digit in a radix-2^m. Hence, the radix-2^m multiplier architecture follows the basic multiplication operation of numbers represented in radix-2^m. The radix-2^m operation in 2's complement representation is given by Eq. (4). This operation is illustrated in Fig. 4.

$$AxB = A'xB' - A'b_{W-1}b_{\frac{W}{m}-1}2^{W-m} - a_{W-1}a_{\frac{W}{m}-1}\sum_{j=0}^{\frac{W}{m}-1} b_j 2^{W-m+j} \quad (4)$$

For the W-m least significant bits of the operands unsigned multiplication can be used. The partial product modules at the left and bottom of the array need to be different to handle the sign of the operands.

For this architecture, three types of modules are needed. *Type I* are the unsigned modules used in the previous section. *Type II* modules handle the m-bit partial product of an unsigned value with a 2's complement value. Finally, *Type III* modules that operate on two signed values. Only one *Type III* module is required for any type of multiplier, whereas $2\frac{W}{m}-2$ *Type II* modules and $(\frac{W}{m}-1)^2$ *Type I* modules are needed. We present a concrete example for W=8 bit wide operands using radix-16 (m=4) in Fig. 5. The modules of the architecture are performed by using Hybrid representation.

Moreover, as can be observed in the dotted lines of the Fig. 5, the sign extension is shown in Hybrid representation (1000 for a negative number).

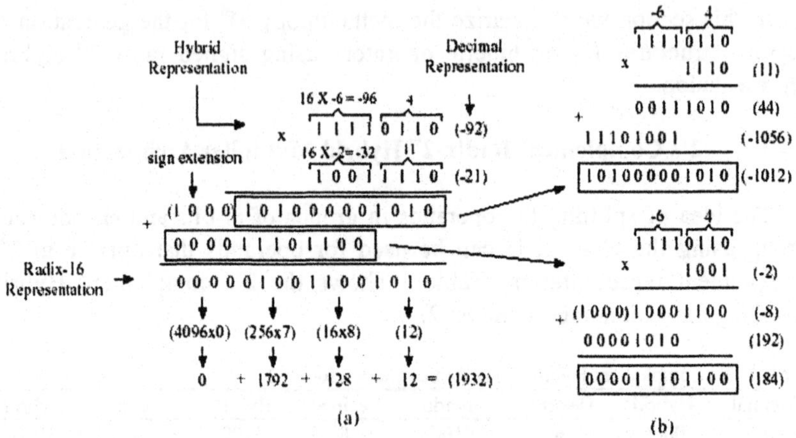

Figure 4. Example of a 2's complement 8-bit wide radix-16 multiplication

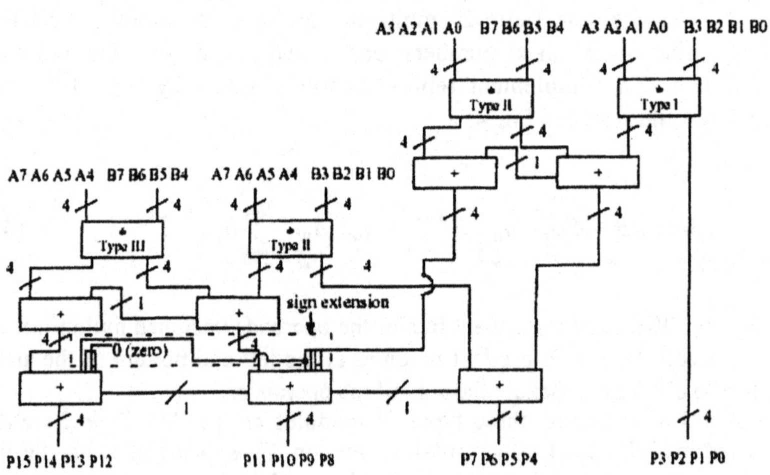

Figure 5. Example of a 8-bit wide 2's complement radix-16 Hybrid multiplier

4.1.2 Modified Booth Multiplier

The radix-4 Booth's algorithm (also called Modified Booth) has been presented in[21]. In this architecture it is possible to reduce the number of partial products by encoding the two's complement multiplier. In the circuit the control signals (0, +X, +2X, -X and -2X) are generated from the multiplier operand for each group of 3-b as shown in the example of Fig. 6 for a 8 bits wide operation. A multiplexer produces the partial product according to the encoded control signal.

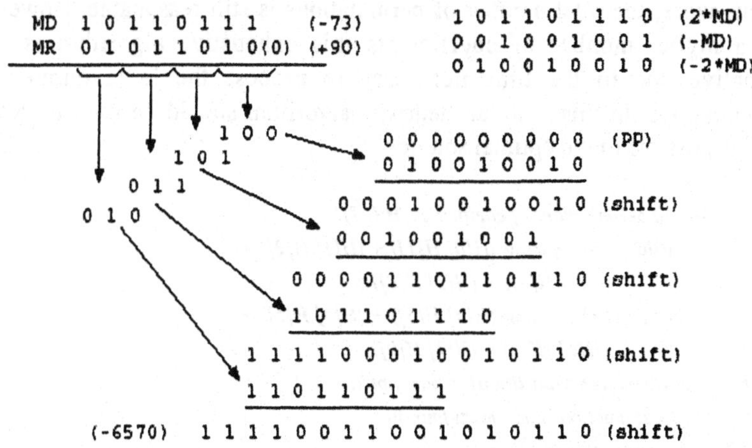

Figure 6. Example of a 8-bit wide Modified Booth multiplication

Common to both architectures is that at each step of the algorithm two bits are processed. However, the basic Booth cells are not simple adders as in the proposed array multiplier, but must perform addition-subtraction-no operation and controlled left-shift of the bits on the multiplicand. Besides taking more area, this complexity also makes it more difficult to increase the radix value in the Booth architecture.

4.2 Coefficient Manipulation

Coefficient ordering can be used as a technique for low power because in a FIR filter computation, the summation operation is both commutative and associative, and the filter output is independent of the order of computing the coefficient product[7]. Coefficient ordering is used in[7] as a technique for low power, where all coefficients are ordered in a Fully-Sequential circuit so as to minimize the transitions in the multiplier input and data bus.

In our work we have experimented an extension of this technique in a Semi-Parallel architecture, where the hardware is duplicated and coefficients are partitioned into groups of coefficients. Thus, the problem is related to finding the best partition for each coefficient by calculating the minimum Hamming distance between the coefficients into each group.

The pseudo-code presented in Fig. 7 describes an example of the algorithm that optimizes the partitioning and ordering of the coefficients. In the example shown in Fig. 7, the cost function is calculated for all the combinations over the coefficients. For the FIR and FFT architectures used in this work, the total number of permutations is still reasonable. However, for a higher number of coefficients this exhaustive algorithm is less attractive due to the time necessary to process the large number of combinations. In this case, an heuristic algorithm should be used to get as near as possible to the optimal solution.

```
for all permutations of coefficients H(0-7){
  partition1=Hamming((H[0],H[1]) + (H[1],H[2]) +
             (H[2],H[3]) + (H[3],H[0]));
  partition2=Hamming((H[4],H[5]) + (H[5],H[6]) +
             (H[6],H[7]) + (H[7],H[0]));
  cost function = partition1 + partition2;
  if(cost function < minimum found){
    save current partition;
    minimum=cost function;
  }
}
```

Figure 7. Pseudo-code of the algorithm for the generation of coefficient partitioning and ordering

5. RESULTS

In This section, we discuss the impact of the proposed low power techniques on dedicated pipelined FIR filter and FFT architectures. Area, delay and power consumption for each architecture are presented. Area is given in terms of the number of literals. Delay values were obtained in SIS environment[22] using the general delay model from the *mcnc* library. This parameter defines the minimum clock period. Power results were obtained with the SLS tool[8] using the general delay model. For the power simulation, we have applied a random pattern signal with 10.000 input vectors represented in 2's complement. For power consumption comparisons, we have chose to compute the power dissipation per sample for the FIR filter and the power dissipation per transform for the FFT.

5.1 Application of the Low Power Arithmetic Operator

In this section, we present results on use of the Hybrid array (m=2) arithmetic operators of Section 4.1 in the FIR and FFT architectures. Area, minimum clock period and power consumption are investigated and compared to the architectures with Modified Booth operator.

5.1.1 Area

Table 2 presents area results for FIR filter and FFT architectures using the Hybrid array (m=2) and Modified Booth operators. As can be observed in this table, there is significant area difference between the architectures with these operators. The Fully-Sequential and Semi-Parallel architectures which use the Hybrid array multiplier operators present more area. This due to the fact that Hybrid array multipliers require more area than Booth circuits.

Table 2. Area results for the pipelined architectures

Architectural Alternative		Operators		Difference (%)
		Booth	Hybrid Array	Hybrid Array vs. Booth
Fully-Sequential	FIR	6427	8035	+25.0
	FFT	24099	32435	+34.6
Semi-Parallel	FIR	10569	13785	+30.4
	FFT	46000	58964	+28.2

5.1.2 Minimum Clock Period

Although FIR filter and FFT architectures with the Hybrid array operators present higher area, these architectures permit a slightly lower clock period than the architectures with Booth operators, as shown in Table 3. This reduction occurs because in the Fully-Sequential and Semi-Parallel FIR and FFT architectures the multiplier circuit is present in the critical path (Fig. 1, Fig. 2 and Fig. 3). For this arithmetic operator, the circuit has a lower delay value[6].

Table 3. Minimum Clock Period results in ns for the pipelined architectures.

Architectural Alternative		Operators		Difference (%)
		Booth	Hybrid Array	Hybrid Array vs. Booth
Fully-Sequential	FIR	260.1	254.8	-2.0
	FFT	355.0	342.6	-3.5
Semi-Parallel	FIR	258.1	252.8	-2.1
	FFT	418.2	414.1	-1.0

5.1.3 Power Dissipation

The Hybrid array and the Modified Booth multiplier applied in this work present reduced power consumption values because of the reduction of the number of partial product lines. In Table 4 we present the power per sample values for the Fully-Sequential and Semi-Parallel FIR architectures in the pipelined version, using the Hybrid array multiplier (m=2) and the Modified Booth multiplier.

Table 4. FIR architecture – Power per sample (μW).

Architectural Alternatives	Modified Booth	Hybrid Array m=2	Difference(%) Hybrid Array vs. Booth
Fully-Sequential	215.4	180.7	-16.1
Semi-Parallel	188.6	158.2	-16.1

As can be observed in Table 4, with the use of the Hybrid array multiplier power per sample savings above 16% are achievable in the Fully-Sequential and Semi-Parallel FIR architectures. This occurs because multiplier circuits are the main responsible for the power consumption in the FIR architectures and the Hybrid array multiplier consumes less power due to the simplest structure and smaller critical path and delay values.

Besides the FIR filter, the FFT architectures also have multiplier circuit in the critical path, as can be observed in Fig. 2 and Fig. 3. For the FFT structure, the higher number of multiplier circuits in the butterfly produces a great amount of glitching activity. Thus, with the use of the Hybrid array multiplier, the FFT architectures become significantly more efficient presenting close to 30% less power consumption per transform, as shown in Table 5. This power reduction is mainly due to the lower logic depth of the array multiplier structure which has a big impact on the reduction of the amount of glitching in the FFT circuits.

Table 5. FFT architecture – Power per transform (mW).

Architectural Alternatives	Modified Booth	Hybrid Array m=2	Difference(%) Hybrid Array vs. Booth
Fully-Sequential	156.6	110.4	-29.5
Semi-Parallel	144.8	92.8	-35.9

5.2 Application of Coefficient Manipulation

In Table 6 we show the power per sample results after using this algorithm in the Pipelined Fully-Sequential FIR filter architecture with Hybrid array (*m*=2 and *m*=4) operators. In this table, it is also shown the power per sample results after applying the ordering algorithm to the Semi-Parallel architecture.

Table 6. FIR architecture – Power per sample (μW)

	Group of Bits	Original Coefficients	Manipulated Coefficients	Difference(%) Manip. vs. Orig.
Fully-Sequential	m=2	180.7	176.9	-3.8
	m=4	228.5	216.0	-12.5
Semi-Parallel	m=2	158.3	151.1	-6.7
	m=4	215.0	201.0	-13.9

As shown in Table 6, there is no significant power per sample reduction in the FIR architectures with *m*=2 Hybrid array operator for the set of coefficients used in this work. However, for groups of *m*=4 bits there is a higher correlation between these coefficients and the architectures with *m*=4 Hybrid array operator present an increasingly power per sample reduction as can be observed in Table 6.

As can be observed in Table 6, for the set of coefficients used in this work, the Semi-Parallel architecture using ordering and partitioning algorithm presents more power per sample reduction compared to Sequential architecture with ordering algorithm. This technique becomes more effective in a set of coefficients with higher correlation (*m*=4).

The manipulation techniques that have been applied to the Fully-Sequential and Semi-Parallel architectures show that the correlation between coefficients can reduce the switching activity in the multipliers input. In the FFT algorithm this aspect becomes more significant due to a higher number of coefficients used in all the stages of the FFT. Thus, we have a higher opportunity for saving power by the manipulation of coefficients. Table 7 shows the power per transform results by the application of the manipulation technique in the Pipelined Fully-Sequential and Semi-Parallel FFT architectures with *m*=2 and *m*=4 Hybrid array multiplier.

In a Semi-Parallel architecture, the coefficients are partitioned into *N*/4 groups at each FFT stage. The aspect of applying the ordering technique in a smaller group of partitioned coefficients increase the proximity between the coefficients. Thus, the *m*=2 and *m*=4 Semi-Parallel architecture presents a higher power per transform reduction compared to the Sequential architecture as can be observed in Table 7.

Table 7. FFT architecture – Power per transform (mW)

	Group of Bits	Original Coefficients	Manipulated Coefficients	Difference(%) Manip. vs. Orig.
Fully-Sequential	m=2	110.8	91.7	-17.2
	m=4	101.1	91.4	-9.6
Semi-Parallel	m=2	93.7	75.5	-19.4
	m=4	87.7	70.1	-20.1

6. CONCLUSIONS

In this work, low power arithmetic operators were experimented in the FIR and FFT architectures. Performance comparisons for pipelined architectures using the array (m=2) and Modified Booth operators were investigated and the results showed that, despite higher area shown by the architectures with the Hybrid array operators, these architectures can present less minimum clock period and power consumption. Due to the characteristics of the FIR and FFT algorithms, which are performed by the product of input data with appropriate coefficients, the best ordering of these coefficients to minimize the power consumption of the implemented architectures was also investigated. The results showed that the FFT architectures can present more power reduction due to the higher opportunity of using the coefficients manipulation technique.

7. REFERENCES

1. M. Mehendale, S. Sherlekar, and G. Venkatesh, G., Low-Power Realization of FIR Filters on Programmable DSP's. *IEEE Transactions on Very Large Scale Integration (VLSI) Systems*, 6(4):546-553, (1998).
2. A. Erdogan, and T. Arslan, High Throughput FIR Filter Design for Low Power SOC Applications. In *13th Annual IEEE International ASIC/SOC Conference*, pp. 21-24, (2000).
3. M. Baas, A Low-Power, High-Performance, 1024-Point FFT Processor. *IEEE Journal of Solid-State Circuits*, 34(3):380-387, (1999).
4. K. Parhi, Algorithms and Architectures for High-Speed and Low-Power Digital Signal Processing. In *Proceedings of 4th International Conference on Advances in Communications and Control*, (1993).
5. E. Mussol, and J. Cortadella, Low-Power Array Multipliers with Transition-Retaining Barriers. *PATMOS*, pp. 227-235. 2002.
6. E. da Costa, J. Monteiro, and S. Bampi, A New Architecture for 2's Complement Gray Encoded Array Multiplier. In *Proceedings of the XV Symposium on Integrated Circuits and Systems Design*, pp. 14-19, (2002).

7. M. Mehendale, S. Sherlekar, and G. Venkatesh, Algorithmic and Architectural Transformations for Low Power Realization of FIR Filters. In *Eleventh International Conference on VLSI Design*, pp. 12-17, (1998).
8. A. Genderen, SLS: An Efficient Switch-Level Timing Simulator Using Min-Max Voltage Waveforms. In *Proceedings of the International Conference on Very Large Scale Integration*, pp. 79-88, (1989).
9. A. Oppenheim, and R. Schafer, *Discrete-Time Signal Processing*. London: Prentice Hall Signal International, (1989).
10. P. Kumhom, J. Johnson, and P. Nagvajara, Design and Implementation of a Universal FFT Processor. *In 13th Annual IEEE International ASIC/SOC Conference*, pp. 182-186, (2000).
11. S. He, and M. Torkelson, Design and Implementation of a 1024-point Pipeline FFT Processor. In *IEEE CICC*, pp. 131.134, (1998).
12. S. Douglas, and et al., 1998, A Pipelined LMS Adaptive FIR Filter Architecture without Adaption Delay. *IEEE Transactions on Signal Processing*, 46(3), (1998).
13. S. Yu, and E. Swartzlander, A New Pipelined Implementation of the Fast Fourier Transform. In *Thirty-Fourth Asilomar Conference on Signals, Systems and Computers*, pp. 423-427, (2000).
14. K. Muhammad, R. Staszewski, and P. Balsara, Speed, Power, Area, and Latency Tradeoffs in Adaptive FIR Filtering for PRML Read Channels. *IEEE Transactions on VLSI Systems*, 9(1):42-51, (2001).
15. M. Mehendale, S. Sherlekar, and G. Venkatesh, Techniques for Low Power Realization of FIR Filters. In *Design Automation Conference*, pp. 404-416, (1995).
16. A. Chandrakasan, and R. Brodersen, *Low Power Digital CMOS Design*. Kluwer Academic Publishers, (1995).
17. M. Stan, and W. Burleson, Low-Power Encodings for Global Communication in CMOS VLSI. *IEEE Trans. on VLSI Systems*, (1997).
18. M. Stan, and W. Burleson, Limited-Weight Codes for Low-Power I/O. *IEEE International Workshop on Low Power Design*, (1997).
19. M. Stan, and W. Burleson, Bus-Invert Coding for Low-Power I/O. *IEEE Transactions on VLSI Systems*, (1995).
20. P. Ramos, and A. Oliveira, Low Overhead Encodings for Reduced Activity in Data and Address Buses. In *IEEE International Symposium on Signals, Circuits and Systems*, pp. 21-24, (1999).
21. I. Khater, A. Bellaouar, and M. Elmasry, Circuit Techniques for CMOS Low-Power High-Performance Multipliers. *IEEE Journal of Solid-State Circuits*, 31:1535-1546, (1996).
22. E. Sentovich, and et al., SIS: A System for Sequential Circuit Synthesis, Technical report, (1992).

STUCK-AT-FAULT TESTABILITY OF SPP THREE-LEVEL LOGIC FORMS

Valentina Ciriani
Dip. di Tecnologie dell'Informazione
Universita' degli Studi di Milano, 26013 Crema, Italy
ciriani@dti.unimi.it

Anna Bernasconi
Department of Computer Science
University of Pisa, 56127 Pisa, Italy
annab@di.unipi.it

Rolf Drechsler
Institute of Computer Science
University of Bremen, 28359 Bremen, Germany
drechsler@informatik.uni-bremen.de

Abstract Recently introduced, three-level logic Sum of Pseudoproducts (SPP) forms allow the representation of Boolean functions with much shorter expressions than standard two-level Sum of Products (SOP) forms, or other three-level logic forms. In this paper the testability of circuits derived from SPPs is analyzed. We study testability under the Stuck-At Fault Model (SAFM). For SPP networks several minimal forms can be considered. While full testability can be proved for some classes, others are shown to contain redundancies. Experimental results are given to demonstrate the efficiency of the approach.

Keywords: Reliability and Testing, Testability, Design for Testability, SPP Three-Level Network.

1. Introduction

An important aspect of logic synthesis is the problem of deriving high-quality design from the initial specifications. A given Boolean function may be realized by a large variety of circuits, very different in terms of structure. In this framework the selection of a logic network, out of all possible known models (e.g., SOP [Coudert, 1994], ESOP [Koda and Sasao, 1995,

Please use the following format when citing this chapter:
Ciriani, Valentina, Bernasconi, Anna, Drechsler, Rolf, 2006, in IFIP International Federation for Information Processing, Volume 200, VLSI-SOC: From Systems to Chips, eds. Glesner, M., Reis, R., Indrusiak, L., Mooney, V., Eveking, H., (Boston: Springer), pp. 299-313.

Sasao, 1993, Sasao, 1996], EXSOP [Chattopadhyay et al., 1997, Debnath and Sasao, 1999, Dubrova et al., 1999], OR-AND-OR [Debnath and Vransic, 2003], SPP [Ciriani, 2003b, Luccio and Pagli, 1999], ESPP [Ishikawa et al., 2002]), is critical and depends on multiple factors. Moreover it is very difficult to define a theoretical model that captures the problem in its generality. Thus the objective is to synthesize a circuit that optimizes a cost function involving different factors. In particular we are interested in several features like:

1 the size of the algebraic expression, in order to estimate the area occupied by the logic gates;
2 the number of levels in the network, in order to estimate the delay of the longest path through the gates;
3 the implementability of the network in the current technologies;
4 the existence of efficient minimization algorithms;
5 the testability properties of the network;
6 the power consuming of the network.

The standard synthesis is performed with Sum of Products (SOP) minimization procedures, leading to two-level circuits. More-than-two level minimization is much harder, but the size of the circuits can significantly decrease. In many cases three-level logic is a good trade-off among circuit speed, circuit size, and the time needed for the minimization procedure [Sasao, 1989]. Algorithms for exact minimization have worst case exponential complexity, hence the time to attain minimal forms may become huge for increasing size of the input.

In this paper we focus on a special three-level network called *Sum of Pseudoproducts* (*SPP*) and on the more general *Sum of k-Pseudoproducts* (*k-SPP*). This choice is motivated by the fact that SPP networks often satisfy the above mentioned properties:

SPP expressions, introduced in [Luccio and Pagli, 1999], can be seen as a direct generalization of SOP expressions using EXOR gates. An SPP form consists of the OR of *pseudoproducts*, where a pseudoproduct is the AND of EXOR factors (i.e., EXOR of literals). In the recent paper [Ishikawa et al., 2002] a modified version of SPP networks, called ESPP and consisting of an EXOR of pseudoproducts, has been proposed. Among three-level networks, SPP forms are particularly compact [Ciriani, 2003a, Ciriani, 2003b]. However SPP forms have two major disadvantages: *(i)* they require large computational effort for the minimization; *(ii)* they have been originally defined for EXOR gates with unbounded fan-in, but in most technologies, EXOR gates with many inputs are slow, expensive and often not easily implementable [Weste and Eshraghian, 1993]. Therefore, in recent studies [Ciriani, 2003a, Ciriani, 2003b, Ciriani and Bernasconi, 2002], *k-SPP forms* with a fixed maximum number of literals (k) in the EXOR factors have been introduced.

Experimental results [Ciriani, 2003a, Ciriani, 2003b, Ciriani and Bernasconi, 2002] show that the size of the k-SPP minimal forms is not significantly larger than the one for unbounded fan-in, but the computational effort drastically decreases, especially when $k = 2$. Thus, 2-SPP forms are reasonable upper bounds of the exact SPP forms, and are a good trade-off between the compactness of SPP forms and the efficiency of SOP minimization. Furthermore 2-SPP forms require a reduced number of different EXOR gates and are more practicable for the current technology. Moreover, preliminary results on multipliers indicate that SPP networks are also low power consuming [Ciriani et al., 2003].

Beside the synthesis aspect, testability is a major aspect of the design process. Up to 40% of the overall design costs are due to testing. For this, aspects of testability should be considered from the very beginning [Williams and Parker, 1982]. For several two-level forms detailed studies on testability have been performed. But, to the best of our knowledge, for three-level networks testability has not been considered so far.

In this paper the testability of 2-SPP and SPP forms is studied from a theoretical and practical point of view under the Stuck-At Fault Model (SAFM).

The classical stuck-at fault model (SAFM) is well-known and used throughout the industry [Breuer and Friedman, 1976]. In SAFM it is assumed that a defect causes a basic cell input or output to be fixed to either 0 or 1. Thus, all failures with this effect will be detected by tests for stuck-at faults.

The investigations with respect to the SAFM are usually based on the single fault assumption, i.e., one assumes that there is at most one fault in the circuit. Under this model it is proved that general SPP networks, minimized with respect to the number of literals, are free of redundancies by construction. Whereas it can be shown by counter-examples that SPPs, minimized with respect to the number of products, are not fully testable. The same result holds for the specific class of 2-SPPs. Experimental results are given to demonstrate the efficiency of the approach.

The paper is structures as follows: In Section 2 notation and definitions are given. The stuck-at fault model is introduced and basics on SPP networks are reviewed. The testability results are presented in Section 3. In Section 4 details on the experimental setup and the practical results are given. Finally, the results are discussed in Section 5.

2. Preliminaries

2.1 Stuck-at Fault Model (SAFM)

Let C be any combinational logic circuit over a fixed library. A fault in the SAFM [Breuer and Friedman, 1976] causes exactly one input or output pin of a node in C to have a fixed constant value (0 or 1) independently of the values applied to the primary inputs of the circuit. More precisely we have the following

DEFINITION 1 *A* stuck-at fault *with fault location v is a tuple $(v[i], \epsilon)$ or $([i]v, \epsilon)$, where $v[i]$ ($[i]v$) denotes the i-th input (output) pin of v, and $\epsilon \in \{0, 1\}$ is the fixed constant value.*

For brevity, in the following we simply speak of stuck-at-0 or stuck-at-1 (s-a-0, s-a-1) faults, if the context is clear.

DEFINITION 2 *An input t to C is a* test *for a fault F, iff the primary output values of C on applying t in the presence of F are different from the output values of C in the fault free case.*

A fault is *testable*, iff there exists a test for this fault. The goal of any test pattern generation process is a *complete* test set for the circuit under test in the considered fault model, i.e. a test set that contains a test for each testable fault.

The construction of complete test sets requires the determination of the faults which are not testable (= *redundant*), even though it is easy to see that in general the detection of redundancies is *coNP-complete*. Redundancies have further unpleasant properties: they may invalidate tests for testable faults and often correspond to locations of the circuit where area is wasted [Breuer and Friedman, 1976]. For this, synthesis procedures which result in non-redundant circuits are desirable.

A node v in C is called *fully testable*, if there does not exist a redundant fault with fault location v. If all nodes in C are fully testable, then C is *fully testable*.

EXAMPLE 3 *Consider the circuit in Figure 1. A s-a-0 fault at the output of the gate $(x_1 \oplus x_2)$ can be tested by setting inputs x_3 and x_4 to 1.*

This is needed to ensure the propagation along the upper AND-gate. Since the EXOR of x_3 and x_4 then becomes 0, the output of the lower AND-gate becomes also 0, ensuring the propagation of the faulty value along the OR-gate at the output. The test is independent of the value of input x_1.

2.2 2-SPP and SPP Networks

In this section we recall some basic definitions from [Ciriani, 2003a, Ciriani, 2003b, Ciriani and Bernasconi, 2002].

In a Boolean space $\{0, 1\}^n$ described by n variables $x_1, x_2, \ldots, x_n$, a 2-*EXOR factor* is an EXOR with at most 2 variables, one of which possibly complemented (an EXOR with just one literal corresponds to the literal itself).

Given two Boolean variables x_1, x_2, all the possible 2-EXOR factors are essentially $x_1, \overline{x}_1, x_2, \overline{x}_2, (x_1 \oplus x_2)$ and $(x_1 \oplus \overline{x}_2)$ (in fact, $\overline{x}_1 \oplus x_2 = x_1 \oplus \overline{x}_2$, and $\overline{x}_1 \oplus \overline{x}_2 = x_1 \oplus x_2$).

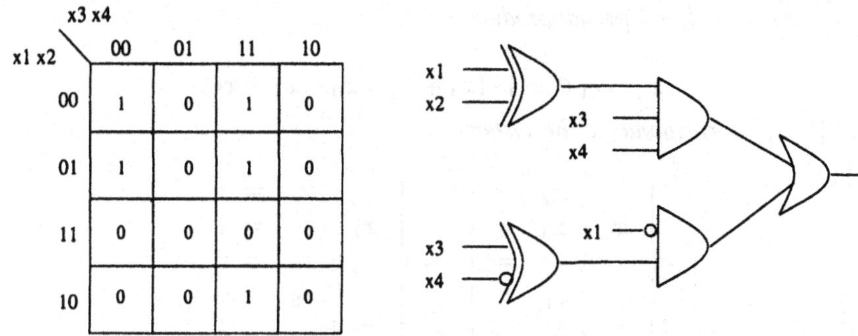

Figure 1. *Karnaugh map of function f with a 2-SPP cover $(x_1 \oplus x_2)x_3x_4 + \overline{x}_1(x_3 \oplus \overline{x}_4)$, minimal with respect to the number of 2-pseudoproducts, and the corresponding 2-SPP circuit representation.*

DEFINITION 4 *A* 2-pseudoproduct *is a product of* 2*-EXOR factors, and a* 2-SPP form *is a sum of* 2*-pseudoproducts.*

A 2-pseudoproduct P of a Boolean function f is *prime* iff no other 2-pseudoproduct P' of f exists such that $P \subseteq P'$.

DEFINITION 5 *A set of points whose characteristic function can be represented as a* 2*-pseudoproduct is a* 2-pseudocube.

This is a generalization of the concept of cubes. A SOP form is a particular 2-SPP form where each EXOR factor contains only one literal.

In the space $\{0,1\}^n$ the number of different 2-EXOR factors with exactly 2 literals is $2 \cdot \binom{n}{2} = n(n-1)$. Thus in the worst case, 2-SPP forms require a quadratic number of different 2-EXOR gates.

The 2-SPP synthesis problem can be stated as: *given a set of points in the Boolean space* $\{0,1\}^n$, *find its minimal cover composed of* 2*-pseudocubes*, where a minimal cover is represented by a sum of 2-pseudoproducts with a minimal number of literals or with a minimal number of 2-pseudoproducts.

EXAMPLE 6 *For the function f represented by the Karnaugh map in Figure 1, the following 2-SPP cover is a minimal expression with respect to 2-pseudoproducts:* $(x_1 \oplus x_2)x_3x_4 + \overline{x}_1(x_3 \oplus \overline{x}_4)$.

The 2-SPP circuit representation is on the left side of the figure. On the other hand, a 2-SPP form minimal with respect to the number of literals is $\overline{x}_2x_3x_4 + \overline{x}_1(x_3 \oplus \overline{x}_4)$.

Finally, a minimal SOP form of such function is $\overline{x}_2x_3x_4 + \overline{x}_1\overline{x}_3\overline{x}_4 + \overline{x}_1x_3x_4$.

We can observe that a 2-pseudoproduct corresponds to a system of linear equations, and a 2-pseudocube corresponds to the set of solutions of such a system.

EXAMPLE 7 *The 2-pseudoproduct*

$$x_2 \cdot (x_1 \oplus x_3) \cdot (x_3 \oplus \overline{x}_5) \cdot \overline{x}_6 \cdot (x_7 \oplus x_8)$$

in $\{0,1\}^9$ *corresponds to the system*

$$\begin{cases} x_2 &= 1 \\ x_1 \oplus x_3 &= 1 \\ x_3 \oplus \overline{x}_5 &= 1 \\ \overline{x}_6 &= 1 \\ x_7 \oplus x_8 &= 1 \end{cases} = \begin{cases} x_2 &= 1 \\ x_1 \oplus x_3 &= 1 \\ x_3 \oplus x_5 &= 0 \\ x_6 &= 0 \\ x_7 \oplus x_8 &= 1 \end{cases}$$

When the 2-pseudocube is actually a cube, the system has only one variable in each equation.

A 2-pseudocube can be represented with different 2-pseudoproducts corresponding to different linear systems. For example, the three 2-pseudoproducts $x_1 \cdot x_1 \cdot (x_2 \oplus x_3) \cdot (x_2 \oplus x_4)$, $x_1 \cdot (x_2 \oplus x_3) \cdot (x_2 \oplus x_4) \cdot (x_3 \oplus \overline{x}_4)$, and $x_1 \cdot (x_2 \oplus x_3) \cdot (x_2 \oplus x_4)$ represent the same set of points (i.e., 2-pseudocube): $\{1011, 1100\}$. Of course the most convenient representation is the third one. The corresponding linear systems are:

$$\begin{cases} x_1 = 1 \\ x_1 = 1 \\ x_2 \oplus x_3 = 1 \\ x_2 \oplus x_4 = 1 \end{cases} = \begin{cases} x_1 = 1 \\ x_2 \oplus x_3 = 1 \\ x_2 \oplus x_4 = 1 \\ x_3 \oplus x_4 = 0 \end{cases} = \begin{cases} x_1 = 1 \\ x_2 \oplus x_3 = 1 \\ x_2 \oplus x_4 = 1 \end{cases}$$

Observe that only the third system has maximum rank, i.e. its equations are linearly independent, and indeed it corresponds to the smaller 2-pseudoproducts. Therefore minimal 2-SPP forms are sums of 2-pseudoproducts whose systems have maximum rank.

In [Ciriani and Bernasconi, 2002] a 2-SPP minimization algorithm is proposed. As in the Quine-McCluskey approach for the SOP forms, the generation of prime 2-pseudoproducts is performed in steps by successive unions of 2-pseudoproducts. A minimal 2-SPP form is generated by choosing a minimal subset of prime 2-pseudoproducts that covers the original function (this is the classical set covering step of Quine-McCluskey optimization).

The *SPP forms*, proposed and studied in [Ciriani, 2003a, Ciriani, 2003b, Luccio and Pagli, 1999], are a direct generalization of 2-SPP expressions, where the EXOR factors can have an unbounded number of literals.

3. Testability in the SAFM

In this section we study the testability of 2-SPP and SPP networks under the SAFM. As observed in Section 2.2 there exist two different notions of cost function for the minimization of 2-SPP (SPP) forms:

1 The cost function is the total number of 2-pseudoproducts (pseudoproducts) in the form.

2 The cost function is the total number of literals in the form.

In both cases, the minimal forms are prime and irredundant. The full testability of 2-SPP and SPP forms is guaranteed only in the second case, as proved below, while forms minimized with respect to the number of pseudoproducts may contain redundancies.

3.1 2-SPP Networks

We first consider 2-SPP forms minimal w.r.t. the number of 2-pseudoproducts.

THEOREM 8 *2-SPP expressions minimal with respect to the number of 2-pseudoproducts are not fully testable.*

Proof.

We provide a counter-example. Consider the function $f = \{0101, 0111, 1001, 1010, 1101, 1110\}$. There are three prime 2-pseudoproducts for f: $(x_1 \oplus x_2)(x_3 \oplus x_4)$, $x_2(x_3 \oplus x_4)$, and $x_1(x_3 \oplus x_4)$. The sum of any couple of them provides a 2-SPP form, prime and irredundant, and minimal w.r.t. the number of 2-pseudoproducts. In fact we have:

$$\begin{aligned} f &= (x_1 \oplus x_2)(x_3 \oplus x_4) + x_2(x_3 \oplus x_4) \\ &= (x_1 \oplus x_2)(x_3 \oplus x_4) + x_1(x_3 \oplus x_4) \\ &= x_1(x_3 \oplus x_4) + x_2(x_3 \oplus x_4). \end{aligned}$$

Let us choose the form

$$f = (x_1 \oplus x_2)(x_3 \oplus x_4) + x_2(x_3 \oplus x_4).$$

Suppose that there is a s-a-0 at the input x_2 of the gate $(x_1 \oplus x_2)$. In this case the output of the 2-pseudoproduct $(x_1 \oplus x_2)(x_3 \oplus x_4)$ is identical to the output of $x_1(x_3 \oplus x_4)$. Therefore the faulty network is equivalent to

$$x_1(x_3 \oplus x_4) + x_2(x_3 \oplus x_4),$$

that is exactly the original function f. ∎

We now consider 2-SPP forms minimal w.r.t. the number of literals. We first need a preliminary result. Recall that 2-SPP networks are composed of three levels of logic: a level of 2-EXORs whose inputs are the variables; a level of ANDs whose inputs are the outputs of the EXOR layer; and an OR among the outputs of the AND layer.

LEMMA 9 *All possible values can be applied to the inputs of the AND layer of a minimal 2-SPP network.*

Proof. Recall that a 2-pseudoproduct can be seen as a linear system. In a minimal 2-SPP form each 2-pseudoproduct contains a number of 2-EXOR factors equal to the rank of its system. In other words the equations in the corresponding system are linearly independent. This means that the outputs of the EXOR gates are independent, i.e., the inputs to the AND layer have all the possible values.

We can now prove the full testability of minimal 2-SPP networks.

THEOREM 10 *2-SPP forms minimal with respect to the number of literals are fully testable.*

Proof. Since 2-SPP forms are prime and irredundant, the proof of the full testability for AND and OR gates is the same as for SOP forms. As proved in Lemma 9, the inputs to the AND gates are directly controllable, i.e., all possible values can be applied. We are then left only with the case of s-a-fault at inputs of EXOR gates. We prove by contradiction that any fault can be tested.

Let $(x_i \oplus x_j) \cdot p + s$ be a representation of f in 2-SPP form minimal w.r.t the number of literals, where p is a 2-pseudoproduct and s is the rest of the minimal 2-SPP form.

Let us consider the case $x_i \equiv 0$, i.e., s-a-0 in x_i. Then the network computes the faulty function $f_F = x_j \cdot p + s$. By contradiction suppose that $f_F \equiv f$, then

$$\begin{aligned} x_j \cdot p + s &\equiv (x_i \oplus x_j) \cdot p + s \\ x_j x_i \cdot p + x_j \overline{x}_i \cdot p + s &\equiv \overline{x}_j x_i \cdot p + x_j \overline{x}_i \cdot p + s \\ x_j x_i \cdot p + s &\equiv \overline{x}_j x_i \cdot p + s\,. \end{aligned}$$

Since $x_j x_i \cdot p \cap \overline{x}_j x_i \cdot p = \emptyset$, we have that

$$x_j x_i \cdot p \subseteq s \quad and \quad \overline{x}_j x_i \cdot p \subseteq s,$$

which implies that $x_i \cdot p \subseteq s$. Therefore f contains $(x_i \oplus x_j) \cdot p$ and $x_i \cdot p$.

We now observe that

$$x_i \cdot p + (x_i \oplus x_j) \cdot p = x_i \cdot p + x_j \cdot p\,.$$

We have

$$\begin{aligned} x_i \cdot p + (x_i \oplus x_j) \cdot p &= x_j x_i \cdot p + \overline{x}_j x_i \cdot p + x_j \overline{x}_i \cdot p \\ &= x_j x_i \cdot p + \overline{x}_j x_i \cdot p + x_j x_i \cdot p + x_j \overline{x}_i \cdot p \\ &= x_i \cdot p + x_j \cdot p\,. \end{aligned}$$

Therefore we reach a contradiction to the minimality w.r.t. the number of literals of the 2-SPP form for f. The minimal 2-SPP form for f would be $x_j \cdot p + s$ instead of $(x_i \oplus x_j) \cdot p + s$.

The case of negated variables is identical. An analogous proof holds for a s-a-1 fault.

3.2 SPP Networks

SPP networks have an unbounded number of literals in the EXOR gates. If we consider forms minimal w.r.t. the number of products, then we have the same result as for 2-SPP networks, since the counter-example given in the proof of Theorem 8 still holds.

Consider now SPP forms minimal w.r.t. the number of literals. The result is analogous to the one for 2-SPP forms:

THEOREM 11 *SPP forms minimal with respect to the number of variables are fully testable.*

Proof.

Following the proof for 2-SPP forms we now have to prove the testability of general EXOR gates. Let $(x_i \oplus h) \cdot p + s$ be a representation of f in SPP form minimal w.r.t the number of literals, where h is an EXOR factor, not including x_i, p is a pseudoproduct and s is the rest of the minimal SPP form.

Let us consider the case $x_i \equiv 0$, i.e. s-a-0 in x_i. Then the network computes the faulty function $f_F = h \cdot p + s$. By contradiction suppose that $f_F \equiv f$, then

$$\begin{aligned} h \cdot p + s &\equiv (x_i \oplus h) \cdot p + s \\ hx_i \cdot p + h\overline{x}_i \cdot p + s &\equiv \overline{h}x_i \cdot p + h\overline{x}_i \cdot p + s \\ hx_i \cdot p + s &\equiv \overline{h}x_i \cdot p + s \,. \end{aligned}$$

Since $hx_i \cdot p \cap \overline{h}x_i \cdot p = \emptyset$, we have that

$$hx_i \cdot p \subseteq s \quad and \quad \overline{h}x_i \cdot p \subseteq s,$$

which implies that $x_i \cdot p \subseteq s$. Therefore f contains $(x_i \oplus h) \cdot p$ and $x_i \cdot p$. Observe that

$$x_i \cdot p + (x_i \oplus h) \cdot p = x_i \cdot p + h \cdot p \,.$$

We have

$$\begin{aligned} x_i \cdot p + (x_i \oplus h) \cdot p &= hx_i \cdot p + \overline{h}x_i \cdot p + h\overline{x}_i \cdot p \\ &= hx_i \cdot p + \overline{h}x_i \cdot p + hx_i \cdot p + h\overline{x}_i \cdot p \\ &= x_i \cdot p + h \cdot p \,. \end{aligned}$$

Therefore we reach a contradiction to the minimality w.r.t. the number of literals of the SPP form for f.

Indeed a minimal form for f would be $h \cdot p + s$ instead of $(x_i \oplus h) \cdot p + s$. An analogous proof holds for the s-a-1 fault. 76

However, in practice SPP networks are defined once a variable ordering is fixed. In this case the above theorem, which refers to SPP forms minimal with respect to any possible variable ordering, does not hold any more. Moreover, as shown below, the SPP forms minimal w.r.t. a fixed variable ordering are no longer fully testable.

Let us consider minimal SPP forms depending on a variable ordering (for more details on SPP networks, see [Ciriani, 2003a, Ciriani, 2003b, Luccio and Pagli, 1999]). For example, consider the Boolean function $f = \{0011, 0100, 1000, 1111\}$, and the variable ordering $o = x_1 < x_2 < x_3 < x_4$.

The function f is indeed a pseudocube, and its minimal SPP network, w.r.t. the variable ordering o, is $(x_1 \oplus x_2 \oplus x_3)(x_1 \oplus x_2 \oplus x_4)$. Meanwhile if we choose the variable ordering $x_3 < x_1 < x_2 < x_4$, then a minimal SPP form is $(x_3 \oplus x_1 \oplus x_2)(x_3 \oplus \overline{x}_4)$, which contains less literals than the former form.

In the case of 2-SPP networks, the number of literals in the minimal forms is instead independent of the variable ordering (see [Ciriani and Bernasconi, 2002] for more details); for this reason the testability theorem holds in any case.

If we fix an ordering, then the proof of testability given above cannot be applied anymore, as the following counter-example shows. Consider the function $f = \{00011, 00100, 00110, 01001, 01011, 01110, 10001, 10011, 10110, 11011, 11100, 11110\}$. Once the variable ordering $o = x_1 < x_2 < x_3 < x_4$ is fixed, there are eleven prime pseudoproducts for f.

A minimal form for f in the variable ordering o is:

$$f = (x_1 \oplus x_2 \oplus x_3 \oplus x_4)(x_3 \oplus x_5) + x_4(x_3 \oplus x_5)\,.$$

Suppose that there is a s-a-0 at the input x_4 of the gate $(x_1 \oplus x_2 \oplus x_3 \oplus x_4)$. In this case the faulty function is:

$$f_F = (x_1 \oplus x_2 \oplus x_3)(x_3 \oplus x_5) + x_4(x_3 \oplus x_5)\,.$$

It is easy to verify that $f \equiv f_F$ but the pseudoproduct $p_F = (x_1 \oplus x_2 \oplus x_3)(x_3 \oplus x_5)$ is not represented in the order o. Therefore it is not in the set of eleven prime pseudoproducts used to form the minimal expression. In this case the fault cannot be detected because f is indeed in minimal form w.r.t. the variable ordering o and $f \equiv f_F$. Of course, if we do not fix a variable ordering then

$$(x_1 \oplus x_2 \oplus x_3 \oplus x_4)(x_3 \oplus x_5) + x_4(x_3 \oplus x_5)$$

is not a minimal form for f.

In summary, we get:

THEOREM 12 *SPP forms minimal with respect to the number of literals in a fixed variable ordering are not fully testable.*

4. Experimental Results

In this section experimental results for the SAFM are reported. The methods described above have been implemented in C. The experiments have been run on a Pentium III 450MHz CPU with 128 MByte of main memory.

The three-level forms have been optimized using the tools described in [Ciriani and Bernasconi, 2002] and the generated networks have been written as BLIF files. The verification of the correctness of the synthesis process and the testability analysis have been carried out in SIS [Sentovich et al., 1992]. The benchmarks are taken from LGSynth93 [Yang, 1991].

In a first series of experiments the quality of SPP forms (optimized by different criteria) are compared to two-level approaches. By this, an impression on the quality of the approaches is provided for a set of benchmarks. To this end we count the number of literals and gates (AND and EXOR) of an expression.

In the multi-level context the cost function is the total number of literals in all gates (see [Eggerstedt et al., 1993, Hachtel and Somenzi, 1996]). The problem is that in many technologies EXOR and OR (or AND) gates have different costs.

In [Hachtel and Somenzi, 1996] the authors consider a 2-input EXOR gate as $x \oplus y = x \cdot y + \overline{x} \cdot \overline{y}$. Thus the cost in literals of a 2-input EXOR gate is 4, while the cost of the 2-input OR and AND gates is 2. This is also proportional to the number of transistors used for the CMOS technology mapping (i.e., 4 transistors for AND/OR gates and 8 transistors for the EXOR gate).

More in general, by the associative property of the EXOR operator, we can always see a k-input EXOR gate as the composition of $(k-1)$ 2-input EXOR gates. Therefore, we can use a function μ where a k-input EXOR gate costs $4(k-1)$, and k-input OR/AND gates cost k. This cost function corresponds to the CMOS cost described in [Eggerstedt et al., 1993].

Table 1 compares the costs of minimal 2-SPP, SOP and SPP forms (2-SPP and SPP networks are minimized with respect to the number of literals in the expressions). In the first column the *name* of the benchmark is given. In the next column the costs are given for 2-SPP, SOP and SPP forms. Here, μ is the cost for the 2-SPP network, while μ' is the cost for the SOP network. The cost for the SPP network is μ''. #E is the number of different EXOR gates in 2-SPP and SPP forms. The star * indicates that the SPP algorithm did not terminate after 172800 seconds (corresponding to 2 CPU days).

The minimization algorithms are designed for exact synthesis of 2-SPP and SPP forms. Indeed the set of prime 2-pseudoproducts (pseudoproducts) is ex-

Table 1. Costs for benchmark functions in 2-SPP, SOP and SPP forms

	2-SPP		SOP		SPP		
name	μ	#E	μ'	μ/μ'	μ''	#E	μ/μ''
9sym	168	18	588	0.29	188	30	0.89
addm4	694	34	1407	0.49	*	*	*
adr4	105	5	415	0.25	118	10	0.89
clip	402	26	769	0.52	*	*	*
dist	471	26	879	0.54	636	50	0.74
f51m	232	19	402	0.58	243	23	0.95
life	180	16	756	0.24	180	16	1.00
m4	735	28	1214	0.61	835	48	0.88
max512	620	35	1032	0.60	*	*	*
mlp4	500	25	869	0.58	524	32	0.95
newcond	161	11	239	0.67	*	*	*
radd	105	5	415	0.25	118	10	0.89
rd53	64	6	175	0.37	66	7	0.97
rd73	212	11	903	0.23	187	15	1.13
root	281	21	376	0.75	366	31	0.77
squar5	101	6	120	0.84	112	8	0.90
xor5	24	2	96	0.25	18	1	1.33
z4	91	6	311	0.29	100	10	0.91

actly computed. Since we used some heuristics [Fiorenzo-Catalano and Malucelli, 2001, Tebboth and Daniel, 2001] in solving the set covering problem, the number of literals in the expressions in Table 1 are upper bounds for the minimal solutions.

The corresponding minimization times are given in Table 2. We note that 2-SPP and SPP forms are much more compact than the corresponding SOP expressions, 2-SPP minimization is also faster than SPP minimization with the exceptions of $9sym$ and $xor5$. This is due to the fact that the SPP minimization algorithm takes advantage of some regularities of functions (see [Bernasconi et al., 2003]), which cannot be exploited by the 2-SPP synthesis.

For all forms, the number of redundancies under the SAFM are given in Table 3. If SOPs are minimized, i.e. they are prime and irredundant, the corresponding networks are also fully testable. But compared to 2-SPP forms they are significantly larger in size (see above). Corresponding to the theoretical results in Section 3, it can be observed that 2-SPPs are fully testable in the SAFM (see Theorem 10), while SPPs may contain redundancies. Indeed the redundancies in SPP networks are due to the heuristic used for their synthesis, and to the fact that the variable ordering in the minimization algorithms is fixed (see Theorem 12).

Table 2. Minimization times (in seconds)

name	2-SPP	SOP	SPP
9sym	242.67	5.32	147.58
addm4	50.96	0.87	*
adr4	6.69	0.10	88.22
clip	1662.27	0.38	*
dist	924.10	0.14	8196.00
f51m	64.00	0.23	443.00
life	120.40	0.03	262.00
m4	890.94	0.67	9929.40
max512	341.24	0.53	*
mlp4	339.51	1.62	1423.74
newcond	1485.01	0.01	*
radd	15.20	0.08	144.00
rd53	0.10	0.01	0.20
rd73	24.10	0.03	114.00
root	272.32	0.08	1597.70
squar5	0.42	0.01	0.64
xor5	0.05	0.01	0.02
z4	5.30	0.04	6.75

In summary, the experiments have shown that 2-SPP forms provide a very good compromise between compact representation, complexity of the minimization process and testability. Beside being more efficient than SOP regarding number of literals, they are so far the only three-level form that ensures full testability of the resulting circuit by construction.

5. Conclusion

Several approaches for three-level synthesis have recently been proposed. The resulting circuits have small delay but are more compact than two-level forms. The algorithmic complexity of the minimization algorithms are moderate. This makes them a promising candidate for synthesis.

In this paper we studied for the first time the testability of the resulting networks. For specific classes, i.e. 2-SPPs and SPPs minimal w.r.t. the number of literals in any variable ordering, full testability has been proved for the SAFM, while for other classes counter-examples were provided. Experimental results demonstrated the efficiency of the approach.

It is focus of current work to study more complex fault models, that allow to model dynamic behavior, like e.g. path-delay faults.

Table 3. Number of redundancies

name	original	2-SPP	SOP	SPP
9sym	0	0	0	0
addm4	24	0	0	*
adr4	24	0	0	0
clip	0	0	0	*
dist	0	0	0	0
f51m	56	0	0	0
life	0	0	0	0
m4	22	0	0	3
max512	4	0	0	*
mlp4	24	0	0	2
newcond	0	0	0	*
radd	0	0	0	0
rd53	0	0	0	0
rd73	0	0	0	0
root	0	0	0	1
squar5	12	0	0	1
xor5	0	0	0	0
z4	12	0	0	0

References

[Bernasconi et al., 2003] Bernasconi, A., Ciriani, V., Luccio, F., and Pagli, L. (2003). Three-Level Logic Minimization Based on Function Regularities. *IEEE Transactions on TCAD*, 22(8):1005–1016.

[Breuer and Friedman, 1976] Breuer, M.A. and Friedman, A.D. (1976). *Diagnosis & reliable design of digital systems.* Computer Science Press.

[Chattopadhyay et al., 1997] Chattopadhyay, S., Roy, S., and Chaudhuri, P. (1997). KGPMIN: An Efficient Multilevel Multioutput AND-OR-XOR Minimizer. *IEEE Transaction on CAD*, 16(3):257–265.

[Ciriani, 2003a] Ciriani, V. (2003a). Synthesis of SPP Three-Level Logic Networks using Affine Spaces. *IEEE Transactions on TCAD*, 22(10):1310–1323.

[Ciriani, 2003b] Ciriani, V. (2003b). *Three-Level Logic Synthesis: Algebraic Approach and Minimization Algorithms.* PhD thesis, Dipartimento di Informatica, University of Pisa.

[Ciriani and Bernasconi, 2002] Ciriani, V. and Bernasconi, A. (2002). 2-SPP: a Practical Trade-Off between SP and SPP Synthesis. In *5th International Workshop on Boolean Problems (IWSBP2002)*, pages 133–140.

[Ciriani et al., 2003] Ciriani, V., Luccio, F., and Pagli, L. (2003). Synthesis of Integer Multipliers in Sum of Pseudoproducts Form. *Integration - the VLSI journal*, 36(3):103–118.

[Coudert, 1994] Coudert, O. (1994). Two-Level Logic Minimization: an overview. *INTEGRATION*, 17:97–140.

[Debnath and Sasao, 1999] Debnath, D. and Sasao, T. (1999). Multiple–Valued Minimization to Optimize PLAs with Output EXOR Gates. In *IEEE International Symposium on Multiple-Valued Logic*, pages 99–104.

[Debnath and Vransic, 2003] Debnath, D. and Vransic, Z.G. (2003). A Fast Algorithm for OR-AND-OR Synthesis. *IEEE Transactions on Computer Aided Design*, 22(9):1166–1176.

[Dubrova et al., 1999] Dubrova, E.V., Miller, D.M, and Muzio, J.C. (1999). AOXMIN-MV: A Heuristic Algorithm for AND-OR-XOR Minimization. In *4th Int. Workshop on the Applications of the Reed Muller Expansion in circuit Design*, pages 37–54.

[Eggerstedt et al., 1993] Eggerstedt, M., Hendrich, N., and von der Heide, K. (1993). Minimization of Parity-Checked Fault-Secure AND/EXOR Networks. In *IFIP WG 10.2 Workshop on Applications of the Reed-Muller Expansion in Circuit Design*, pages 142–146.

[Fiorenzo-Catalano and Malucelli, 2001] Fiorenzo-Catalano, M. S. and Malucelli, F. (2001). Parallel Randomized Heuristics For The Set Covering Problem. *International Journal of Computer Research*, 10(4).

[Hachtel and Somenzi, 1996] Hachtel, G. and Somenzi, F. (1996). *Logic Synthesis and Verification Algorithms*. Kluwer Academy Publishers.

[Ishikawa et al., 2002] Ishikawa, R., Igarashi, T., Hirayama, T., and Shimizu, K. (2002). Pseudocube-based expressions to enhance testability. In *IEEE Asia-Pacific Conference on Circuits and Systems*, volume 2, pages 305–310.

[Koda and Sasao, 1995] Koda, N. and Sasao, T. (1995). An Upper Bound on the Number of Products in Minimum ESOPs. In *IFIP WG 10.5 Workshop on Applications of the Reed-Muller Expansions in Circuit Design*.

[Luccio and Pagli, 1999] Luccio, F. and Pagli, L. (1999). On a New Boolean Function with Applications. *IEEE Transactions on Computers*, 48(3):296–310.

[Sasao, 1989] Sasao, T. (1989). On the Complexity of Three-Level Logic Circuits. In *Int. Workshop on Logic Synthesis*.

[Sasao, 1993] Sasao, T. (1993). AND-EXOR Expressions and their Optimization. In Sasao, T., editor, *Logic Synthesis and Optimization*. Kluwer Academic Publisher.

[Sasao, 1996] Sasao, T. (1996). Representation of Logic Functions Using EXOR Operators. In Sasao, T. and Fujita, M., editors, *Representation of Discrete Functions*. Kluwier Academic.

[Sentovich et al., 1992] Sentovich, E., Singh, K., Lavagno, L., Moon, Ch., Murgai, R., Saldanha, A., Savoj, H., Stephan, P., Brayton, R., and Sangiovanni-Vincentelli, A. (1992). SIS: A system for sequential circuit synthesis. Technical report, University of Berkeley.

[Tebboth and Daniel, 2001] Tebboth, J. and Daniel, R. (2001). A Tightly Integrated Modelling and Optimisation Library. *Annals of Operations Research*, 104:313–333.

[Weste and Eshraghian, 1993] Weste, N.H.E. and Eshraghian, K. (1993). *Principles of CMOS VLSI Design*. Addison-Wesley Publishing Company.

[Williams and Parker, 1982] Williams, T.W. and Parker, K.P. (1982). Design for Testability - A Survey. *IEEE Transactions on Computers*, 31(1):2–15.

[Yang, 1991] Yang, S. (1991). Synthesis on Optimization Benchmarks. User guide, Microelectronic Center. Benchmarks available at ftp://ftp.sunsite.org.uk/computing/general/espresso.tar.Z.

Authors Index